California Cuckoo Wasps in the Family Chrysididae (Hymenoptera)

Lynn S. Kimsey

California Cuckoo Wasps in the Family Chrysididae (Hymenoptera)

Lynn S. Kimsey

University of California Press
Berkeley • Los Angeles • London

University of California Press, one of the most distinguished university presses in the United States, enriches lives around the world by advancing scholarship in the humanities, social sciences, and natural sciences. Its activities are supported by the UC Press Foundation and by philanthropic contributions from individuals and institutions. For more information, visit www.ucpress.edu.

University of California Publications in Entomology, Volume 125

University of California Press
Berkeley and Los Angeles, California

University of California Press, Ltd.
London, England

Printed in the United States of America

Library of Congress Cataloging-in-Publication Data

Kimsey, Lynn Siri.
California cuckoo wasps in the family Chrysididae (Hymenoptera) / Lynn S. Kimsey.
p. cm. -- (University of California publications in entomology ; v. 125)
Includes bibliographical references.
ISBN 978-0-520-09857-2 (pbk. : alk. paper)
1. Chrysididae--California--Classification. I. Title.
QL568.C47K55 2007
595.79'8--dc22

2006029139

The paper used in this publication meets the minimum requirements of
ANSI/NIZO Z39.48-1992 (R 1997) (Permanence of Paper).{∞}

CONTENTS

Abstract

Species and genera of the wasp family Chrysididae in California are reviewed and their California and overall distributions mapped. In addition, keys to California genera and species, and discussions of these species are given. Three new synonymies are given, *Chrysis eurekana* Linsenmaier 1994 and *Chrysis angustianalis* Linsenmaier 1994 under *Chrysis nitidula* Fabricius, and *Chrysis antiochicola* Linsenmaier 1994 under *Chrysis schusteri* Bohart 1982.

INTRODUCTION

CHRYSIDID WASPS, also called cuckoo wasps, are nest parasitoids of predatory wasps and sawflies, and walking stick (Phasmatodea) egg parasites. They are best known for the brilliant metallic body coloration seen in most species. In California chrysidids are generally colored in metallic blues, greens and purples. A few may also be brightly marked with gold, coppery red and white; while others may have non-metallic coloration at least in part. Some *Cleptes* species and *Xerochrum rubeum* have non-metallic red, brown or black metasomae. *Microchridium minutum* is entirely non-metallic black. Overall Old World species tend to be much more colorful than those in the Americas.

The family Chrysididae was completely revised from a world standpoint by Kimsey and Bohart (1991). Worldwide the family consists of four subfamilies, Cleptinae, Amiseginae, Loboscelidiinae and Chrysidinae. Only two of these, Cleptinae and Chrysidinae are found in California. The subfamily Chrysidinae is further divided into six tribes, Elampinae, Chrysidini, Elampini, Parnopini, Kimseyini, and Allocoeliini. The first four of these occur in California. Allocoeliini is endemic to South Africa and Kimseyini is represented by a single genus and species in Uzbekistan.

Bohart and Kimsey (1982) reviewed the North American species. There are 227 chrysidid species in America north of Mexico and 166 of these (73%) are in California (Table 1). In North America the greatest diversity occurs west of 100° N longitude. A number of genera are endemic to this western region, including *Minymischa, Pseudolopyga, Microchridium, Xerochrum, Hedychreides, Chrysurissa* and *Argochrysis*. All of these western endemics occur in California.

About 10% of the North American chrysidid species are endemic to California. A number of taxa found elsewhere in North America are noticeably absent in California, including the subfamily *Amiseginae,* and the genera *Muesebeckidium, Pseudospinolia, Neochrysis, Ipsiura* and *Exochrysis*. With the possible exception of *Pseudospinolia* these genera are primarily Neotropical and extend into North America east of 100° N latitude. *Pseudospinolia* species are primarily Palearctic, with one species found in Chile.

This is a distinctive family of insects, with a number of unique structural modifications. The metasoma is highly modified, with two or more terminal segments telescoped internally. Members of the subfamily Cleptinae have a roughly cylindrical metasoma with four external segments in females and five in males. In the Chrysidinae the terga are strongly convex with a discrete laterotergite that wraps ventrally around the sterna. The sterna are flat or concave and generally not visible in lateral view. Male and female Chrysidinae have three visible metasomal segments, except male *Parnopes* have four. The internally retracted metasomal segments form an eversible genital or egg-laying tube. Another distinctive feature is the metallic coloration of most species mentioned above. The female sting has been modified into an egg-laying tube with highly reduced valvulae and poison gland. As a result, unlike most other aculeates, chrysidids cannot sting, and they can be easily handled whether male or female. As is typical for most aculeate wasps, males have thirteen antennal articles and females twelve.

Table 1. Comparison of the number of chrysidid species found in genera occurring in California, among different geographic regions and the number endemic to California.

Taxon	No. World Species	No. North American Species	No. California Species	No. Species Endemic to California
Cleptinae				
Cleptes	64	7	6	2
Chrysidinae				
Argochrysis	15	15	12	2
Caenochrysis	51	6	2	0
Ceratochrysis	27	27	20	4
Elampus	57	6	6	0
Chrysis	1000	77	47	4
Chrysura	48	10	9	1
Chrysurissa	1	1	1	0
Hedychreides	1	1	1	0
Hedychridium	240	31	27	4
Hedychrum	136	9	3	0
Holopyga	91	6	4	0
Microchridium	1	1	1	0
Minymischa	3	3	3	0
Omalus	20	4	3	0
Parnopes	16	7	5	0
Philoctetes	43	8	8	0
Pseudolopyga	2	2	2	0
Pseudomalus	42	5	5	0
Xerochrum	1	1	1	0
TOTAL	1859	227	166	17

Chrysidines are fairly unspecialized parasites and those that are nest parasites will enter host nests with impunity, often encountering the host female in the process. The cuplike metasoma in chrysidines gives them protection while entering host nests. It can be folded up against the head and thorax, almost completely covering the legs and lower head. They are heavily armored and once rolled in a ball are impervious to attack by hosts. Female *Sceliphron caementarium* (Drury) (Sphecidae) can be observed grasping rolled up female *Chrysis angolensis* Radoszkowski in their mandibles and tossing them away from their nest. One odd draw back of this ability to roll into a ball is that a number of specimens end up pinned that way in collections, which makes determining species very difficult.

California has six major entomofaunal provinces according to Powell and Hogue (1979), Sierran, Great Basin, Desert, Central Valley, Californian and Coastal. These provinces can be divided into two basic categories in the state. The lower or Austral Zones (Lower and Upper Sonoran) include the Central Valley, deserts and interior valleys and the southern coastal plain. Boreal zones (Transition, Canadian and Hudsonian) characterize those areas with pine forests. The largest amount of species endemism appears to be in the Central Valley and Californian Regions. Other chrysidid species found in California can be quite widespread. There are species, such as *Chrysis nitidula*, that occur throughout North America. Others have more limited distributions but their ranges extend outside of the State. Species with Great Basin distributions are typically found in California, Nevada and sometimes Utah. Those found in the Sierran Province often also occur both in the north Coastal Range and into the Cascades of Oregon and Washington. Insects found in the southern Desert Province typically occur in northern Mexico, Arizona and New Mexico.

Chrysidids are relatively uncommon in collections and many collectors rarely see them in the field. In fact these wasps can be locally abundant, and given the right search image can be collected in fairly large numbers. They are not often seen because they tend to run along the ground, searching for host nests in and under vegetation or on fallen logs. Collectors in search of beetles, butterflies or other vagile species will rarely see chrysidid wasps. However, malaise traps and pan traps are quite effective for collecting these wasps.

BIOLOGY

California chrysidid species are all parasitoids of other Hymenoptera, since the walking stick parasitic subfamily Amiseginae does not occur in the state. Members of the Cleptinae parasitize sawflies in the families Diprionidae and Tenthredinidae. They attack prepupal sawflies by tearing a hole in the cocoon and laying an egg directly on the larva or elsewhere within the cocoon (Dahlsten 1961, 1967). Chrysidines are nest parasites of bees, crabronid and sphecid wasps and non-social vespids (Eumeninae and Masarinae). They lay their egg in the host cell, on the prepupal larva or on the pupa. The one exception to this behavior occurs in the genus *Pseudolopyga* (Carrillo and Caltagirone 1970, Parker and Bohart 1968). These wasps actually place their eggs directly on free-living first or second instar lygaeid bugs that are provisioned by the host crabronid wasp. The eggs do not hatch unless the bug is stung by the host and placed in its nest.

Species of *Argochrysis* exhibit an unusual behavior not seen in other chrysidids or parasitoids. Rosenheim, in a series of studies (1987a, b, 1988, 1989, 1993), discovered that more than one *Argochrysis* can emerge from a single host cell, and on occasion it is possible for the host to survive as well.

Chrysidids typically develop in one of two ways. In most species the egg does not hatch until the host larva has consumed all of its provisions and is preparing to pupate. Other species of chrysidid lay their egg in the host cell, or in the case of *Pseudolopyga*, on the host provisions. The egg hatches immediately and the larva feeds both on the host egg, or larva, and the provisions.

Hosts are only known for about 10% of the North American species, so it is difficult to say much about parasitic behavior. Where known, most chrysidines are unspecialized parasitoids. With some exceptions as discussed above, they appear to be host nest specialists rather than host species specific. This lack of specificity may very well account for the amount of size variation seen in some species, for example *Hedychridium fletcheri* adults range from 3-8 mm in length. Different sized hosts would result in a range of chrysidid sizes within a single species.

The majority of chrysidid species are active as adults between the months of May and August. There appears to be one generation per year, and larvae, prepupae or pupae diapause overwinter.

MATERIALS AND METHODS

The synonymic lists in this study are strictly based on species synonymy, not published records. In a few synonymies, such as for *Chrysis angolensis*, the synonymy is limited to names applied in North America or names frequently encountered in the literature. Complete synonymy was given in Kimsey and Bohart (1991). Each entry in the synonymy consists of the current genus, species name, author, date and page of publication, original generic name if different from the original name of description, kind of primary type, type collection data, and type repository in parentheses.

A number of terms are used in the descriptions, which are illustrated in Fig. 1. The midocellus diameter is the breadth of the midocellus measured from side to side in front view. The malar space is the distance from the lower eye margin to the mandibular socket viewed from the side. Subantennal distance is measured from the lower margin of the antennal socket to the edge of the lower clypeal margin directly below it. Brow refers the area of the frons immediately above the scapal basin. Least interocular distance (LID) is measured across the face where the eye margins are closest. Midocellus diameter (MOD) is the width of the midocellus measured from side to side and is used to measure facial dimension. Metasoma is the term used for the apparent abdomen, which begins with abdominal segment two; tergum I therefore refers to metasomal tergum I.

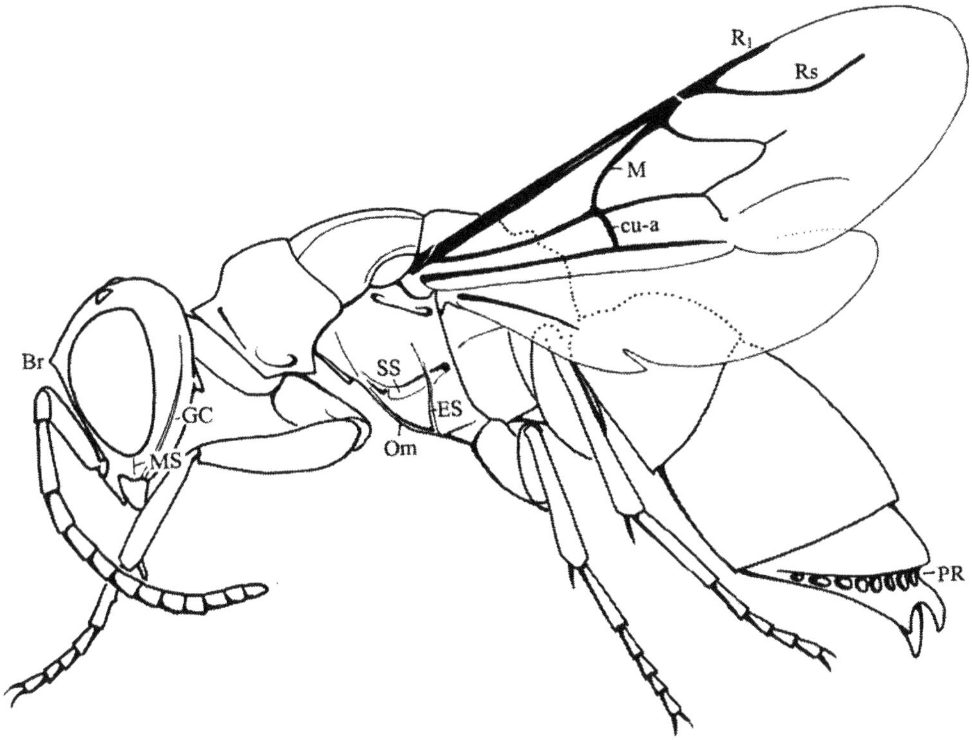

Fig. 1. Lateral view of *Chrysis apontis*, female. Abbreviations: Br = brow, cu-a = cubital-anal cross vein, ES = episternal sulcus, GC = genal carina, M = medial vein, MS = malar space, Om = omaulus, PR = pit row, R_1 = first radial vein, Rs = radial sector, SS = scrobal sulcus.

Male genitalia are used in species distinctions in many groups. However, for the purposes of this study descriptions of genitalic features are minimized to make the keys and descriptions more accessible, without requiring examination of the genitalia, which need to be slide mounted for study. Genitalic features are not needed to identify most California species.

It is often important to distinguish males from females in this family as the sexes sometimes differ in diagnostic features. Sex determination is not always easy in this group, with some obvious exceptions. In the Parnopini and Cleptinae males have four visible metasomal segments and females only three. Female *Elampus* can be recognized by the presence of a regular dense fringe of short setae along the genal margin. The remaining genera are more problematic. Females sometimes have the narrow tip of the ovipositor tube protruding beyond metasomal sternum III. However, the male genital capsule when it protrudes slightly can be easily confused with the ovipositor tube. Male *Chrysura* and *Ceratochrysis* sometimes have metasomal sternum VIII protruding slightly beyond VII. In the remainder of Chrysidines females generally (but not always) have metasomal sternum III convex medially or with a roughly subtriangular medial swelling. In males sternum III is always flat. In *Omalus*, *Philoctetes*, *Pseudomalus* and *Diplorrhos* the only reliable way to determine the sex of a specimen is to pull the genitalia. In any case, the most reliable albeit time-consuming way to determine the sex of a specimen is to extrude the genitalia.

The following are abbreviations used for the repositories for primary types and for specimen material used in this study. Curators and other individuals who loaned specimens for this study are indicated below at the end of each institutional name.

ANSP Department of Entomology, Academy of Natural Sciences, ANSP, Pennsylvania, USA – D. Azuma.

BMNH Department of Entomology, The Natural History Museum, BMNH, England – S. Lewis.

CAS Department of Entomology, California Academy of Sciences, San Francisco, USA – N. Penny, W. Pulawski

CDFA Plant Protection and Pest Management Division, California Department of Food and Agriculture, Sacramento – R. Gill, T. Tyler

CNC Hymenoptera Section, Canadian National Collection, Agriculture Canada, Ottawa – L. Masner.

CU Department of Entomology, Cornell University, Ithaca, New York, USA – J. Liebherr.

ISZP Institute of Systematic Zoology, Polish Academy of Sciences, Krakow – M. Dylewska, J. Razowski.

LACM Natural History Museum of Los Angeles County, Los Angeles, California, USA – R. R. Snelling.

LU Laval University, Quebec City, Quebec, Canada – J. M. Perron.

MCSN Museo Civico di Storia Naturale, Genoa, Italy – R. Poggi.

MCZ Museum of Comparative Zoology, Harvard University, Cambridge, Massachusetts, USA – S. Shaw, P. Perkins.

MHNG	Museum of Natural History, Geneva, Switzerland – C. Besuchet.
MNHN	Muséum National d'Histoire Naturelle, Laboratoire d'Entomologie, Paris, France – J. C. Weulersse.
MNHU	Zoologisches Museum, Humboldt-Universität, Berlin, Germany – F. Koch.
MZLU	Zoologiska Institutionen, Zoologiska Museet, Lund, Sweden – R. Danielsson.
NHMV	Naturhistorisches Museum, Vienna, Austria – M. Fischer.
NML	Natur-Museum, Luzern, Switzerland – L. Reser.
RNH	Rijksmuseum van Natuurlijke Historie, Leiden, Netherlands – C. van Achterberg.
SMEK	Snow Museum of Entomology, University of Kansas, SMEK, USA – R. W. Brooks.
TMB	Zoological Department, Hungarian Natural History Museum, Budapest – J. Papp, L. Zombori.
UCDC	Bohart Museum of Entomology, University of California, BME, USA – S. L. Heydon.
UCR	Entomology Research Museum, University of California, Riverside – S. Frommer, D. Yanega.
USNM	U. S. National Museum of Natural History, Washington, D. C., USA – K. V. Krombein.
USU	Department of Entomology, Utah State University, Logan, USA – T. Griswold.
WSU	Department of Entomology, Washington State University, Pullman, USA – R. S. Zack.
ZMUC	Zoologisk Museum, Universitetsparken, Copenhagen, Denmark – R. Meier.

SYSTEMATICS

Key to the Subfamilies and Tribes in the Family Chrysididae

1 Metasoma with five segments in males and four in females; sternum strongly convex (Fig. 2); face above antennal sockets convex, without indication of scapal basin (as in Fig. 8); pronotum narrowed submedially and transected by transverse groove (as in Figs. 9, 10)..Subfamily Cleptinae

- Metasoma with four or fewer segments in males and three in females; sternum flat or concave; face above antennal sockets flat or concave, usually with well-developed scapal basin; pronotum not narrowed submedially and without transverse groove, instead with narrow anterior collar followed by subquadrate disk; Subfamily Chrysidinae..2

2 Tongue elongate and tubular, as long as head or longer (Fig. 7); metasoma with four external segments in male and three in female, apical tergum apically broadly V-shaped and denticulate, with two large lateral subapical foveae (Fig. 97)..Tribe Parnopini

- Tongue not elongate nor tubular, fitting closely in oral fossa; metasoma with three external segments in both sexes (as in Figs. 1, 5); apical tergum apically rounded, medially notched or with one to six well-developed teeth or lobes (as in Figs. 54-81), without two large lateral subapical foveae...3

3 Tarsal claws edentate; apical tergum with subapical pit row (sometimes obsolescent); mesopleuron with horizontal scrobal sulcus; occiput with transverse welt or carina above foramen, often ending in one small tooth ....Tribe Chrysididini

- Tarsal claws with one or more subsidiary teeth (edentate in *Microchridium* and *Xerochrum*) (as in Figs. 29-37); mesopleuron with oblique scrobal sulcus; apical tergum without subapical pit row; occiput without welt or carina above foramen...Tribe Elampini

SUBFAMILY CLEPTINAE

Genus *Cleptes* Latreille

Generic Diagnosis. Head as wide as long or longer; eyes small, following head contour, not bulging; midocellus equal to or smaller than antennal socket; malar space usually more than 1 MOD long; mandibles with three or more apical teeth; pronotum saddled in profile, with transverse submedial impression; mesopleuron with scrobe, scrobal sulcus and omaulus present in some species; metanotum weakly obtuse or flat in profile; propodeal tooth triangular and generally shorter than broad; tarsal claws with one perpendicular submedial tooth; forewing discoidal cell indicated by stained vein remnants.

Hosts. These wasps are parasites of tenthredinid and diprionid sawflies.

Discussion. The subfamily contains two genera, *Cleptes* and *Cleptidea*; both occur in the New World but only *Cleptes* is found in North America. Kimsey (1981) reviewed the Western Hemisphere species of *Cleptes*.

Key to the California Species of *Cleptes*

1 Metasoma concolorous green, blue, purple, black or dark brown.....................2

\- Metasoma bicolored, two or more terga contrastingly reddish orange to yellow....4

2 Head, thorax and metasoma concolorous, entirely purple to blue; malar space shorter or equal to 1 MOD..*purpuratus* Cresson

\- Head and thorax concolorous metallic green or blue, metasoma bicolored with basal segments black or dark brown, with some metallic highlights laterally, remaining segments green or blue; malar space longer than 1 MOD...............3

3 Both sexes with head and thorax blue, blue-green to purple, or at least head and scutum purple..*alienus* Patton

\- Male head and thorax bright green to bronzy, female head and thorax bronzy to coppery, with green and gold tints......................................*speciosus* Aaron

4 Pronotum without two large subapical pits or subapical pit row (as in Fig.9); head and thorax dark greenish blue to purple, often becoming blackish dorsally in females..*blaisdelli* Bridwell

\- Pronotum with two large subapical pits or subapical pit row (as in Fig. 10); head and thorax bluish green, green or bronze..5

5 Forewing with anterior veinlet (towards costal margin) of discoidal cell faintly indicated or absent; femora red to yellow............................*rufifemur* Kimsey

- Forewing with anterior veinlet of discoidal cell strongly indicated; femora blue or green..6

6 Malar space long, 2 MOD long; femora and coxae dark brown, without metallic highlights...*humboldti* Moczár

- Malar space short, 1.0-1.5 MOD long; femora and coxae metallic green to blue..*rufigaster* Kimsey

***Cleptes alienus* Patton**

Map 1

Cleptes alienus Patton 1879:66. Holotype ♀; USA: Wyoming (lost?).

Cleptes americanus Provancher 1881:304. Holotype ♀; Canada: LU; (lost, invalidly labeled male type, LU; invalid female lectotype designated by Móczár 1962, TMB). Nec Cresson 1879.

Cleptes provancheri Aaron 1885:212. Replacement name for *americana* Provancher 1881.

Geographic Range (Map 1). Western USA, Florida and Georgia, and Alberta, Canada; 81 specimens were studied.

California Records. Collected May through August; **Alpine Co**.: Hope Valley; Woodfords; Ebbetts Pass; **Amador Co**.: 3 km n Silver Lake; **El Dorado Co**.: Echo Lake; Greenwood; Pine Hill near Rescue; **Kern Co**.: Mill Potrero; **Lake Co**.: Upper Lake; **Lassen Co**.: Hallelujah Junction; **Modoc Co**.: 24 mi. ssw Alturas; Cedar Pass campground; Adin Pass; **Nevada Co**.: Sagehen Creek; Lang Crossing; Hobart Mills; **Plumas Co**.: Aden Pass; 1 mi. w Greenville; Little Long Valley 6 mi. e Spring Garden; Spencer Lake; Nelson Creek Rd., Plumas/Sierra Co. line; **Shasta Co**.: Shingletown; Upper Lake; **Sierra Co**.: Sierraville; Sierra Valley; Gold Lake; Weber Lake; Sierra City; **Siskiyou Co**.: Horsethief Creek; Little Grayback; **Tulare Co**.: Ash Mt.; **Tuolumne Co**.: Strawberry; Dardanelles; Leland Meadow; Yosemite National Park; **Yolo Co**.: Davis; Rumsey.

Hosts. Unknown.

Discussion. Males are similar in appearance to those of *speciosus* and there are apparently no structural differences between males of these two species. However, female *alienus* consistently differ in color from those of *speciosus*; they are blue to purple, and those of *speciosus* are green to brassy green. The head and scutum in both sexes of *alienus* are blue to purple.

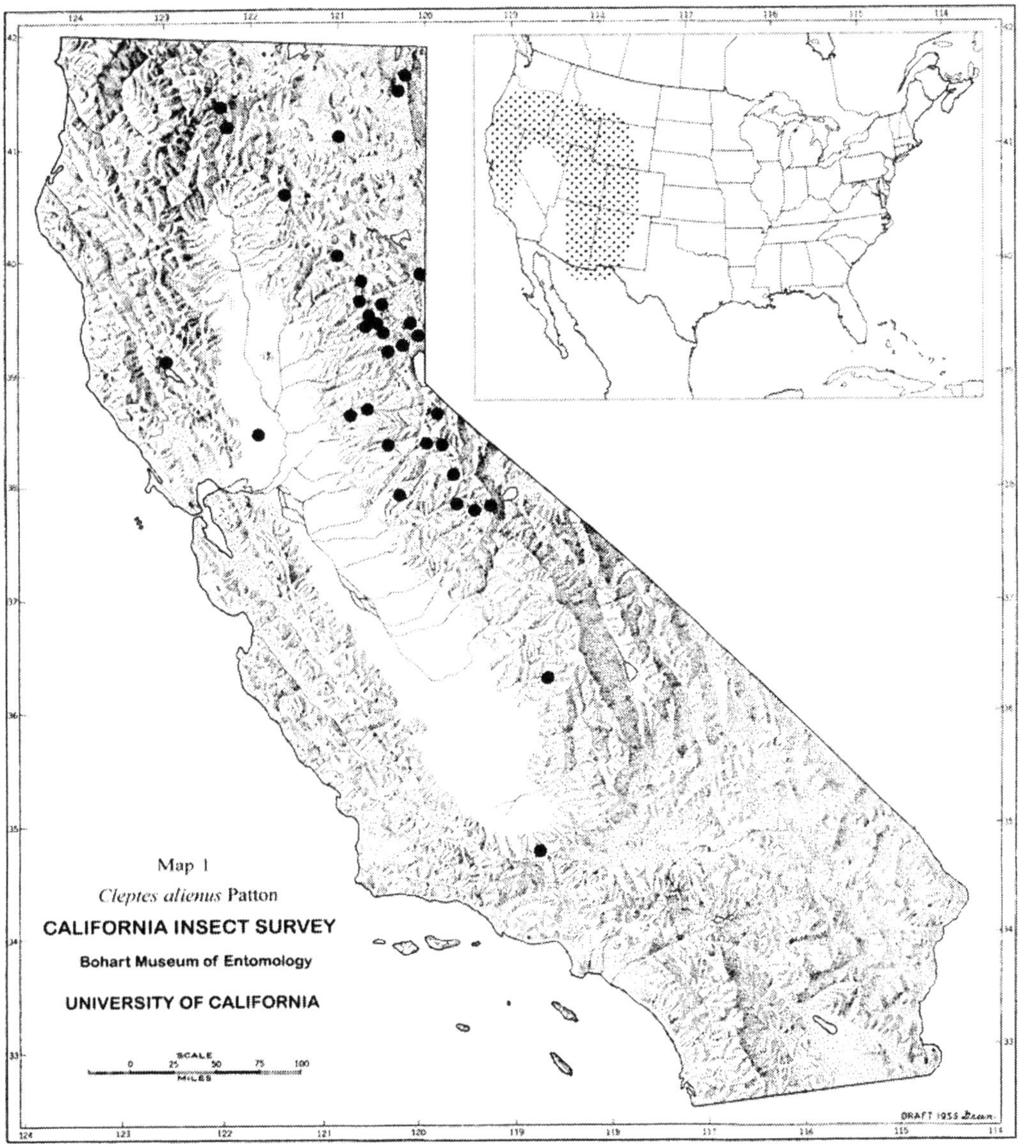

Map. 1. California distribution of *Cleptes alienus*. Inset: overall distribution.

Cleptes blaisdelli Bridwell

Fig. 9, Map 2

Cleptes blaisdelli Bridwell 1919:37. Holotype ♂; USA: California, Poway (USNM).

Geographic Range (Map 2). Southern California to northern Mexico (Sonora), in the Upper and Lower Sonoran Zones; 9 specimens were seen.

California Records. Collected March through June; **Riverside Co**.: Box Springs Mts.; Riverside; Herkey Creek; San Jacinto Mts.; Keen Camp; **Santa Barbara Co**.: Santa Inez; **San Bernardino Co**.: 7mi e Barton Flat; **San Diego Co**.: Borrego Valley; **Tulare Co**.: Woodlake.

Hosts. Unknown.

Discussion. The coloration of *blaisdelli* is similar to that of *rufifemur, humboldti* and *rufigaster,* but *blaisdelli* can be immediately distinguished by the lack of pits or groove along the posterior margin of the pronotum. In addition, the forewing discoidal cell is obsolescent and male femora are blue in this species.

Cleptes humboldti Moczár

Cleptes humboldti Moczár 1996:158. Holotype ♂; USA: California, Humboldt Co., Blair's Ranch, Redwood Creek (USNM).

Geographic Range. Northern California; one paratype was seen.

California Records. Collected in June; **Humboldt Co**.: Blair's Ranch; Redwood Creek.

Hosts. Unknown.

Discussion. *Cleptes humboldti* most closely resembles *rufigaster*, differing primarily in the length of the malar space, making the face quite long compared to *rufigaster*, and the non-metallic coxae and femora. This species is only known from two males and one female from the type locality.

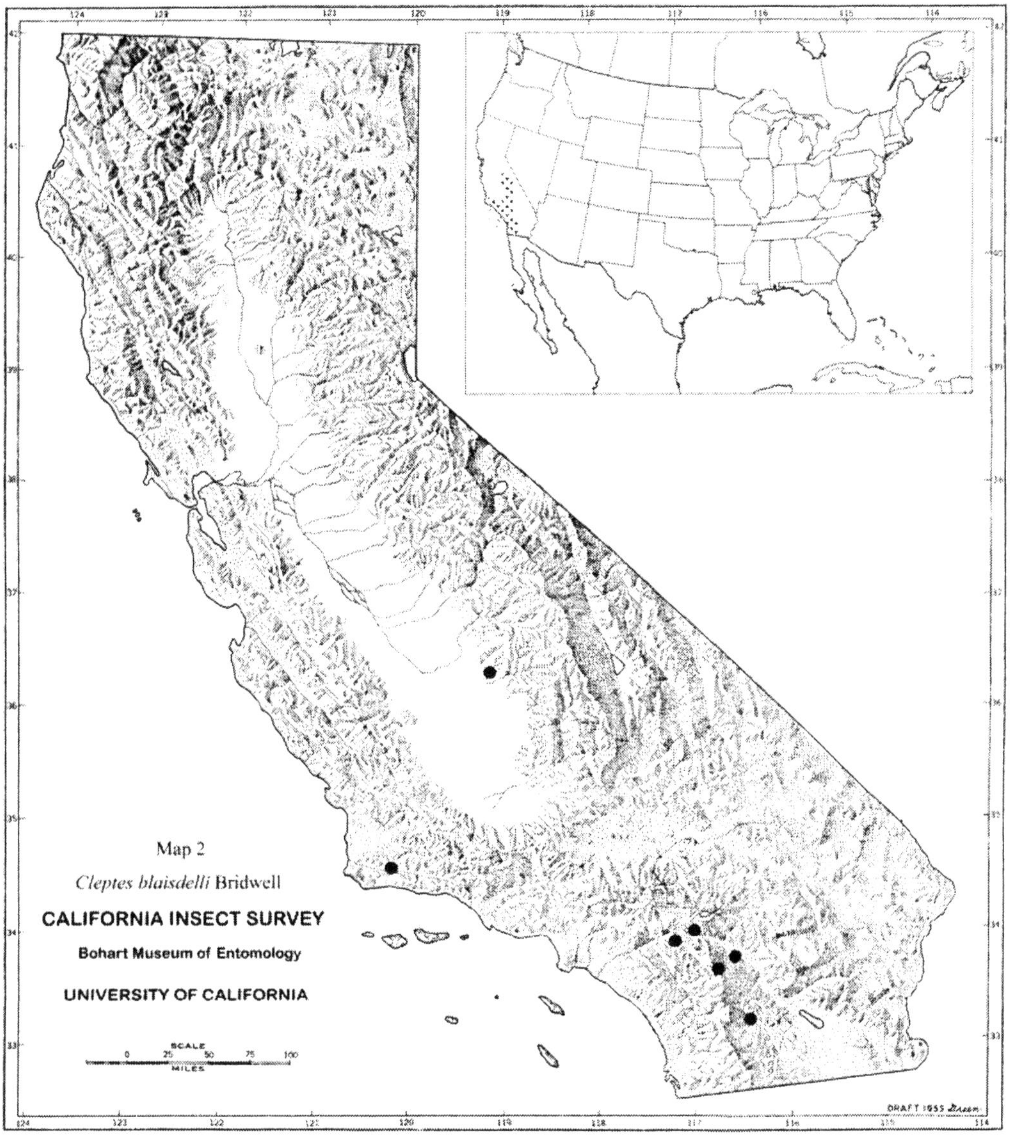

Map. 2. California distribution of *Cleptes blaisdelli*. Inset: overall distribution.

Cleptes purpuratus Cresson

Fig. 8, Map 3

Cleptes purpuratus Cresson 1879:x. Holotype ♀; Canada: British Columbia, Vancouver (ANSP).

Cleptes americanus Cresson 1879:x. Holotype ♀; USA: Nevada (ANSP).

Cleptes insperatus Aaron 1885:212. Holotype ♀; USA: Montana (ANSP). Synonymized by Kimsey 1991 (in Kimsey and Bohart 1991).

Geographic Range (Map 3). Western USA and Florida; an abundant Sierran species; 392 specimens were studied.

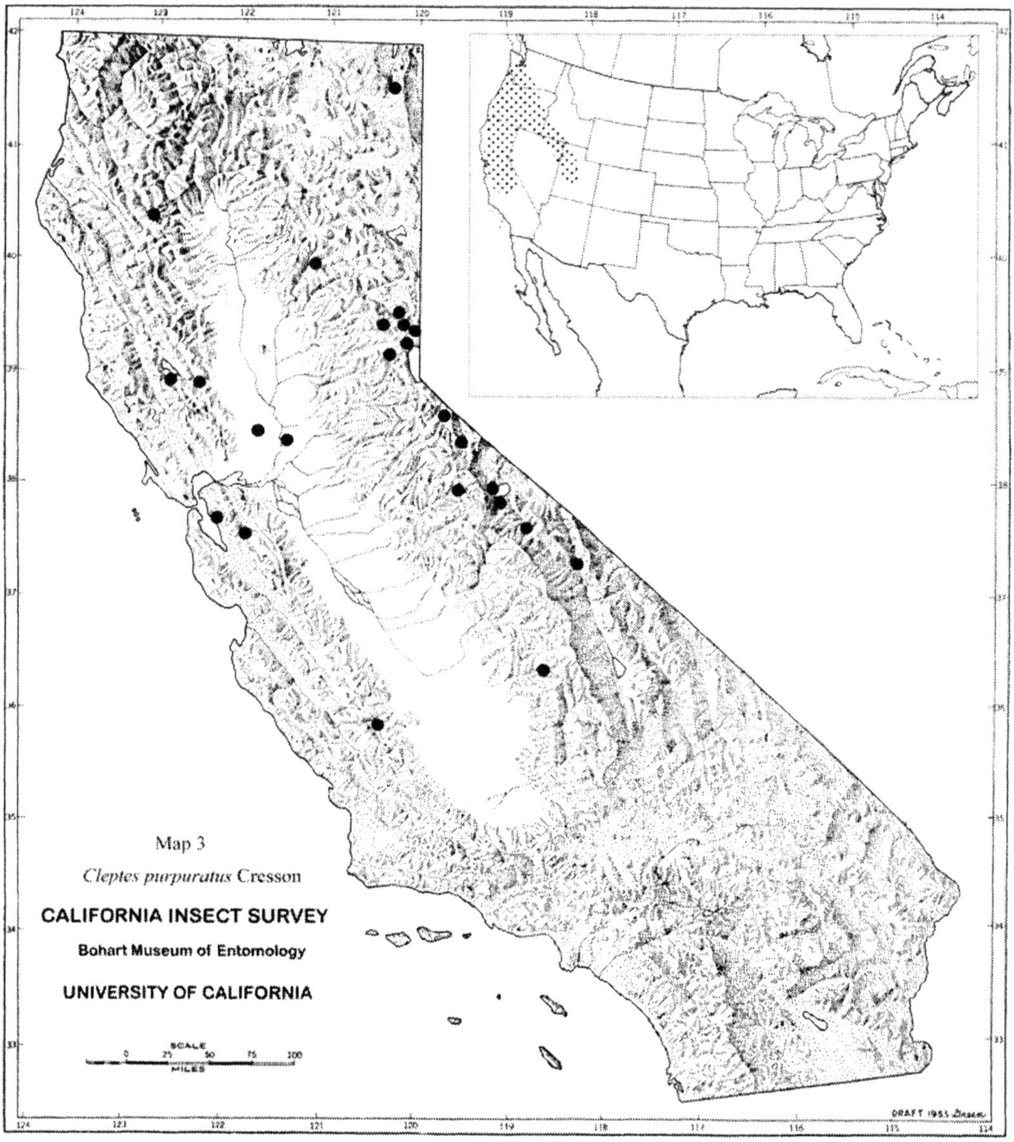

Map 3. California distribution of *Cleptes purpuratus*. Inset: overall distribution.

California Records. Collected April through September; **Alameda Alameda Co.**: Alameda; 1 mi. e Mission Peak; **Alpine Co.**: Hope Valley; **El Dorado Co.**: Carnelian Bay and Pope Beach, Lake Tahoe; **Inyo Co.**: 20 mi. n Bishop; **Lake Co.**: north fork

Cache Creek; Kelseyville; **Modoc Co.**: 24 mi. sw Alturas; **Mono Co.**: Leavitt Meadows; **Monterey Co.**: Arroyo Seco; **Nevada Co.**: Sagehen Creek; Boca; Hobart Mills; **Placer Co.**: Emigrant Gap; **Plumas Co.**: 4 mi. w Quincy; **Sacramento Co.**: Sacramento; **Sierra Co.**: Sattley; Independence Lake; **Siskiyou Co.**: 12 mi. w Coffee Creek; Horsethief Creek; **Trinity Co.**: Trinity River Camp; **Tulare Co.**: Ash Mountain, Sequoia National Park; **Tuolumne Co.**: Leland Meadows; Sonora Pass; **Yolo Co.**: Davis.

Hosts. Dahlsten (1961, 1967) reared *purpuratus* from *Neodiprion* sp. (Diprionidae).

Discussion. The most distinctive feature of *purpuratus* is the all blue to purple head, thorax and metasoma. As with *blaisdelli*, the pronotum lacks pits or a groove along the posterior margin. This species can be locally abundant.

Cleptes rufifemur Kimsey

Fig. 10, Map 4

Cleptes rufifemur Kimsey 1981:814. Holotype ♂; USA: California, Yolo Co., Davis (UCDC).

Geographic Range (Map 4). Central California to northwestern Mexico; 22 specimens were studied.

California Records. Collected April through August; **Alameda Co.**: Berkeley; Oakland; Albany; **Lake Co.**: Borax Lake; **Napa Co.**: Monticello Dam; **Santa Barbara Co.**: Santa Cruz Island; **Yolo Co.**: Davis; Winters.

Hosts. Unknown.

Discussion. Diagnostic features of *rufifemur* include red legs in both sexes and the obsolescent forewing discoidal cell.

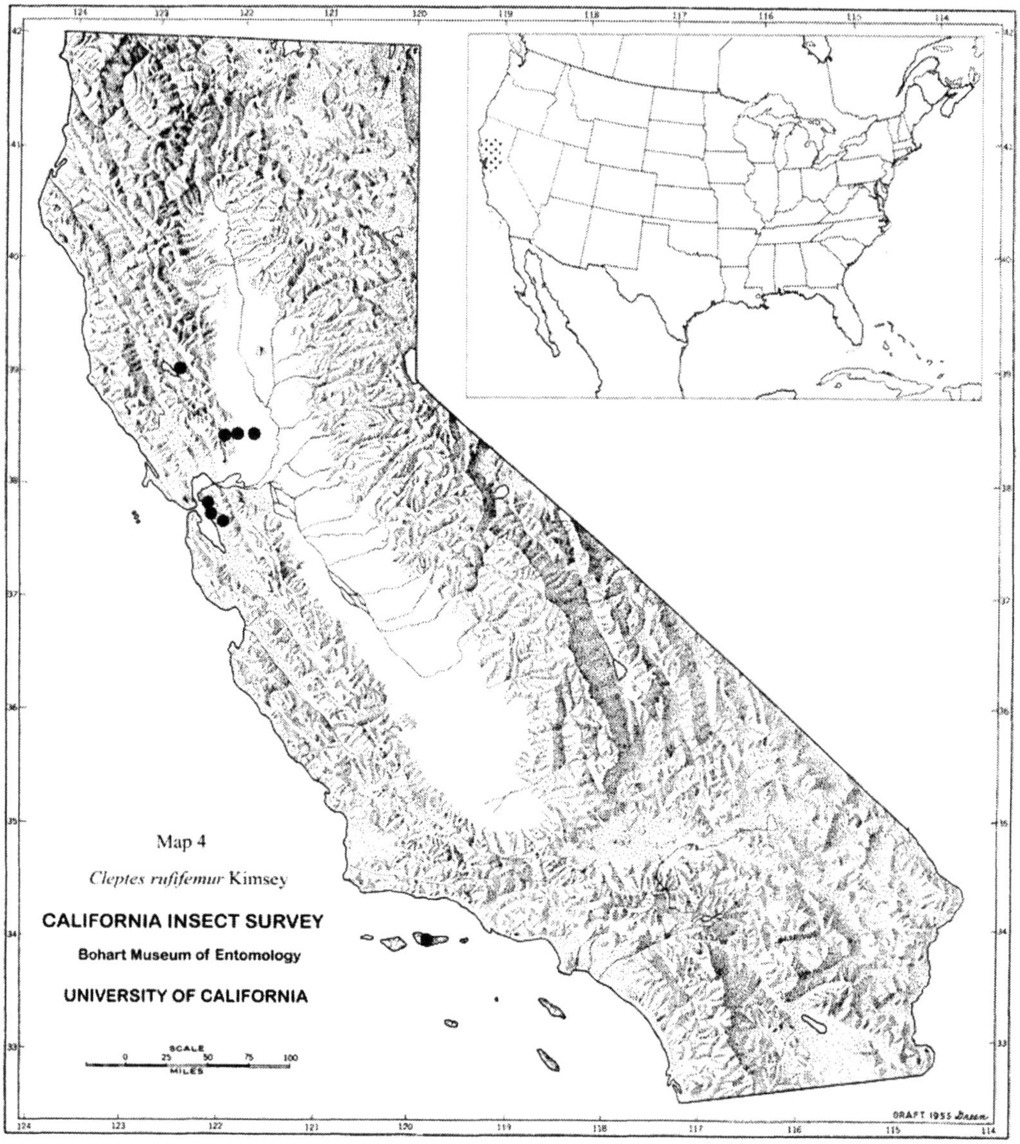

Map 4. California distribution of *Cleptes rufifemur*. Inset: overall distribution.

Cleptes rufigaster Kimsey

Map 5

Cleptes rufigaster Kimsey 1981:814. Holotype ♂; USA: California: Placer Co., Rocklin (UCDC).

Geographic Range (Map 5). This species is endemic to central and southern California; occurring in the Upper Sonoran Zone; 7 specimens were examined.

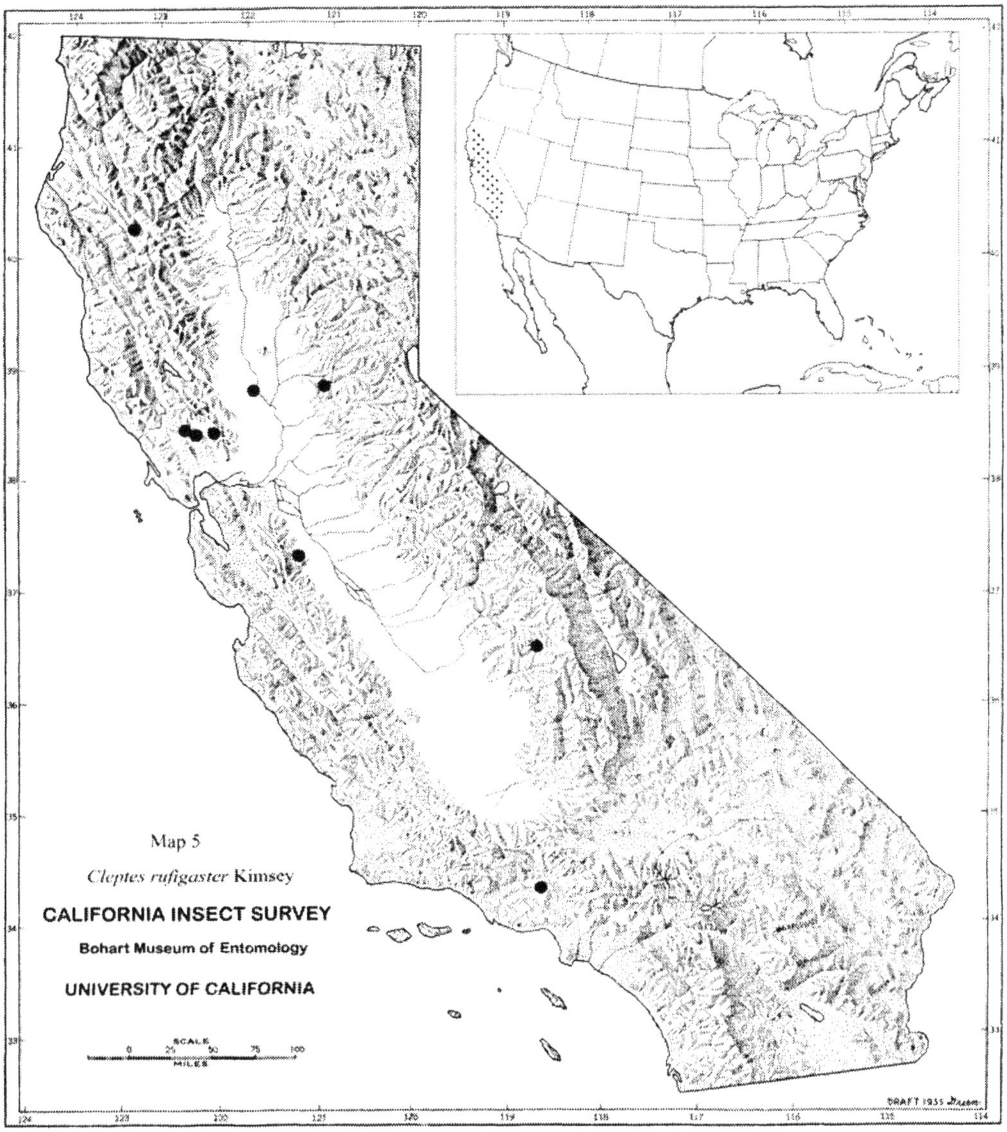

Map 5. California distribution of *Cleptes rufigaster*. Inset: overall distribution.

California Records. Collected April through August; **Los Angeles Co.**: Tanbark Flat; **Mendocino Co.**: Hopland Field Station; **Monterey Co.**: Paraiso Springs; **Napa Co.**: Samuel Springs (beneath Lake Berryessa); Mt. St. Helena; **Placer Co.**: 6 mi. ne Auburn; 4 mi. s Rocklin; **Riverside Co.**: Palm Springs; **Stanislaus Co.**: Del Puerto

Canyon; **Trinity Co.**: Island Mountain; **Tulare Co.**: Ash Mountain; **Yolo Co.**: Knights Landing; Putah Canyon; Davis.

Hosts. Unknown.

Discussion. This species is similar to *rufifemur* but can be distinguished by having blue to black femora, and the anterior veins of the discoidal cell clearly indicated.

Cleptes speciosus Aaron

Fig. 2; Map 6

Cleptes speciosus Aaron 1885:212. Lectotype ♀ (designated by Cresson 1928); USA: Montana (ANSP).

Geographic Range (Map 6). USA and Canada as far east as North and South Dakota, and one record from New York; Alberta, British Columbia and Saskatchewan, and Baja California, Mexico; in transition and Canadian Zones; 231 specimens were seen.

California Records. Collected April through September. **Alameda Co.**: **Alpine Co.**: 2 mi. n Blue Lakes; Hope Valley; **Lake Co.**: Upper Lake; **Lassen Co.**: Lassen National Park; **Los Angeles Co.**: Tanbark Flat; Claremont; Griffith Park; **Modoc Co.**: Cedarville; **Mono Co.**: Tioga Pass; **Monterey Co.**: Bryson; Arroyo Seco; **Napa Co.**: 14 mi. w Monticello Dam; **Nevada Co.**: Sagehen Creek; Grass Valley; **Orange Co.**: Fullerton; **San Luis Obispo Co.**: Atascadero; **Santa Barbara Co.**: Santa Cruz Island; **Santa Clara Co.**: Stanford University; **Shasta Co.**: Redding; **Sierra Co.**: Yuba Pass; **Tehama Co.**: Redding; Lassen Volcanic National Park; **Tulare Co.**: Mineral King; **Tuolumne Co.**: 4 mi. e Sonora; Leland Meadows; **Yolo Co.**: 5 mi. e Guinda; Davis.

Hosts. R. Smith (1962) reared *speciosus* (identified as *provancheri*) from *Neodiprion* sp.

Discussion. This species is close to *purpuratus*, both species have a weakly sculptured pronotum. *C. speciosus* can be distinguished from *purpuratus* by the bronzy coloration and from *alienus* by the greener head and thorax.

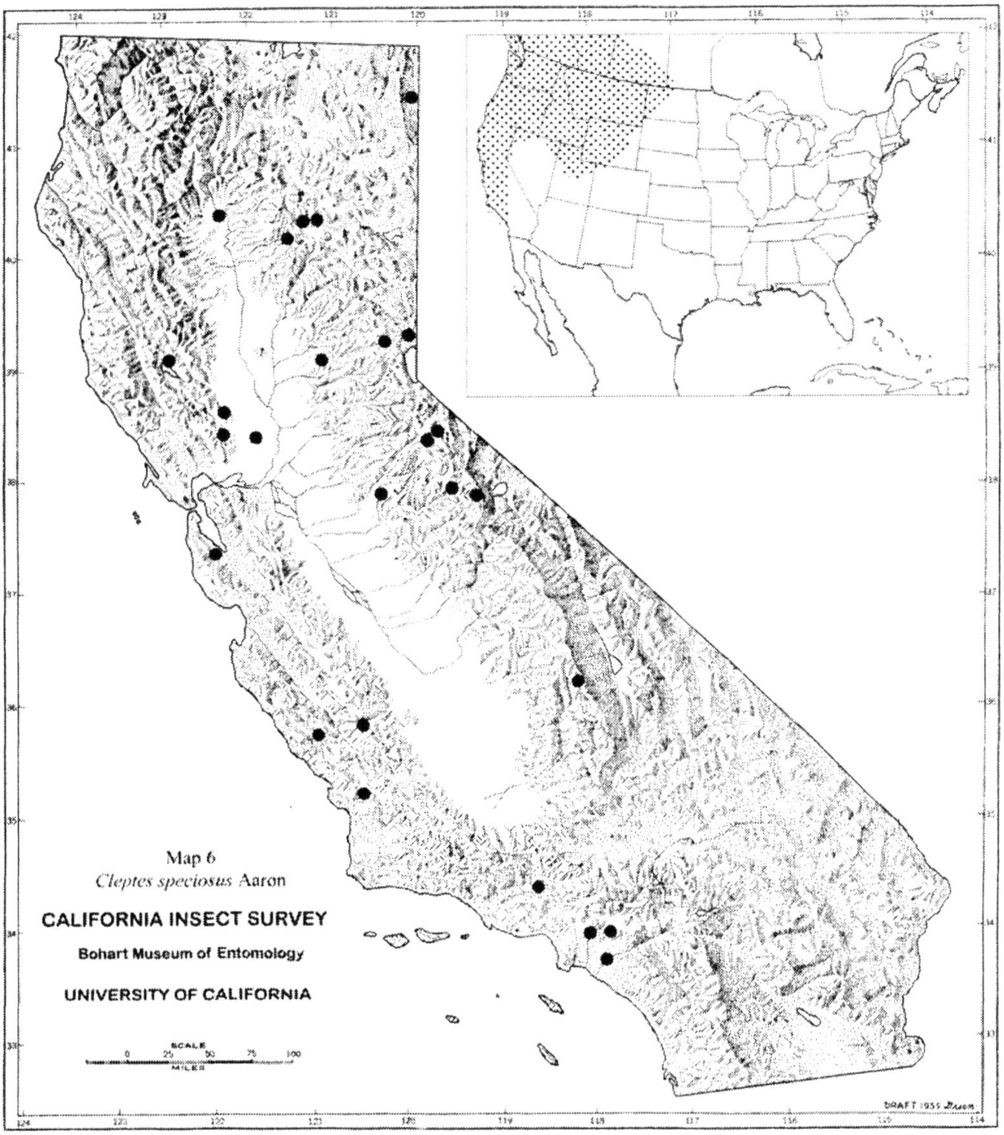

Map 6. California distribution of *Cleptes speciosus*. Inset: overall distribution.

SUBFAMILY CHRYSIDINAE

This is the largest subfamily in the Chrysididae, containing about 3,000 species worldwide. Chrysidines are characterized by having only three metasomal terga visible externally in both sexes (except male parnopines have four), the terga are strongly convex, laterotergites fold sharply ventrally, the venter is flat to concave, and the sterna are not usually visible in side view. Members of this group are parasites of crabronid wasps, bees and eumenine vespids.

Tribe Elampini

Elampini is one of three tribes in the Chrysidinae, which also includes the Chrysidini, Kimseyini and Parnopini. The monotypic tribe Kimseyini is found only in Central Asia. Elampines are characterized by having tarsal claws with one or more subsidiary teeth (except *Xerochrum*), the forewing Rs stub is sclerotized for less than one-half of the marginal cell length, no discoidal cell, and tergum III without subapical grooves or pit row.

Key to the Genera of Elampini

1 Hindtarsal claw edentate (Fig. 35), metasoma non-metallic red...*Xerochrum* Bohart

- Hindtarsal claw dentate with 1-5 subsidiary teeth (as in Figs. 29-34, 34, 37), metasoma brightly metallic-colored in greens, blues, purples etc., or non-metallic black..2

2 Forewing Rs vein reduced to stub, less than one-fifth length of medial vein (as in Fig. 25)..3

- Forewing Rs vein one-half or more as long as medial vein (as in Fig. 24)...........4

3 Forewing venation restricted to basal third of wing (Fig. 25); pronotum without lateral carina; face evenly roughened and somewhat striate, without discrete scapal basin (Fig. 11)..*Microchridium* Bohart

- Forewing venation extending half or more of wing length (Fig. 26); pronotum with lateral carina; face smooth with scattered punctures, scapal basin clearly indicated, polished and impunctate...............................*Minymischa* Kimsey

4 Hindtarsal claw with one perpendicular, submedial tooth (Fig. 31); face flat or slightly concave medially, with narrow to large zone of fine, dense cross-ridging in scapal basin (as in Fig. 17)...........................*Hedychridium* Abeille de Perrin

- Hindtarsal claw with single subparallel subsidiary tooth or more than one subsidiary tooth; face usually strongly concave medially, scapal basin smooth and polished, finely punctate or marked by coarse cross-ridging.......................5

5 Scapal basin with dense tiny punctures covered with dense appressed silvery pubescence; tarsal claw with one subparallel tooth (Fig. 30); forewing medial vein arising at cu-a (Fig. 24)....................................*Hedychreides* Bohart

- Scapal basin either smooth and impunctate or with at least narrow zone of cross-ridging, without appressed silver pubescence; tarsal claw dentition various; forewing medial vein usually arising before or after cu-a............................6

6 Forewing medial vein straight or weakly curved medially, hindtarsal claw with one subsidiary tooth (except in some males) (Figs. 29, 32, 37)..........................7

- Forewing medial vein strongly arched, often forming right angle, hindtarsal claw usually with two or more subsidiary teeth (as in Fig. 36)..............................8

7 Head with sharp tubercle at base of oral fossa; male hindfemur enlarged and non-metallic black or brown; female sternum III with transverse subbasal sulci and often apicomedial lobe or tubercle (Fig. 20); mid and hindtibia usually with discrete pit on inner surface (as in Fig. 23)...........................*Hedychrum* Latreille

- Head without tubercle at base of oral fossa; male hindfemur not enlarged nor non-metallic; female sternum III unmodified; mid and hindtibia without pit on inner surface...*Pseudolopyga* Krombein

8 Forewing medial cell setose; mesopleuron strongly angulate, with scrobal carina, verticaulus and omaulus meeting in ventrally pointing angle (Fig. 3) ....*Holopyga* Dahlbom

- Forewing medial cell asetose; mesopleuron usually not strongly angulate and scrobal carina, verticaulus and omaulus not meeting in ventrally pointing angle 9

9 Metanotum medially mucronate, flat dorsally (Fig. 4); scutum coarsely and irregularly punctate; tergum III with produced and snout-like apically (as in Figs. 4, 38-40), scapal basin flat or shallowly concave, with U-shaped cross-ridges or wrinkles...*Elampus* Spinola

- Metanotum rounded or conical; tergum III usually without truncate snout; scutum with sparse punctures clumped along or between notauli, or entirely absent; scapal basin usually deeply concave and smooth....................................10

10 Scutum medially impunctate or with small, evenly scattered punctures (Figs. 13, 15); mesopleuron with scrobal sulcus foreshortened; omaulus elongate and extending nearly horizontally to scobal sulcus; tergum III usually with transparent rim and usually without apicomedial notch..............*Omalus* Panzer

- Scutum with large punctures, particularly between or along notauli (as in Figs. 12, 16); mesopleuron with scrobal sulcus and omaulus subequal in length and extending at an oblique angle ventrally; tergum III usually without transparent rim and often with apicomedial notch (as in Figs. 41-44)............................11

11 Scutum with large punctures clumped along notauli (Fig. 12); malar space not divided by genal carina..................................*Philoctetes* Abeille de Perrin

- Scutum with large, shallow punctures clumped posteriorly between notauli (Fig. 16); malar space divided by genal carina...................*Pseudomalus* Ashmead

Genus *Elampus* Spinola

Generic Diagnosis. Facial setae scattered, erect; scapal basin nearly flat; malar space less than 1 MOD long; head cupped posteriorly, carinate and angulate behind eyes; scutum and scutellum regularly, densely punctate; metanotum with large flat medial projection or mucro; mesopleuron strongly projecting anterolaterally, with well-developed omaulus and scrobal carina forming sharp ventral angle; forefemur with ventral carina and often subbasally angulate; tarsal claws with 1-5 subsidiary teeth; forewing medial vein strongly arched, arising at or slightly before cu-a, stigma short, broad and apically rounded; tergum III produced into apical, more or less membrane-filled snout, lateral margin sinuate; digitus and cuspis long and slender.

Hosts. *Elampus* parasitize ground-nesting Crabronidae, including species of *Mimumesa* and *Hoplisoides*.

Discussion. Unlike most elampines, the sexes of *Elampus* are relatively easy to distinguish. Males have long, irregularly positioned genal bristles, whereas females have a regular, short dense genal fringe. Species are generally distinguished by color and the configuration of the apical metasomal snout. The genus was reviewed by Huber and Pengelly (1978).

Key to the California Species of *Elampus*

1 Tergum III apical truncation round or oval, almost completely filled with membrane (Fig. 40); hindtarsal claw with one or two subsidiary teeth, metanotal projection subtruncate apically.......................................*viridis* Cresson

- Tergum III apical truncation forming half circle, truncation membrane considerably emarginate medially (as in Figs. 38, 39), hindtarsal claw variously dentate, metanotal projection evenly rounded apically.......................................2

2 Wing membrane water clear (untinted), hindtarsal claw with one or two subsidiary teeth...*hyalinus* (Aaron)

- Wing membrane brown-tinted, hindtarsal claw with three or four subsidiary teeth (rarely fewer)..3

3 Thoracic color, particularly propodeum, sharply contrasting with metasomal color; metasomal terga bright green with strong coppery red highlights...*nitidus* (Aaron)

- Thorax and metasoma concolorous, metasomal terga without coppery red highlights..4

4 Tergum III apical truncation plus narrow anterior band bright (usually) brownish red, truncation membrane indented less than half truncation height (Fig. 38) ....*versicolor* Norton

- Tergum III apical truncation brown to black without anterior light-colored band, truncation membrane various..5

5 Tergum III barely drawn out posteriorly, truncation in side view with flat posterior surface forming right angle with longitudinal axis of tergum, truncation forming less than half circle and membrane indented at least half truncation height (Fig. 39)...*viridicyaneus* Norton

- Tergum III drawn out posteriorly, truncation in side view with flat posterior surface projecting more dorsally than ventrally, truncation usually forming slightly less than half circle and membrane indented less than half truncation height...*marginatus* (Patton)

***Elampus hyalinus* (Aaron)**

Fig. 4, Map 7

Notozus hyalinus Aaron 1885:218. Lectotype ♀ (designated by Cresson 1928); USA: Montana (ANSP).

Geographic Range (Map 7). This species is widespread in the USA, southern Canada (Ontario) and northern Mexico (Chihuahua, Jalisco), but is not particularly common in California; 24 specimens were examined.

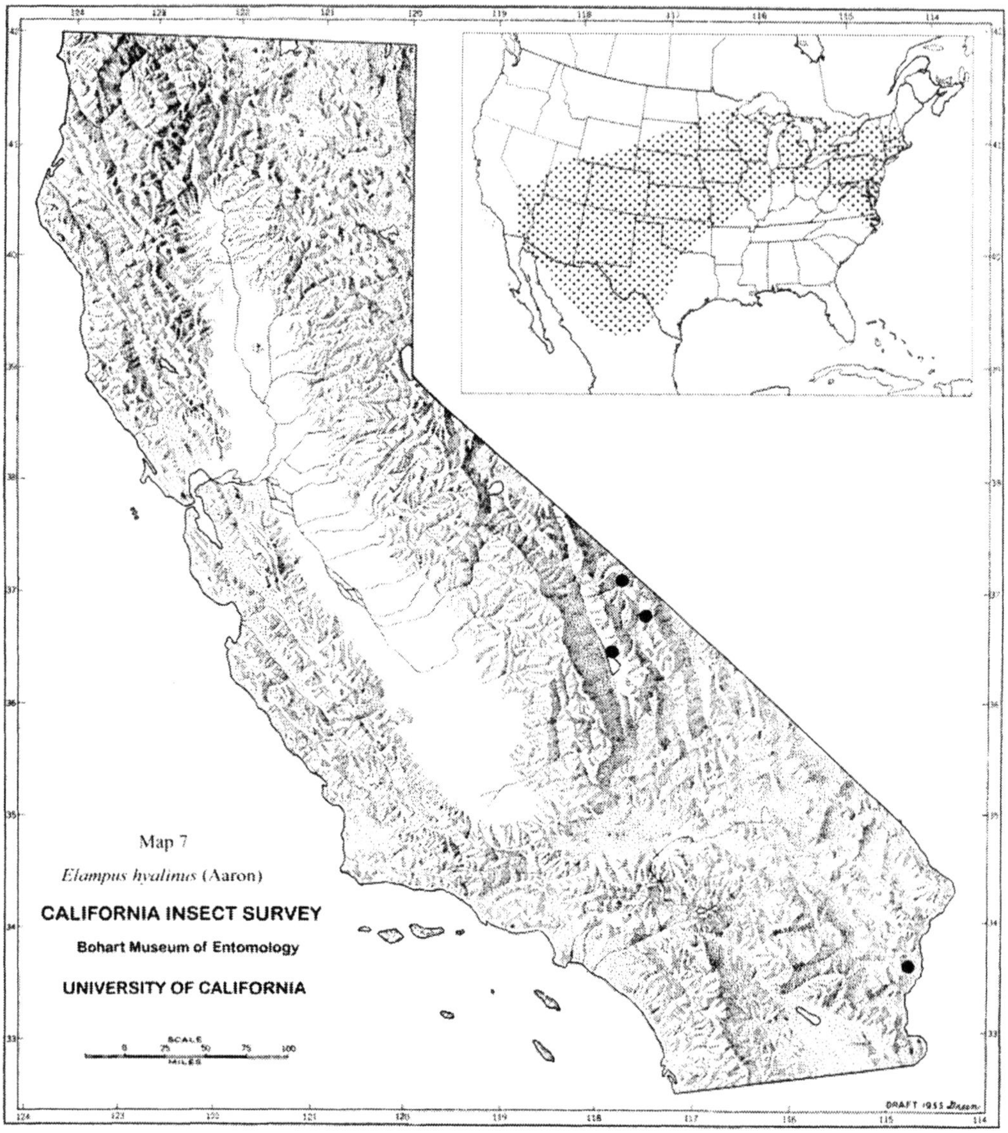

Map 7. California distribution of *Elampus hyalinus*. Inset: overall distribution.

California Records. Collected April through August; **Inyo Co**.: Deep Springs; Owens Lake; Eureka Valley dunes; **Riverside Co**.: 18 mi. w Blythe.

Hosts. Unknown.

Discussion. The most distinctive features of *hyalinus* are the clear, untinted wing membrane, tarsal claws with only one or two subsidiary teeth, and the apical metasomal truncation crescentic and only partly membrane-filled.

Elampus marginatus (Patton)

Map 8

Notozus marginatus Patton 1879:66. Neotype ♀ (designated by Huber and Pengelly 1977); USA: Connecticut, E. Hartford (CORNELL).

Geographic Range (Map 8). Widespread in USA, southern Canada (British Columbia, Manitoba, Saskatchewan, Alberta) and in northern Mexico (Sonora, Chihuahua, Baja California); 322 specimens were studied.

California Records. Collected April through October; **Alpine Co**.: Hope Valley; **Amador Co**.: 6 mi. w Tragedy Springs; Ione; Volcano; **El Dorado Co**.: Meyers; Silver Lake; South Lake. Tahoe; **Imperial Co**.: Seeley; **Inyo Co**.: Deep Springs Lake; **Lassen Co**.: Bridge Creek Camp; Susan River Camp; Spauldings; Hallelujah Junction; **Los Angeles Co**.: Claremont; **Madera Co**.: Big Sandy Flat; **Merced Co**.: Dos Palos; **Modoc Co**.: 5.5 mi. e Cedarville; **Mono Co**.: Cottonwood Creek; White Mountains; **Nevada Co**.: Sagehen Creek; Boca; Hobart Mills; Fuller Lake; Jackson Lake; **Placer Co**.: Brockway Summit; Carnelian Bay, Lake Tahoe; **Plumas Co**.: Blairsden; Bucks Lake; Onion Valley; **Riverside Co**.: Palm Springs; Thousand Palms; **Sacramento Co**.: Carmichael; **Shasta Co**.: Old Station; Moose Camp; **Sierra Co**.: Sierra Valley; Gold Lake; Sattley; Independence Lake; Sierraville; Yuba Pass; **Trinity Co**.: Big Flat, Coffee Creek; Coffee Creek 10 mi. w Trinco; **Tulare Co**.: Woodlake; **Tuolumne Co**.: Kennedy Meadow; Dardanelles; Pinecrest; **Yolo Co**.: Davis.

Hosts. Unknown.

Discussion. Diagnostic features of *marginatus* include the dark color, tarsal claws with three or more subsidiary teeth, and a well-developed, extensively membrane-filled apical truncation, with membrane brown (black in *viridicyaneus*). This species is difficult to separate from *viridicyaneus*; *marginatus* is generally smaller, 3.5-4.0 mm long, versus 4.5-6.0 mm long in *viridicyaneus*.

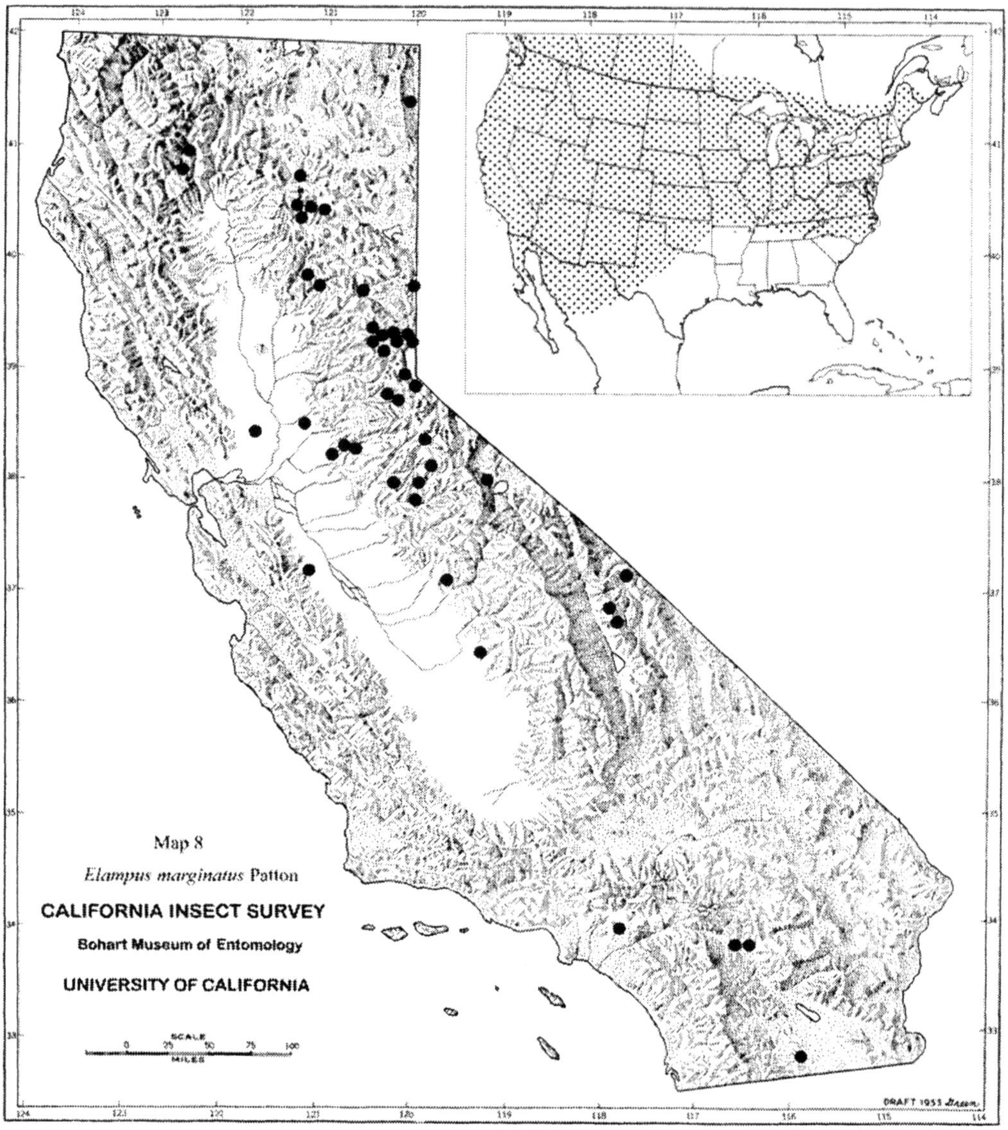

Map 8. California distribution of *Elampus marginatus*. Inset: overall distribution.

Elampus nitidus (Aaron)

Map 9

Notozus nitidus Aaron 1885:218. Lectotype ♂ (designated by Cresson 1928); USA: Montana (ANSP).

Elampus mexicanus Mocsáry 1889:72. Lectotype ♂ (designated by Huber and Pengelly 1977); Mexico: Anganguco (GENEVA).

Elampus nitidus californicus Huber and Pengelly 1978:97 Holotype ♂; USA: California, "Cala" (ANSP).

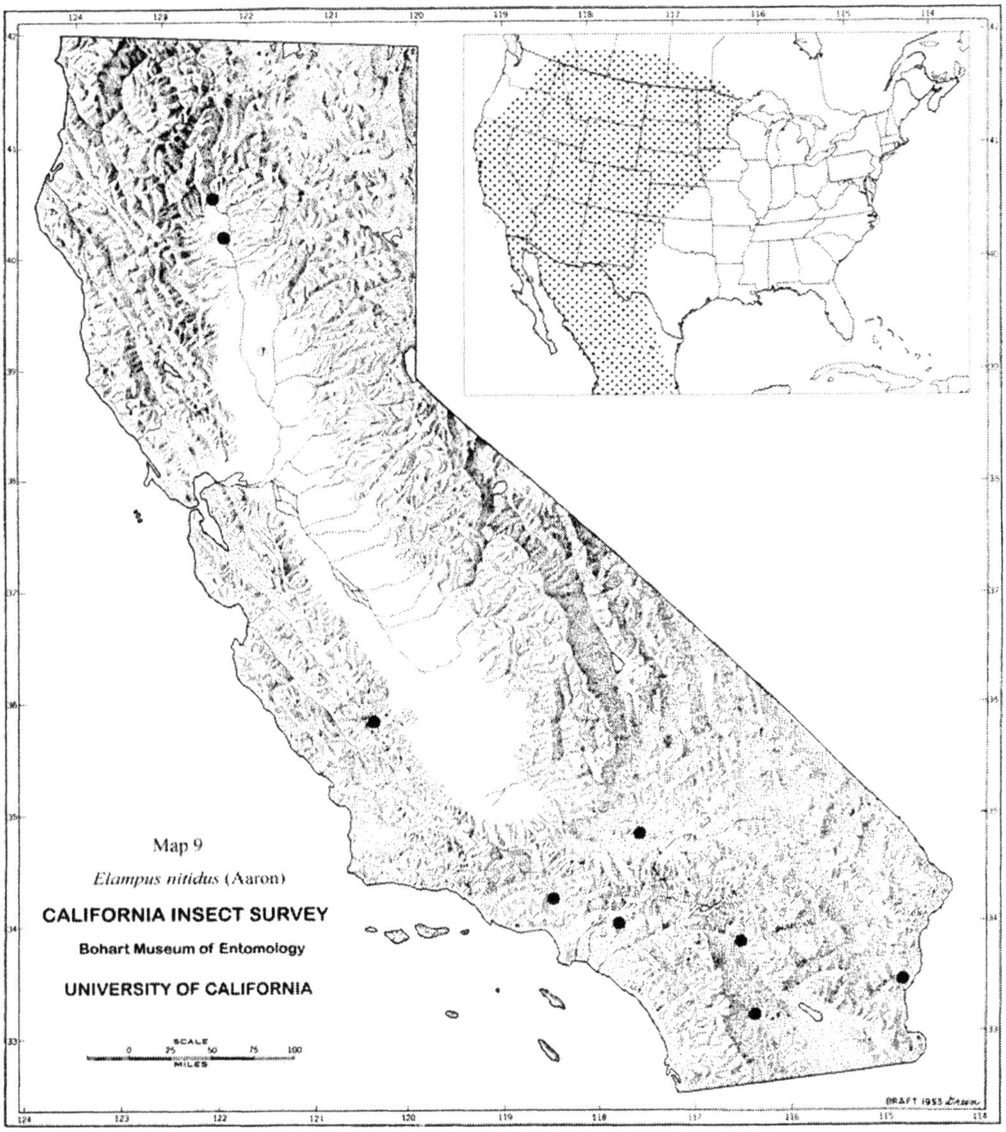

Map 9. California distribution of *Elampus nitidus*. Inset: overall distribution

Geographic Range (Map 9). Widespread in the USA and southern Canada (Alberta, Saskatchewan, Manitoba), reaching as far south as El Salvador; 86 specimens were seen.

California Records. Collected March through August; **Imperial Co**.: Bard; **Los Angeles Co.**: Claremont; Newhall; **Monterey Co**.: Arroyo Seco; **Riverside Co**.: Thousand Palms; **San Bernardino Co**.: 6 mi. nw Adelanto; **San Diego Co**.: Borrego Valley; **Shasta Co**.: Redding; **Tehama Co**.: Red Bluff.

Hosts. Unknown.

Discussion. *E. nitidus* is the most brightly colored of the North American *Elampus* species. It superficially resembles *Omalus telfordi* and *Minymischa deserti*. In all three species the thoracic color is blue to green and strongly contrasts with the coppery red metasoma.

Elampus versicolor Norton

Fig. 38, Map 10

Elampus versicolor Norton 1879:235. Holotype ♂; USA: "Dacota" (ANSP).

Notozus productus Aaron 1885:219. Lectotype ♂ (designated by Cresson 1928); USA: Montana (ANSP). Nec Dahlbom 1845.

Ellampus aaroni Bodenstein 1951:719. Replacement name for *productus* Aaron 1885.

Geographic Range (Map 10). Scattered across the western and midwestern USA and Mexico (Chihuahua, Sonora, Baja California); primarily west of the 100th meridian; 36 specimens were studied.

California Records. Collected April through October; **Imperial Co**.: Bard; Palo Verde; 12 mi. e Herber; Calexico; 3.5 mi. w Gordon's Well; **Plumas Co**.: 4 mi. w Quincy; **Riverside Co**.: Thermal; **San Bernardino Co**.: Mesquite Springs; **Yolo Co**.: Davis.

Hosts. Unknown.

Discussion. Similar to *marginatus*, *versicolor* can be distinguished from that species by the distinctive red-colored truncation membrane and orange tergal ring just anterior to it. The punctation toward the apex of tergum II is fine, but more closely placed than in *marginatus*.

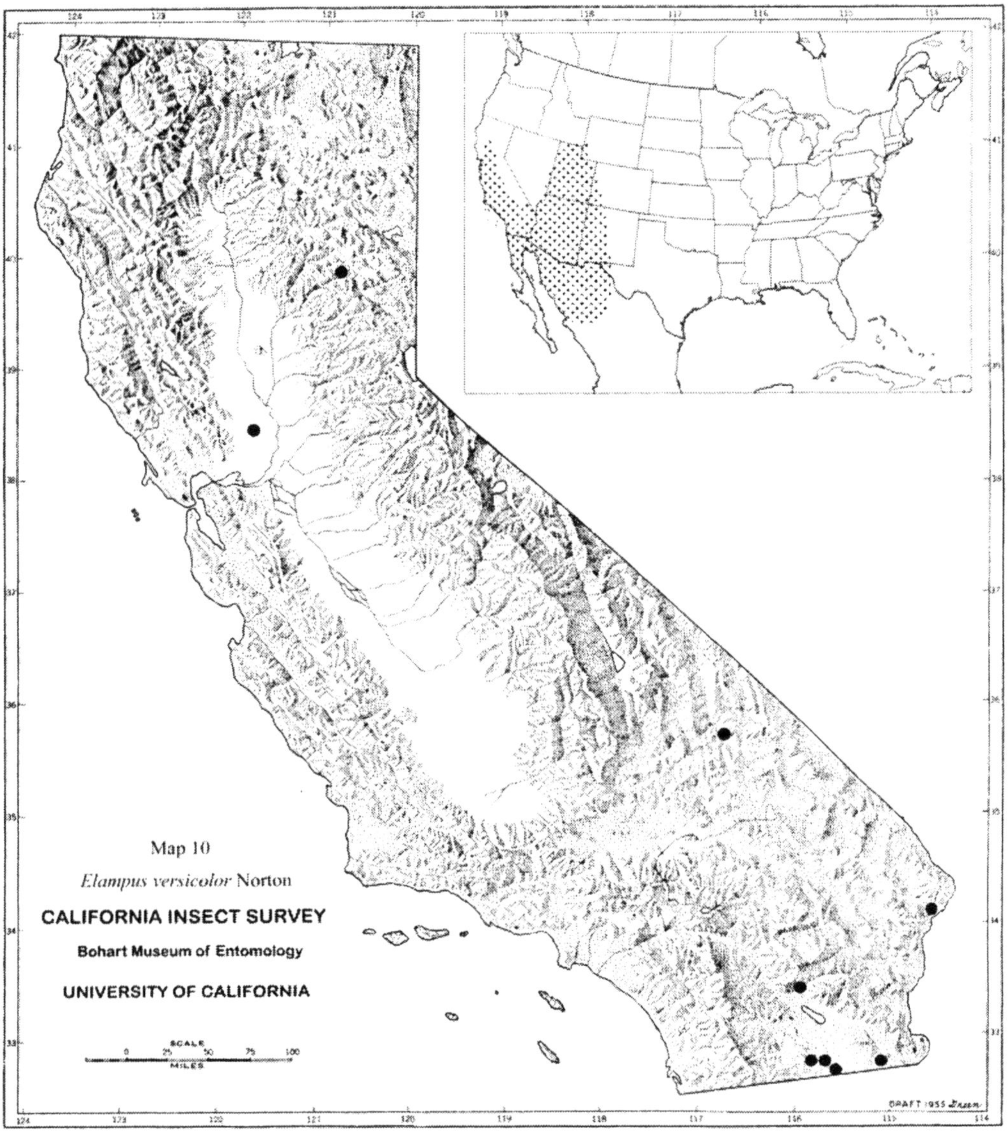

Map 10. California distribution of *Elampus versicolor*. Inset: overall distribution

***Elampus viridicyaneus* Norton**

Fig. 39, Map 11

Elampus viridicyaneus Norton 1879:235. Holotype ♀; USA: Massachusetts (ANSP).

Elampus spinosus Provancher 1881:302. Holotype ♀; Canada: Quebec (LU).

Notozus connexus Viereck 1906:192. Holotype ♂; USA: Kansas, Clark Co. (SMEK).

Geographic Range (Map 11). Widespread in the USA and in southern Canada (Alberta, Ontario, Saskatchewan); 162 specimens were studied.

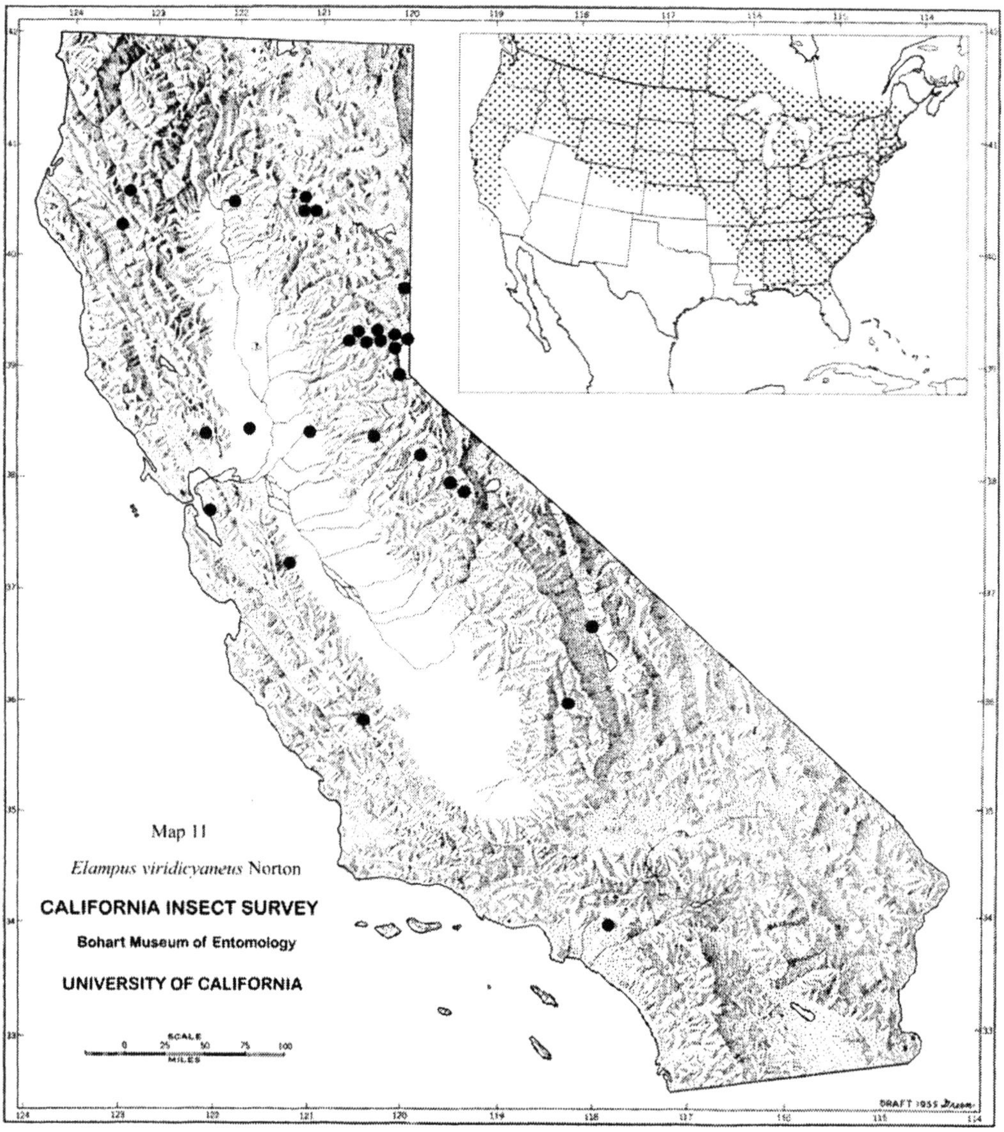

Map 11. California distribution of *Elampus viridicyaneus*. Inset: overall distribution.

California Records. Collected May through October; **Alameda Co**.: Berkeley; **Amador Co**.: 6 mi. w Tragedy Springs; **Inyo Co**.: Lone Pine; **Lassen Co**.: Hallelujah Junction; Bridge Creek Camp; **Los Angeles Co**.: Claremont; **Monterey Co**.: Arroyo Seco; **Napa Co**.: Samuel Springs (beneath Lake Berryessa); **Nevada Co**.: Russell

Valley; Boca; Sagehen Creek; 4 mi. s Graniteville; **Placer Co.**: Brockway; Carnelian Bay, Lake Tahoe; **Sacramento Co.**: American River; **Shasta Co.**: Cow Creek near Whitmore; **Sierra Co.**: Sierraville; Sierra Valley; Independence Lake; Sattley; Yuba Pass; **Stanislaus Co.**: Patterson; **Trinity Co.**: Trinity River; Eagle Creek; **Tulare Co.**: Ash Mountain; **Tuolumne Co.**: Strawberry; Yosemite National Park; **Yolo Co.**: Davis.

Hosts. Krombein (1958, 1967) reported this species from *Hoplisodes costalis* (Cresson) nests. Rosenheim and Grace (1987) reared *viridicyaneus* from cells of *Mimumesa mixta* (W. Fox).

Discussion. Similar to *marginatus* as discussed under that species, the form of the apical metasomal truncation is variable in *viridicyaneus* and may be reduced to a narrow strip in some individuals.

Elampus viridis Cresson

Fig. 40, Map 12

Elampus viridis Cresson 1865a:103. Holotype; Cuba (ANSP).

Elampus rotundus Huber and Pengelly 1978:92. Holotype ♂; USA: California, Imperial Co. (UCDC).

Geographic Range (Map 12). Central to western USA south to northern Mexico (Baja California, Sonora, Sinaloa, Tamaulipas), Cuba and Venezuela; 22 specimens were seen.

California Records. Collected August through October; **Imperial Co.**: Bard; **Riverside Co.**: 18 mi. w Blythe.

Hosts. Unknown.

Discussion. *E. viridis* can be immediately distinguished from other North American species by the nearly circular membrane-filled apical metasomal truncation (Fig. 40). This species is rarely collected in California.

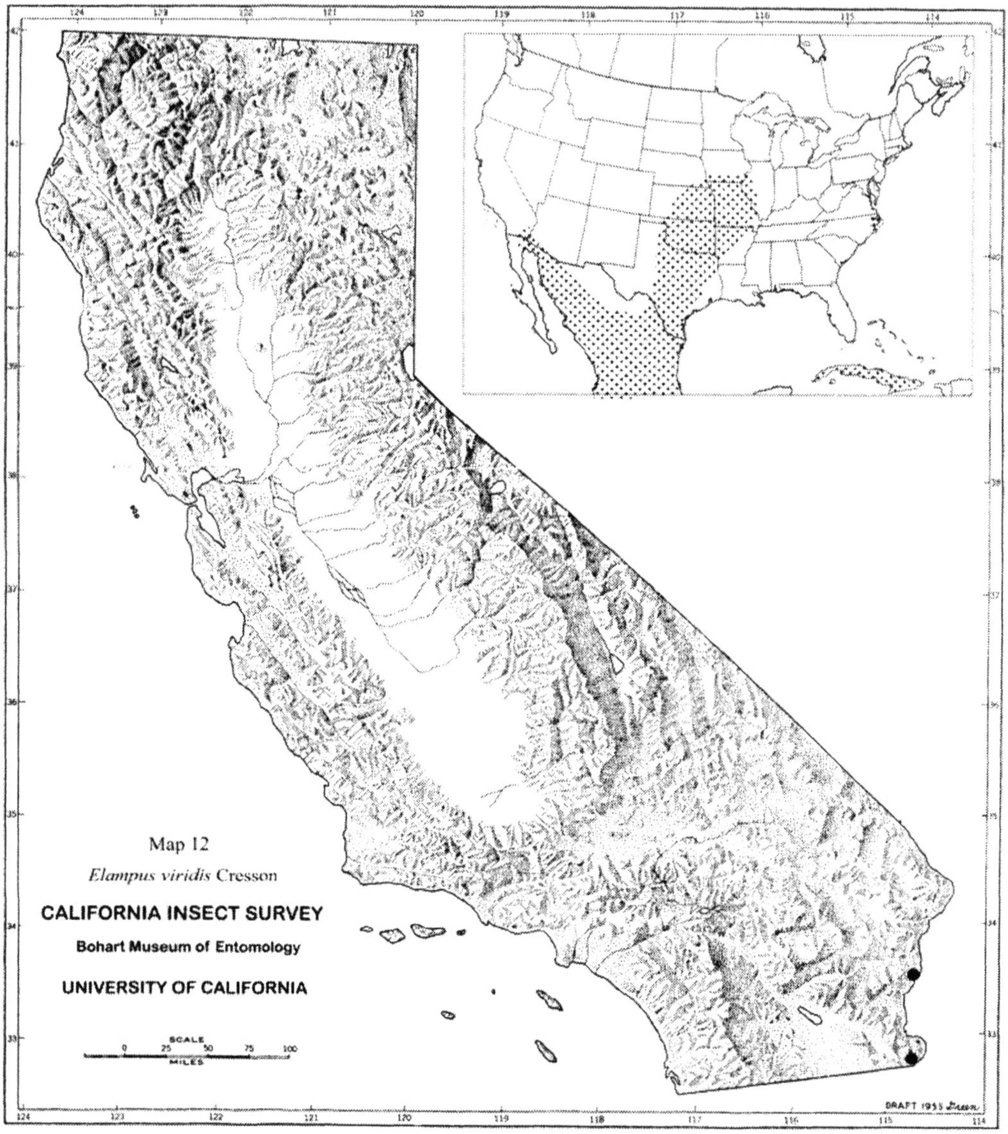

Map 12. California distribution of *Elampus viridis*. Inset: overall distribution.

Genus *Hedychreides* Bohart

Generic Diagnosis. Scapal basin flat or shallowly convex with contiguous fine punctures, covered with dense appressed silver setae; malar space 1 MOD long; pronotal anterior margin with faint anterolateral carina; mesopleuron rounded, without omaulus or scrobal carina; notauli sulciform; scutellar wing fossa without anterior lobe; mid and hindtibia with large, dark, shallow depression on inner surface; male midtarsomeres laterally expanded and flattened, with long curved marginal setae; tarsal claws apically bifid (Fig. 30); forewing medial vein slightly curved, arising at cu-a, stigma slender and apically acute (Fig. 24); tergum III apical margin evenly rounded, slightly reflexed with hyaline rim; volsella with digitus and cuspis

Host. Unknown.

Discussion. *Hedychreides* is a monotypic genus, occurring only in the dry regions of the southwestern United States. It is structurally somewhat intermediate between *Hedychrum* and *Hedychridium*, having the tarsal claw dentition of *Hedychrum* and facial and propodeal sculpturing of *Hedychridium*.

Hedychreides mirum Bohart

Figs. 24, 30, Map 13

Hedychreides mirum Bohart 1980:135. Holotype ♂; USA: California, Inyo Co., Lone Pine (UCDC).

Geographic Range (Map 13). Upper Sonoran Zone of California, Arizona and Nevada; 212 specimens have been seen.

California Records. Collected April through July. **Imperial Co**.: Holtville; 6 mi. w Mt. Signal; **Inyo Co**.: Panamint Springs; 4 mi. e Big Pine; Owens Lake; Deep Springs; Pinto Flat; 2 mi. e Lone Pine; **Kern Co**.: 19 mi. w Shafter; Red Rock Canyon; **Riverside Co**.: Palm Springs; 18 mi. w Blythe; 7 mi. n Palm Springs; **San Bernardino Co**.: 3 mi. s Kramer Junction; Rialto; 11 mi. w Ludlow; **San Diego Co**.: Borrego Valley.

Discussion. This species is easily confused with *Hedychridium amabile* and *H. mirabile*, and all three are often sympatric. They are brightly colored, with coppery red on a greenish background. Both sexes of *mirum* have a V-shaped brow and scalpal basin densely covered in silvery appressed pubescence.

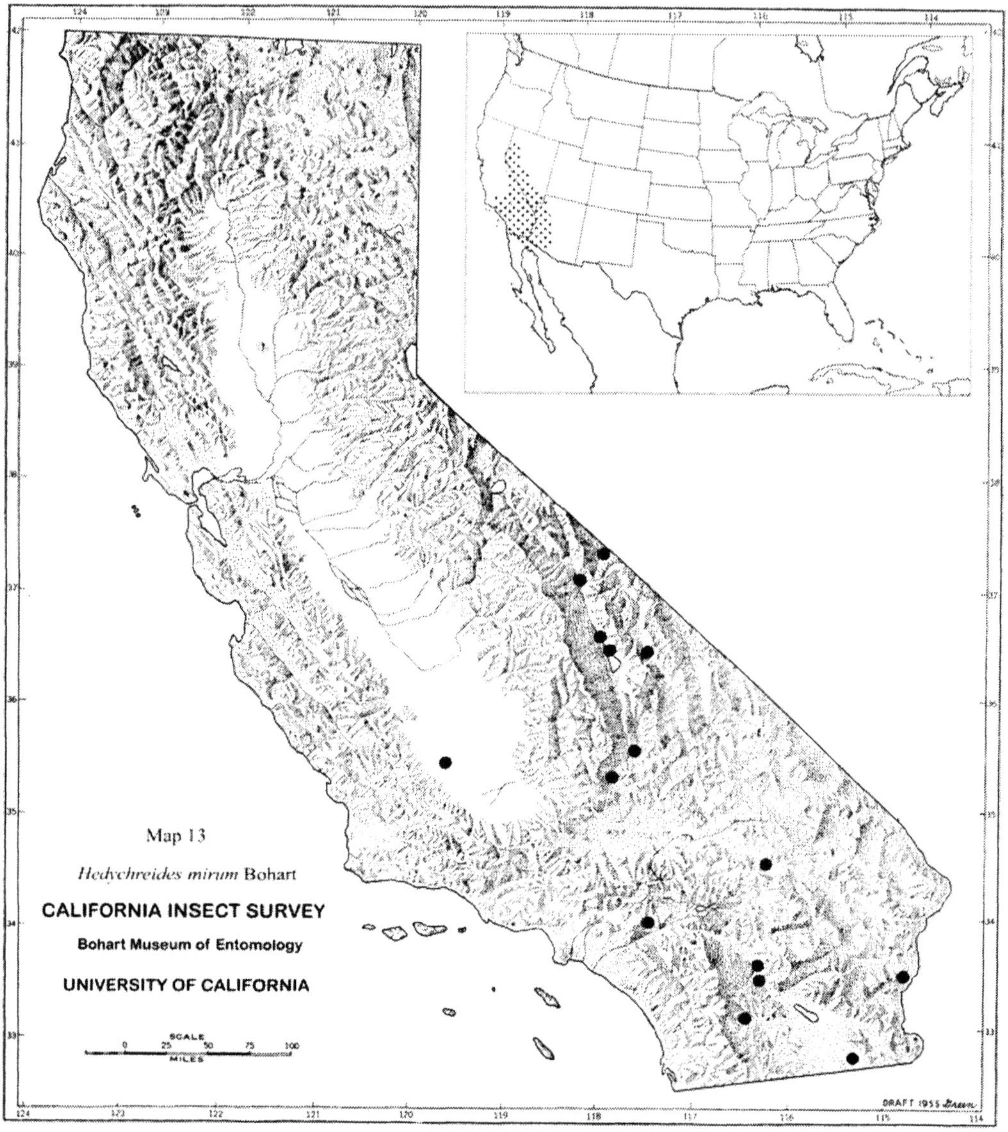

Map 13. California distribution of *Hedychreides mirum*. Inset: overall distribution.

Genus *Hedychridium* Abeille de Perrin

Generic Diagnosis. Scapal basin flat to shallowly concave, usually with dense appressed silvery pubescence; base of oral fossa somewhat elevated; malar space usually less than 1 MOD long; pronotal anterior margin with sublateral carina; mesopleuron rounded without omaulus or scrobal carina; scutellar wing fossa without anterior lobe; mid and hindtibia without pits on inner surface; tarsal claw with one submedial, perpendicular tooth; forewing medial vein straight to strongly arched, arising at cu-a, stigma slender and apically acute; tergum III usually evenly rounded, may be apicomedially drawn out; digitus and cuspis present.

Hosts. Old World species of *Hedychridium* parasitize Crabronidae, Halictidae and Megachilidae. Little is known about the biology of New World species, although they are recorded as nest parasites of ground nesting Crabronidae, including species of *Solierella* and *Tachysphex*.

Distribution. *Hedychridium* has a nearly cosmopolitan distribution, with Australia and South America being the only regions depauperate in species.

Discussion. This is the second largest genus of chrysidids in California. In North America the vast majority of species occur west of the 100th meridian. The genus contains a diverse group of species and as further studies are made there will undoubtedly be more species discovered. *Hedychridium* is characterized by the long slender tarsal claw, with a small submedial, perpendicular tooth, unmodified hindtibia, rounded mesopleuron, and the Rs stub usually longer than the stigma. The American species of *Hedychridium* were reviewed by Bohart and Kimsey (1978).

Key to California Species of *Hedychridium*

1 Flagellomere I more than twice as long as pedicel, sternum III green or blue.......2

- Flagellomere I twice or less as long as pedicel; sternum III variously colored, often brown...4

2 Tergum II strongly thickened toward apex..............................*crassum* Bohart

- Tergum II thinned toward apex..3

3 Tergum III margin rounded apically, flagellomere I less than twice as long as broad, wing membrane whitish.......................................*purum* Kimsey

- Tergum III margin indented apicomedially, flagellomere I more than twice as long as broad, wing membrane smoky or brown tinted.......................*incism* Bohart

4 Forefemur subbasally sharply angled (Fig. 22), sternum III entirely brown or brown with medial green spot...*olene* Kimsey

- Forefemur subbasally rounded, sternum III variously colored.........................5

5 Flagellomere I more than twice as long as broad, head and thorax not closely microreticulate or shagreened..6

- Flagellomere I less than twice as long as broad, and/or head and thorax closely microreticulate or shagreened..15

6 Forewing basal vein reaching MCu at broadly obtuse angle, mesopleuron with microridged area above midcoxa..7

- Forewing basal vein reaching MCu at right angle or nearly so; mesopleuron with polished impunctate spot above midcoxa..11

7 Malar space much shorter than pedicel, tergum II usually blackish medially, pubescence of lower frons various...8

- Malar space subequal to length of pedical, tergum II various, lower frons covered with dense pubescence (viewed from above) (as in Fig. 17)..........................10

8 Pubescence of lower frons dense, whitish or silvery................*maricopae* Bohart

- Pubescence of lower frons sparse, often fulvous...9

9 Tergum II punctures contiguous toward middle, integument appearing dull, sternum II usually brown..................................*solierellae* Bohart and Brumley

- Tergum II punctures separated by one puncture diameter or more, integument appearing shiny overall, sternum II blue or green in male, usually with blue or green highlights in female..*rasile* Bohart

10 Basal vein of forewing as long as Rs stub............................*amabile* Cockerell

- Basal vein of forewing longer than Rs stub.............................*coruscum* Bohart

11 Tergum III rounded over apically and indented before apical rim, apex drawn out into obtuse medial point particularly in female; forefemur ventrally longitudinally grooved, with associated well-developed carina (Fig. 19); frons medially overhanging scapal basin; male sternum II-III green or blue......*dimidiatum* (Say)

- Tergum III thin-edged apically or very slightly thickened, apex broadly rounded or drawn out medially; forefemur at most weakly carinate; other characters various..12

12 Tergum III margin posteromedially drawn out into rounded point, especially in female; pronotal punctures nearly contiguous and evenly spaced, similar or slightly larger than on tergum II; sterna II-III green or blue........*crebrum* Kimsey

- Tergum III margin posteromedially broadly rounded; pronotal punctures coarse, irregularly spaced and much larger than those on tergum II; sterna II-III variously colored...13

13 Tergum II basomedially with at most a small indistinctly edged black area; tegula and male sternum III entirely green or blue; tergum III apex broadly rounded ....*politum* Bohart

- Tergum II basomedially with large diffuse, or medium-sized and distinctly edged black area; male sternum III entirely or mostly brown; other characters various 14

14 Tergum III apicomedially somewhat angulate, especially in female; tegula green; tergum II with definite basomedial bilobate or triangular black spot; forefemur may be weakly longitudinally grooved and keeled ventrally......*bilobatum* Bohart

- Tergum III apicomedially broadly rounded; tegula usually brown; tergum II more than half covered by irregularly margined dark area; forefemur rounded ventrally ....*menkei* Bohart

15 Subantennal distance about equal to pedicel length......................................16

- Subantennal distance measurably less than pedicel length...............................17

16 Pronotal vertical surface with coarse longitudinal ridging; lower frons with inconspicuous pubescence; terga purplish, with dark pubescence ....*nevadae* Kimsey

- Pronotal vertical surface punctate without longitudinal ridging; lower frons with dense fine pubescence arising from fine punctation; terga usually greenish to blue, with pale pubescence..*fletcheri* Bodenstein

17 Pronotal punctures separated by obvious microsculpture: shagreening or micropunctation..18

- Pronotal punctures not separated by obvious microsculpture, polished between 19

18 Pronotum and tergum II with large punctures sparse, microsculpture extensive, with some transverse structure to it; flagellomere V of male longer than broad, flagellum slender..*frugale* Bohart

- Pronotum and tergum II with large punctures separated by micropunctured areas about one macropuncture diameter; flagellomere V of male broader than long, flagellum broadened and flattened......................................*cornutum* Bohart

19 Forewing stigma pale yellow, flagellum reddish, wing membrane water clear...20

- Forewing stigma brown to black, other characters various...............................21

20 Mesopleuron densely covered with silver pubescence, obscuring punctation; most of frons similarly covered with dense sliver pubescence.........*argenteum* Kimsey

- Mesopleural pubescence sparser, punctation obvious; frons moderately covered with...*leucostigma* Bohart

21 Lower frons with punctation coarser or more irregular toward eye margins, pubescence moderate to sparse, often pale fulvous.......................................22

- Lower frons with fine, even punctation extending to eye margins, pubescence forming dense silvery mat on either side of midline.....................................24

22 Basal vein strongly curved; sternum III and tegula brown............*lividium* Bohart

- Basal vein straight, sternum III usually blue or with bluish highlights, tegula bluish to green or brown...23

23 Brow rough overall, punctures nearly contiguous; tegula partly bluish or greenish ....*paulum* Bohart

- Brow smooth overall, punctures separated by small smooth areas; tegula brown ....*milleri* Kimsey

24 Punctures across brow much finer toward eye margin, brow with V-shaped medial swelling...*frontis* Kimsey

- Punctures across brow even in size toward eye margin, brow without V-shaped medial swelling...25

25 Brow well-developed all across, overhanging scapal basin; tegula bright green or blue; Rs stub shorter than basal vein....................................*mirabile* Kimsey

- Brow not well-developed, tegula and Rs stub various..................................26

26 Flagellomere V about as long as broad or slightly longer........*gemmatum* Kimsey

- Flagellum V distinctly broader than long..............................*arietinum* Bohart

***Hedychridium amabile* Cockerell**

Fig. 17, 31; Map 14

Hedychridium amabile Cockerell 1903:262. Holotype ♀; USA: New Mexico: Las Cruces, Mesilla Park (USNM).

Hedychridium azurellum Bohart 1978:601. Holotype ♂; USA: California, Inyo Co., Deep Springs (UCDC).

Geographic Range (Map 14). Western USA, mostly west of the 100th meridian, Mexico (Baja California Norte and Sur, Sonora, Tamaulipas); 276 specimens were examined.

California Records. Collected March through October. **Alpine Co**.: Carson Pass; Hope Valley; **Contra Costa Co**.: Danville; **Imperial Co**.: Glamis; Palo Verde; Pinto Flat; Fish Creek; **Inyo Co**.: Big Pine Creek; Deep Springs; Kearsarge Pass Rd. 8 mi. w Independence; Eureka Valley; Saline Valley; Owens Lake Valley; **Kern Co**.: 19 mi. w Shafter; **Los Angeles Co**.: Tanbark Flat; Huntington Park; **Marin Co**.: Dillon Beach; **Mono Co**.: Bridgeport; 11 mi. n Bridgeport; **Monterey Co**.: Salinas; **Nevada Co**.: Boca; Sagehen Creek; **Placer Co**.: Brockway Summit; Carnelian Bay, Lake Tahoe; **Plumas Co**.: Greagle; **Riverside Co**.: Thousand Palms; 18 mi. w Blythe; 7 mi. se Freda; **Sacramento Co**.: Carmichael; Sacramento River; Sacramento River Levee; Grand Island; **San Bernardino Co**.: Needles; 15 mi. w Baker; 5 mi. n Vidal Junction; Apple Valley; **San Diego Co**.: Borrego Valley; **Sacramento Co**.: Lone Mountain; **San Luis Obispo Co**.: Oso Flaco Lake; **Sierra Co**.: Sattley; Independence Lake; **Ventura Co**.: Foster Park; **Yolo Co**.: Davis.

Hosts. Unknown.

Discussion. Most individuals of *amabile* have bright gold coppery markings, although a few may be all blue to green instead. This species most closely resembles *coruscum* based on the long flagellomere I (2.8-3.0 times as long as broad), brown tegula, long malar space (1.4-1.6 MOD), medial vein almost straight, and tergum III broadly rounded. *H. amabile* can be distinguished from *coruscum* by the longer Rs stub.

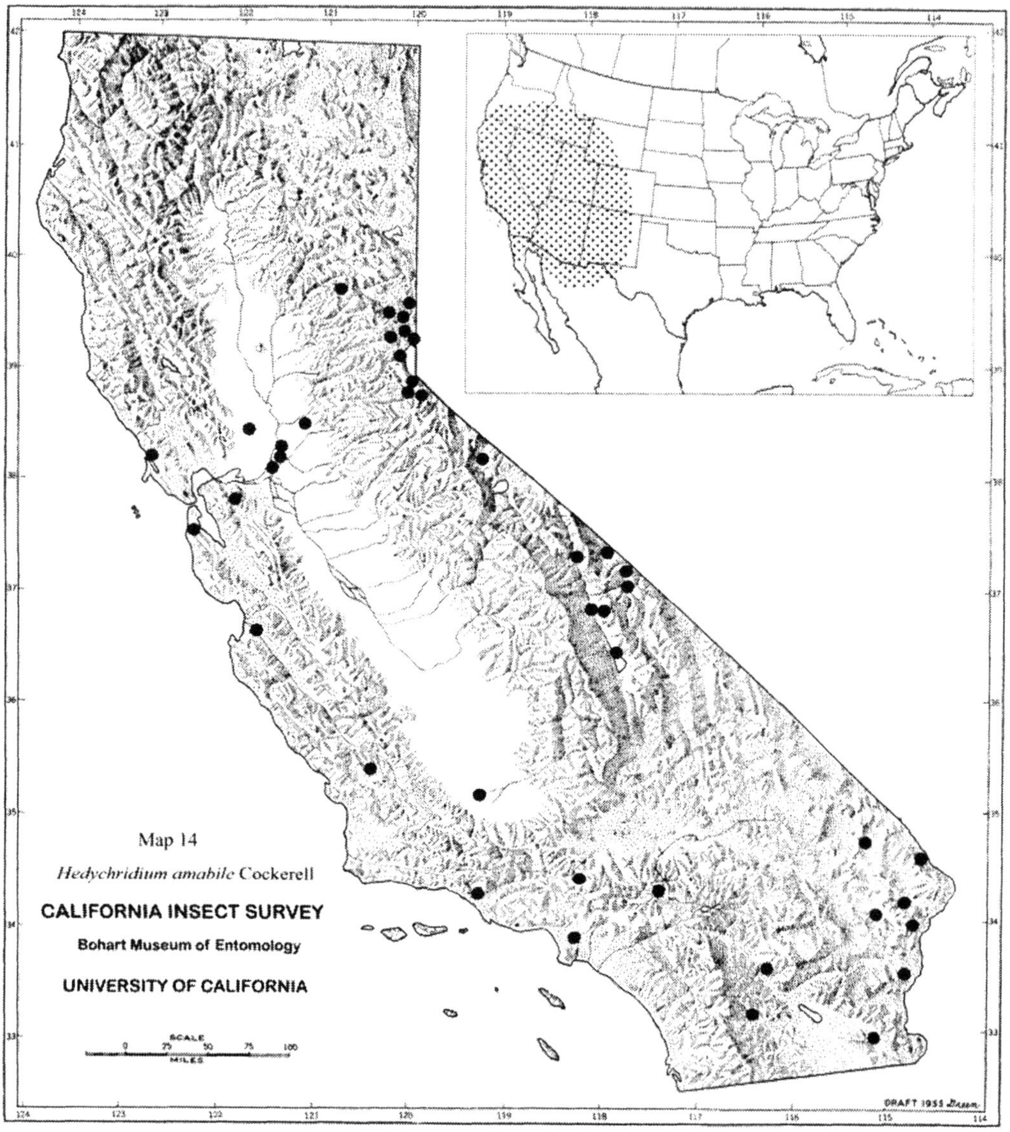

Map 14. California distribution of *Hedychridium amabile*. Inset: overall distribution.

Hedychridium argenteum Kimsey

Hedychridium argenteum Kimsey, in Bohart and Kimsey 1978:599. Holotype ♂; USA: California, Riverside Co., 18 mi. w Blythe (UCDC).

Geographic Range. This species appears to be endemic to eastern Riverside Co.; 3 specimens in the type series were seen.

California Records. Collected in April, October; **Riverside Co**.: 18 mi. w Blythe.

Hosts. Unknown.

Discussion. It would be difficult to confuse *argenteum* with any other species. The body is covered with dense, scale-like whitish pubescence. The wing membrane is untinted, and the veins yellow. In addition, the male flagellum is pale red.

Hedychridium arietinum Bohart

Map 15

Hedychridium arietinum Bohart, in Bohart and Kimsey 1978:600. Holotype ♂; USA: California, Riverside Co., Thousand Palms (UCDC).

Geographic Range (Map 15). Desert regions of California, Arizona and Nevada; 31 specimens were seen.

California Records. Collected in March, April; **Imperial Co**.: Fish Creek; Glamis; Ogilby Road; **Riverside Co**.: 18 mi. w Blythe; Thousand Palms; **San Bernardino Co**.: Helendale; Apple Valley; **San Diego Co**.: Borrego Valley; Split Mountain.

Hosts. Unknown.

Discussion. This is one of several tiny, blue to green *Hedychridium* species. *H. arietinum* is about 2 mm in length. Diagnostic features include the clear, untinted wings, brown tegula, straight medial vein, broadened flagellum, with flagellomere V twice as long as broad, and the brow weakly developed.

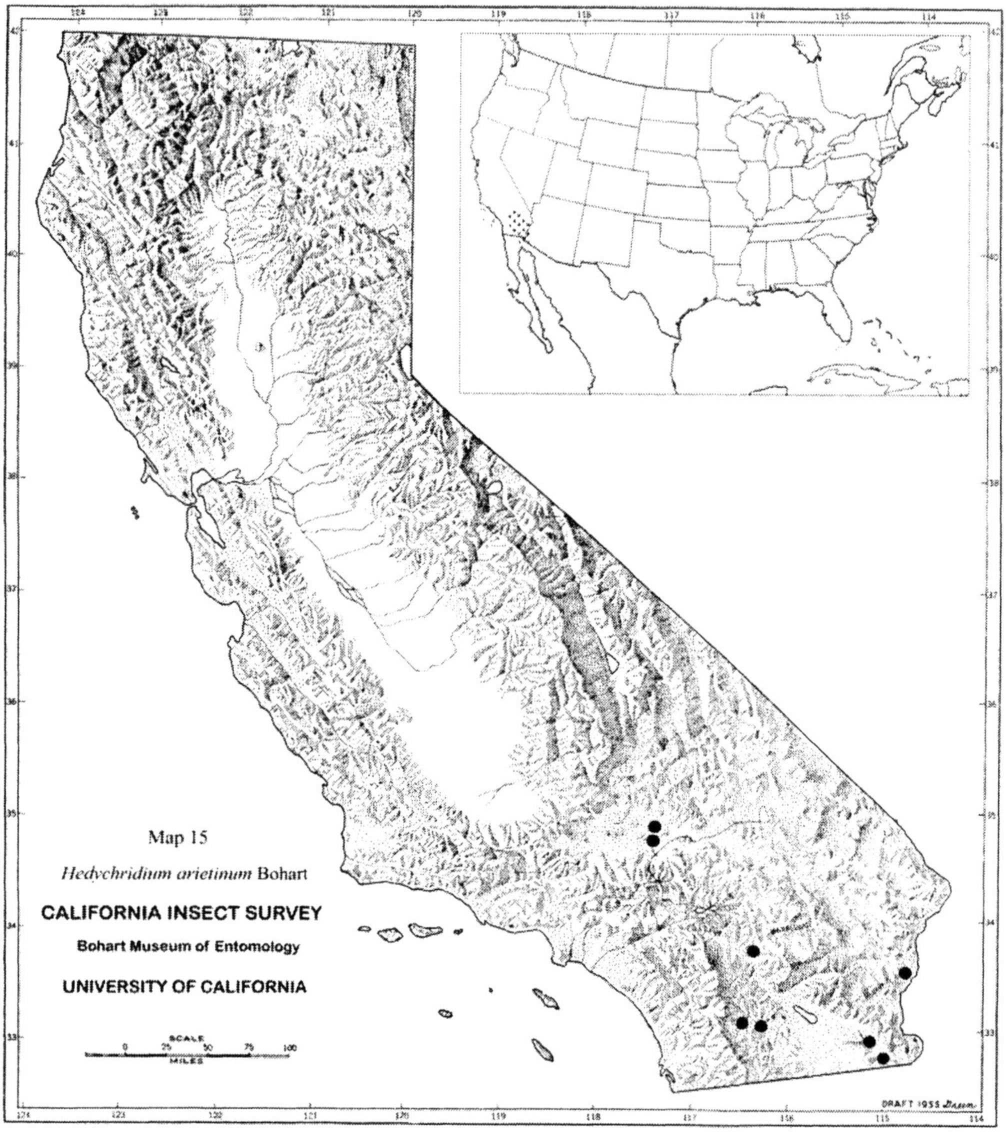

Map 15. California distribution of *Hedychridium arietinum*. Inset: overall distribution.

Hedychridium bilobatum Bohart

Map 16

Hedychridium bilobatum Bohart, in Bohart & Kimsey 1978:602. Holotype ♂; USA: California, Nevada Co., Sagehen Creek (UCDC).

Geographic Range (Map 16). Western USA in the Upper Sonoran and Canadian Zones; 71 specimens were seen.

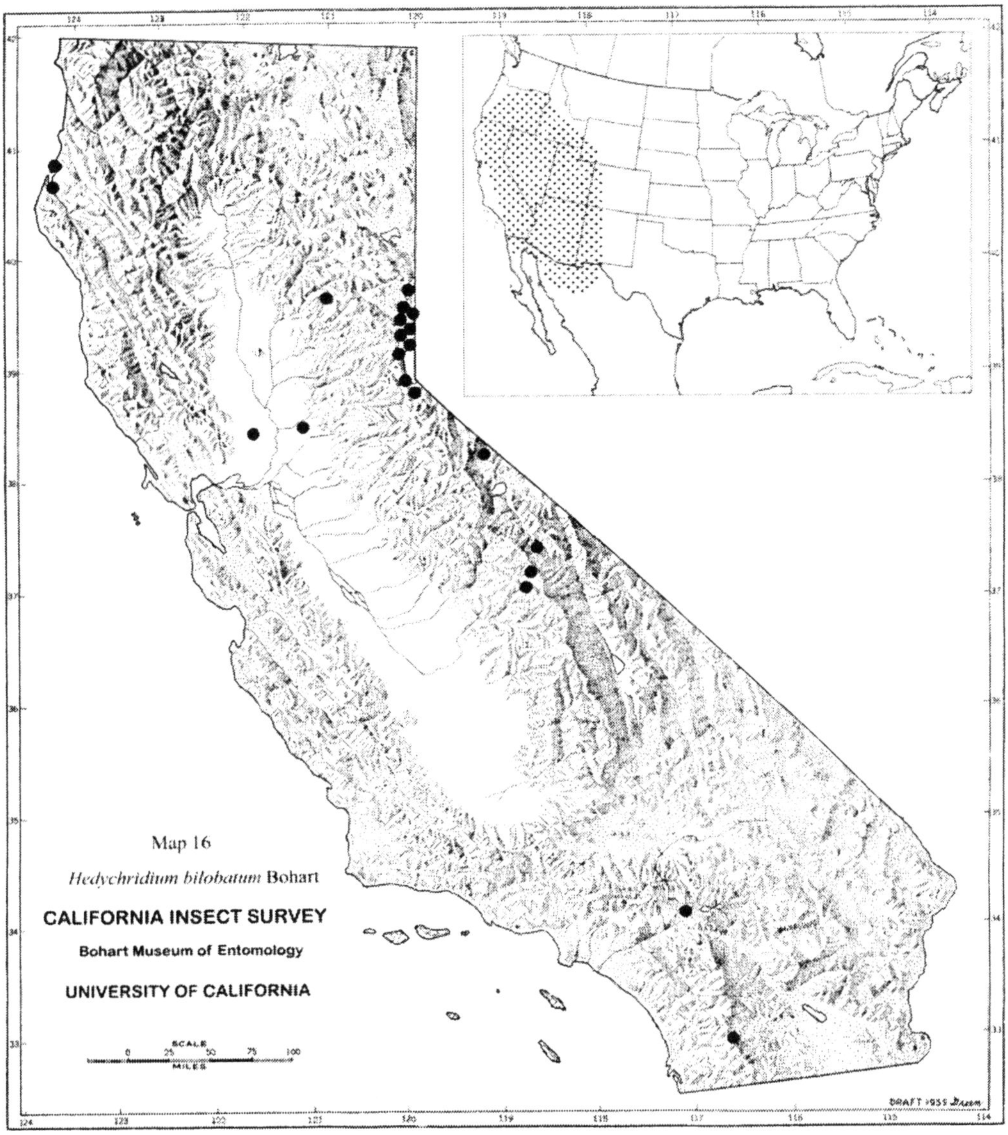

Map 16. California distribution of *Hedychridium biolobatum*. Inset: overall distribution.

California Records. Collected April through September; **El Dorado Co**.: Grass Lake; **Fresno Co**.: Huntington; Salie Keyes Lake; **Humboldt Co**.: 1 mi. s Samoa; Samoa dunes; **Lassen Co**.: Hallelujah Junction; **Mono Co**.: Mill Creek Canyon 7 mi. ne Hwy. 395; 5 mi. nw Paradise Camp; Twin Lake; **Nevada Co**.: Sagehen Creek;

Hobart Mills; **Placer Co.**: Brockway; Carnelian Bay, Lake Tahoe; **Plumas Co.**: Blairsden; **Sacramento Co.**: Carmichael; **San Bernardino Co.**: Dollar Lk Trail San Bernardino Mountains; **San Diego Co.**: Julian; **Sierra Co.**: Independence Lake; Sattley; Sierra Valley; Sierraville; **Yolo Co.**: Davis.

Hosts. Unknown.

Discussion. The name, *bilobatum*, refers to the large, black, often bilobate, spot on tergum II of this species. Other diagnostic features include: flagellomere I 2.5x as long as broad, malar space short, medial vein curved and reaching MCu at a 90° angle, tegula green, and sternum III brown.

Hedychridium cornutum Bohart

Map 17

Hedychridium cornutum Bohart, in Bohart & Kimsey 1978:606. Holotype ♂; USA: California, San Luis Obispo Co., Dune Lakes (CDFA).

Geographic Range (Map 17). This species occurs in central and southern California, Arizona and Baja California, Mexico; 48 specimens were studied.

California Records. Collected May through July; **Contra Costa Co.**: Mt. Diablo; **Los Angeles Co.**: Tanbark Flat; **Marin Co.**: Baltimore Park; **Riverside Co.**: Riverside; **San Bernardino Co.**: 27 mi. n Vidal; **San Diego Co.**: Borrego Valley; **San Luis Obispo Co.**: Dune Lakes 3 mi. s Oceano; Oso Flaco Lake; **Santa Barbara Co.**: Santa Cruz Island; Goleta; **Ventura Co.**: Quatal Canyon; Lockwood.

Hosts. Unknown.

Discussion. Most similar to *frugale*, both *frugale* and *cornutum* have the head and thoracic integument densely microreticulate. *H. cornutum* has closer and more distinct punctation than *frugale*, flagellomere I is shorter, less than 1.7x as long as broad, stouter flagellomere V (the length less than 0.7x breadth), and the forewing medial vein nearly straight. Some specimens are blue to green others have coppery reddish coloration.

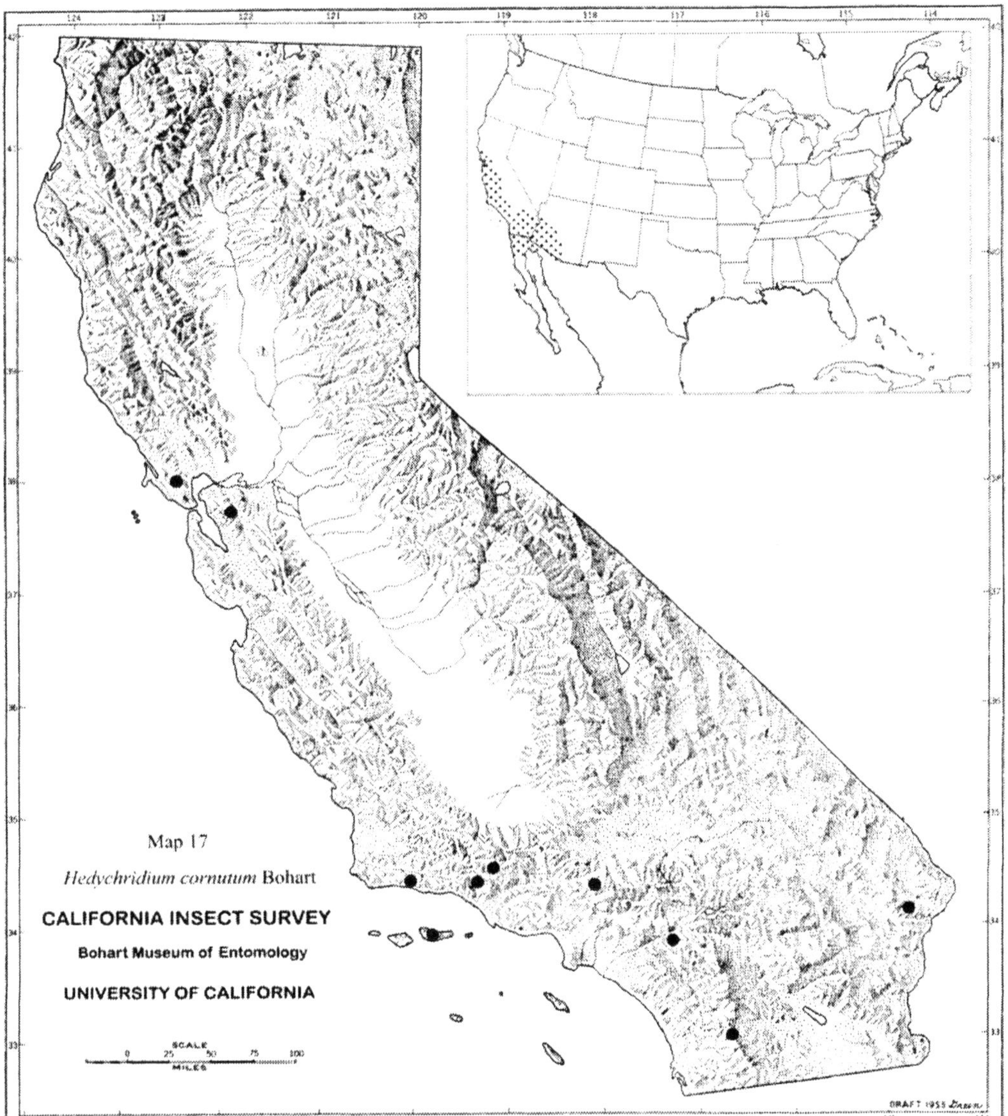

Map 17. California distribution of *Hedychridium cornutum*. Inset: overall distribution.

Hedychridium coruscum **Bohart**

Map 18

Hedychridium coruscum Bohart, in Bohart & Kimsey 1978:607. Holotype ♂; USA: California, Santa Barbara Co., Goleta (UCDC).

Geographic Range (Map 18). California, Arizona, Baja California Sur (Mexico); 25 specimens were seen including the type series.

California Records. Collected June and July; **Los Angeles Co.**: Canipole; **Santa Barbara Co.**: Goleta; **Santa Clara Co.**: Stanford University.

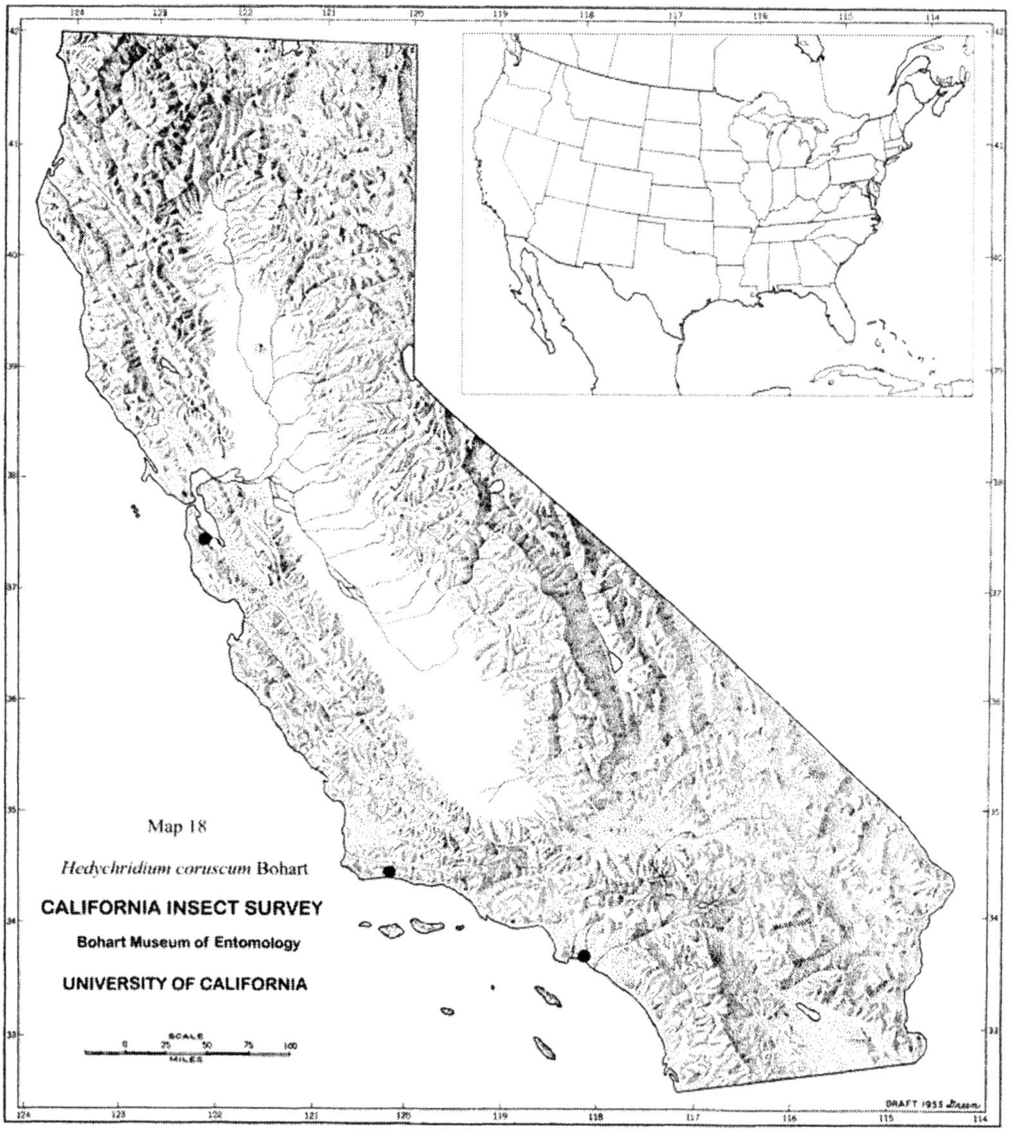

Map 18. California distribution of *Hedychridium coruscum*. Inset: overall distribution.

Hosts. Unknown.

Discussion. Similar in color to *amabile*, *coruscum* has flagellomere I is 2.4x as long as broad and longer than the pedicel, and the Rs stub is shorter than that of *amabile*.

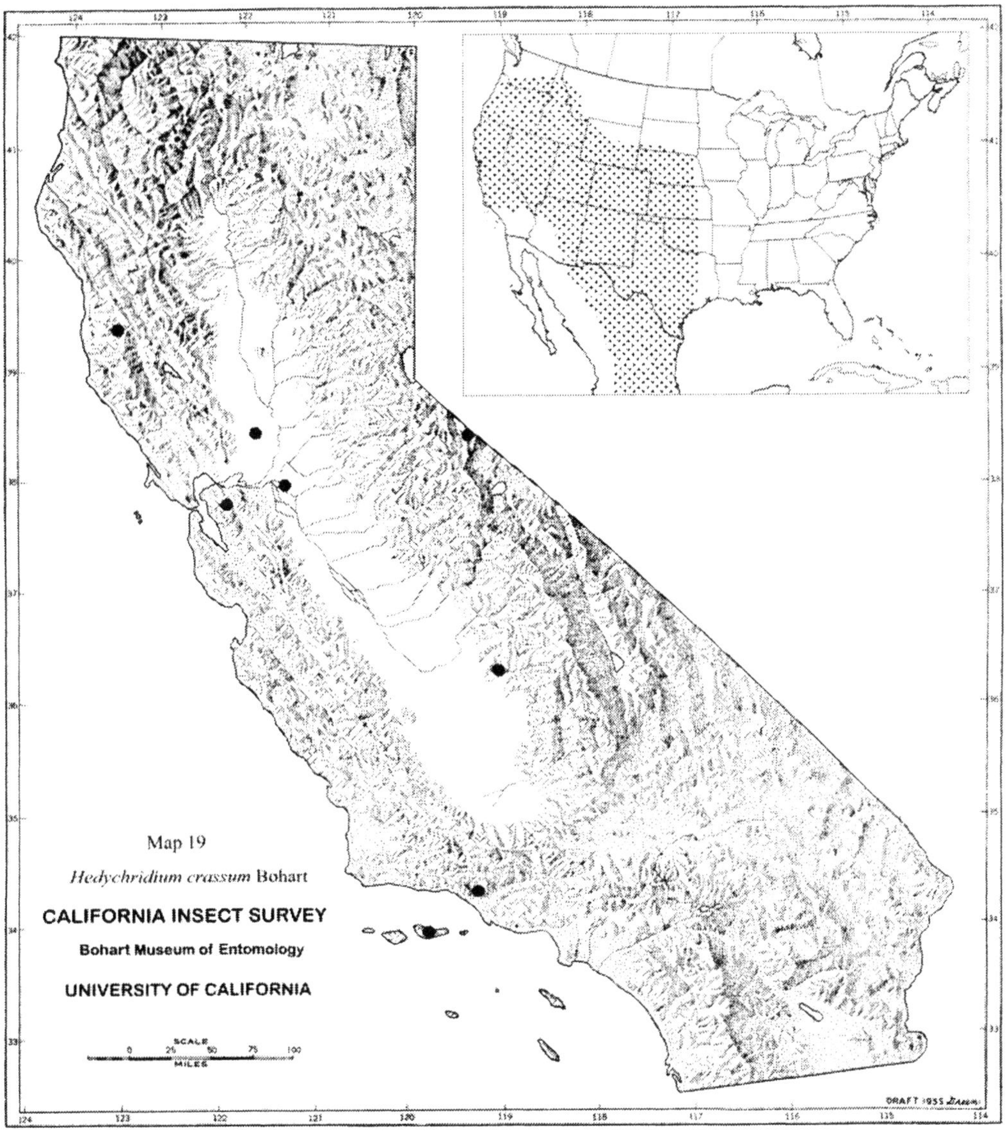

Map 19. California distribution of *Hedychridium crassum*. Inset: overall distribution.

Hedychridium crassum Bohart

Map 19

Hedychridium crassum Bohart, in Bohart & Kimsey 1978:608. Holotype ♂; USA: California, Lassen Co., Hallelujah Junction (UCDC).

Geographic Range (Map 19). Western USA, also Iowa, Nebraska, and western Canada; 85 specimens were studied.

California Records. Collected April through August; **Contra Costa Co**.: Mt. Diablo; **Mendocino Co**.: Willits; **Mono Co**.: Topaz Lake; **San Joaquin Co**.: Stockton; **Santa Barbara Co**.: Santa Cruz Island; **Tulare Co**.: Woodlake; **Ventura Co**.: Foster Park; **Yolo Co**.: Davis.

Hosts. Unknown.

Discussion. This species can be immediately recognized by the large body size (5-6 mm long) and swollen apical margin of tergum II. In addition, flagellomere I is more than twice as long as broad and is longer than the pedicel, the medial vein is curved, the facial punctation is undifferentiated between the brow and scapal basin, and sternum III is blue.

Hedychridium crebrum Kimsey

Map 20

Hedychridium crebrum Kimsey, in Bohart & Kimsey 1978:609. Holotype ♂; USA: California, Mariposa Co., Mariposa Canyon (UCDC).

Geographic Range (Map 20). Western USA (CA, OR, AZ, NM); 31 specimens were seen.

California Records. Collected April through October; **Contra Costa Co**.: Danville; Tassajara; **El Dorado Co**.: Farnham Ridge; **Mariposa Co**.: Mariposa; **Riverside Co**.: Riverside; Thousand Palms; **San Diego Co**.: Julian; Borrego Valley; **Yolo Co**.: Clarksburg; Davis.

Hosts. Unknown.

Discussion. *H. crebrum* is a distinctive species, which particularly in females, can be recognized by the apical margin of tergum III drawn out medially into a rounded angle. In addition, the malar space is short, flagellomere I is 2.8x as long as broad and 1.8x the pedicel length, medial vein is strongly arched, the lower frons is sparsely punctate, and tergum II has a dark basomedial spot.

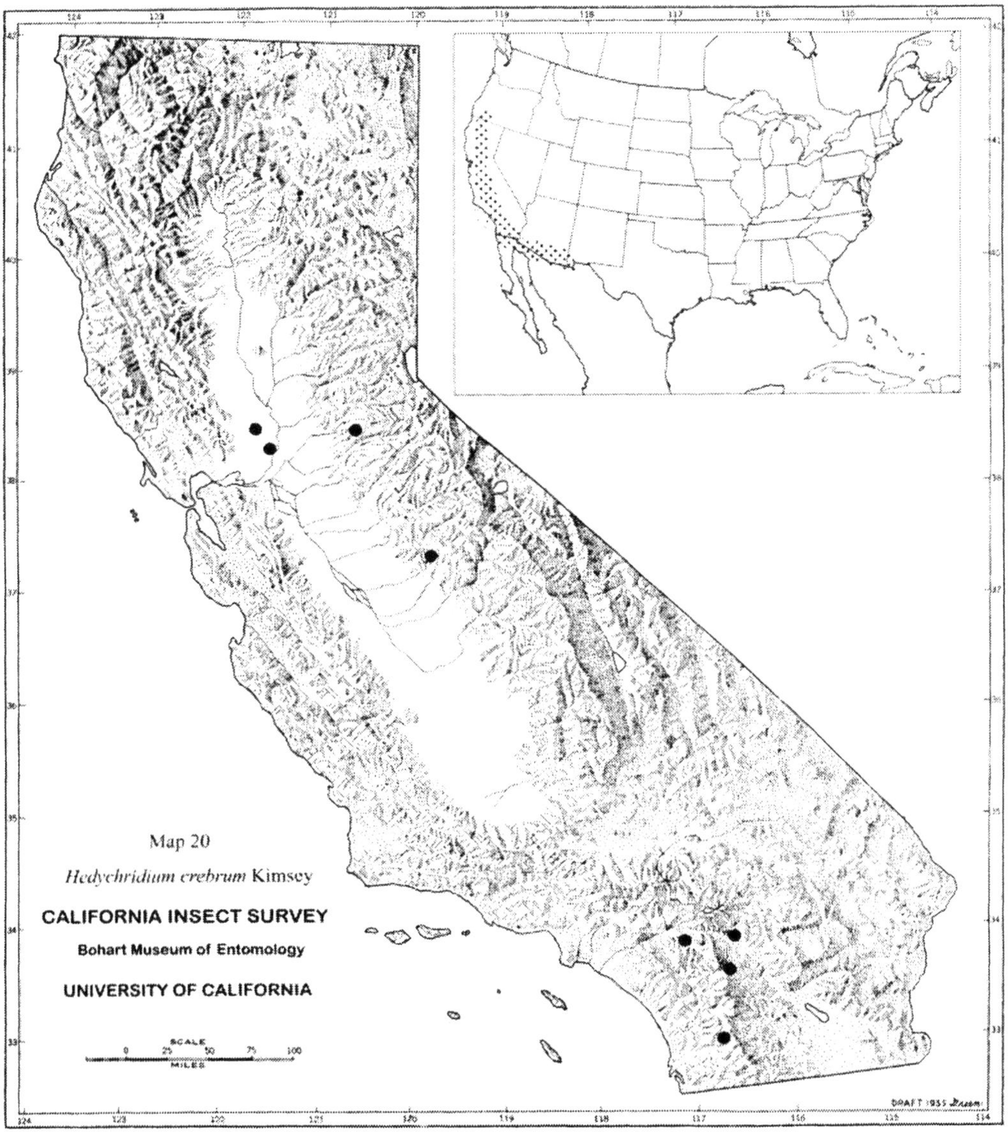

Map 20. California distribution of *Hedychridium crebrum*. Inset: overall distribution.

Hedychridium dimidiatum (Say)

Fig. 19, Map 21

Hedychrum dimidiatum Say 1824:330. Holotype ♀; USA: Pennsylvania (destroyed).

Chrysis mexicanum Cameron 1888:460. Holotype ♀; Mexico: Zacatecas City (BMNH).

Holopyga mexicanum Mocsáry 1911b:448. Holotype ♂; Mexico (TMB). Nec Cameron 1888.

Geographic Range (Map 21). This species is widespread in the Nearctic Region: in the USA, Canada (British Columbia, Manitoba, Ontario, Nova Scotia) and Mexico (Chihuahua, Jalisco, Oaxaca, Veracruz, San Luis Potosi); 692 specimens were studied, including the types of *mexicanum* Cameron and *mexicanum* Mocsáry.

California Records. Collected April through September; **Amador Co**.: Daffodil Hill; **Butte Co**.: Feather Falls; **Contra Costa Co**.: Mt. Diablo; **Inyo Co**.: Antelope Springs; **Lassen Co**.: Hallelujah Junction; **Los Angeles Co**.: Claremont; Tanbark Flat; **Marin Co**.: Mill Valley; Point Reyes Station; **Mono Co**.: Rock Creek campground; Mill Creek Canyon, 7 mi. s Junction Hwy 395; 11 mi. n Bridgeport; Pickel Meadow; **Nevada Co**.: 3 mi. n Boca; Boca; Prosser Dam; Sagehen Creek; Hobart Mills; **Placer Co**.: Carnelian Bay, Lake Tahoe; **Sacramento Co**.: Carmichael; **San Bernardino Co**.: 7 mi. e Barton Flat; **San Diego Co**.: 9 mi. s Warner Springs; Borrego Valley; Olivehain; **Santa Barbara Co**.: Santa Cruz Island; **Shasta Co**.: Cassel; **Sierra Co**.: Independence Lake; Sierra Valley; **Siskiyou Co**.: McBride Spring; Big Flat Campground Salmon Trinity Alps; **Sonoma Co**.: Tolay Creek; **Stanislaus Co**.: Turlock; Del Puerto Canyon; **Tehama Co**.: Mineral King; **Ventura Co**.: Foster Park; **Yolo Co**.: Davis; Elkhorn Ferry; **Yuba Co**.: Marysville.

Hosts. Unknown.

Discussion. This is one of several species with flagellomere I 2.5x as long as broad and twice as long as the pedicel. The most distinctive features of *dimidiatum* include tergum III apical margin angulate and rounded over, and the forefemur with a ventral, longitudinal groove and associated carina (Fig. 19). Other diagnostic features include the short malar space, sparse facial pubescence, strongly curved medial vein, and the tegula usually but not always entirely brown. These are large bodied chrysidids, ranging from 4-8 mm in length.

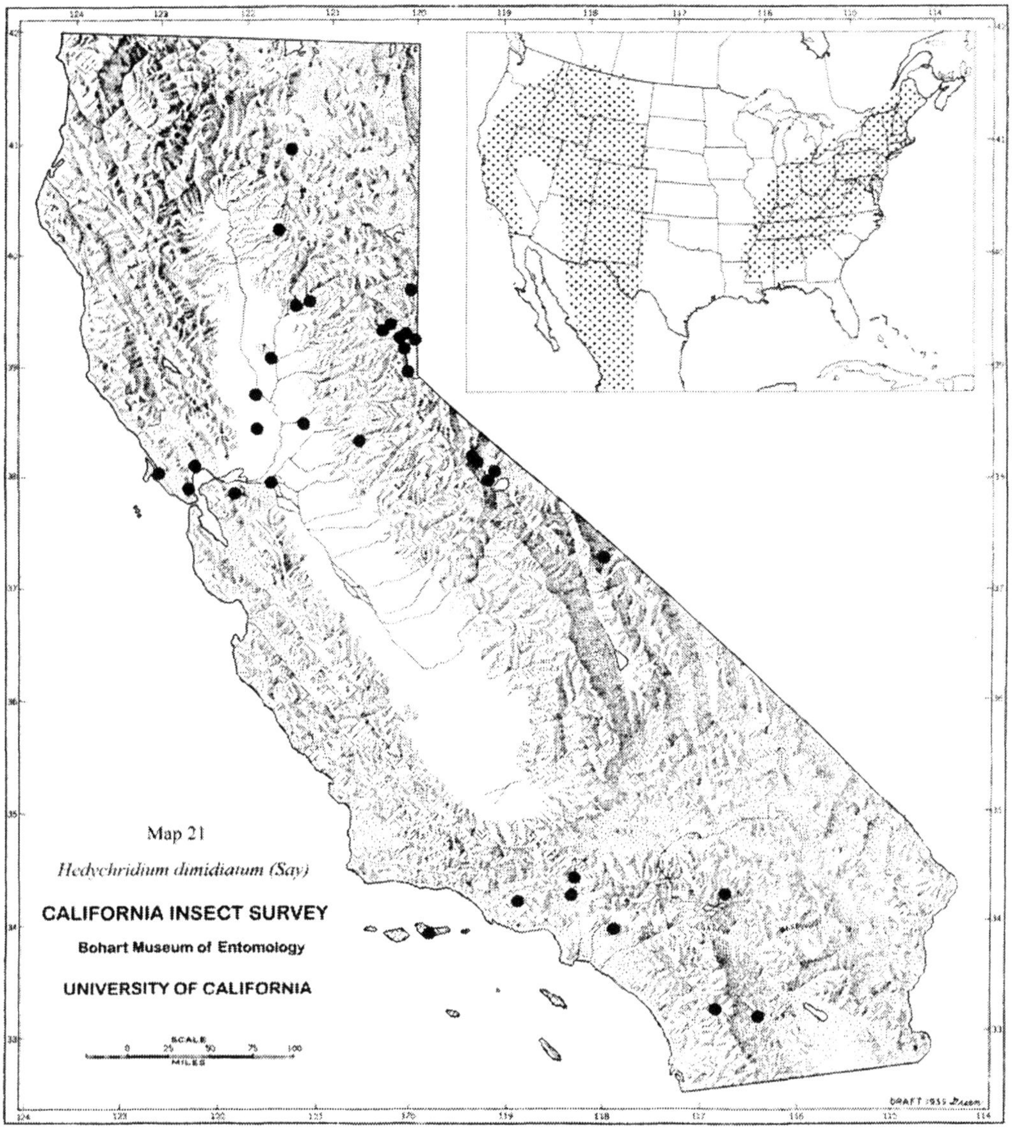

Map 21. California distribution of *Hedychridium dimidiatum*. Inset: overall distribution.

Hedychridium fletcheri Bodenstein

Map 22

Hedychrum viride Cresson 1865b:306. Lectotype ♀ (designated by Cresson 1916); USA: "Colorado Territory" (ANSP). Nec Guérin de Méneville 1842.

Hedychridium fletcheri Bodenstein 1951:720. Replacement name for *viride* Cresson 1865.

Geographic Range (Map 22). This species is widespread in the USA, southern Canada (Alberta, British Columbia, Manitoba, Saskatchewan), and northern Mexico (Baja California, Chihuahua); 2,085 specimens were seen, including the type of *viride* Cresson.

California Records. Collected May to August; **Alpine Co**.: Luther Pass; Carson Pass; Hope Valley; Lake Winnemucca; Wolf Creek; 4 mi. s Monitor Pass; **Amador Co**.: Ione; **Contra Costa Co**.: Danville; Mt. Diablo; Antioch; **El Dorado Co**.: Tallac; 2 mi. off Ice House Rd.; 4 mi. s Wrights Lk; 8 mi. n on Ice House Rd; Blodgett Forest 13 mi. e Georgetown; Echo Lake; El Dorado; Grass Lake; Meyers; South Lake Tahoe; **Fresno Co**.: Rea Lake; Jacalitos Canyon; Pioneer Basin; **Humboldt Co**.: 1 mi. s Samoa; Garberville; Samoa Peninsula; **Inyo Co**.: Big Pine; 7 mi. n Parchers Camp; 2 mi. e Lone Pine; Big Pine Creek; Independence; Lone Pine Creek; Ruby Lake; **Kern Co**.: 19 mi. w Shafter; **Lake Co**.: north fork Cache Creek Hwy 20; **Lassen Co**.: Hallelujah Junction; Summit Camp; Bridge Creek Camp; **Los Angeles Co**.: Tanbark Flat; **Marin Co**.: Lagunitas; McClure Beach, Point Reyes National Seashore; **Mendocino Co**.: 2 mi. n Point Arena; 4 mi. e Point Arena; **Modoc Co**.: Cedar Pass campground; 3 mi. w Alturas; 30 mi. s Tule Lake; **Mono Co**.: Tom's Place; 1 mi. s Saddlebag Lake; 4 mi. e Junction Hwys 120-395; Chalfant; Convict Lake; Cottonwood Creek; Hilton Creek; Rock Canyon; Sardine Creek; White Mountains; **Monterey Co**.: Salinas River; 1 mi. s Soledad; Arroyo Seco; **Napa Co**.: 4 mi. nw Berryessa; Pope Valley; Samuel Springs (beneath Lake Berryessa); **Nevada Co**.: Sagehen Creek; 17 mi. n Truckee; 3 mi. n Boca; Boca; Jackson Lake; Hobart Mills; Russell Valley; **Placer Co**.: Carnelian Bay, Lake Tahoe; Dutch Flat; Dollar Point, 2 mi. e Tahoe City; **Plumas Co**.: Blairsden; Buck's Lake; Graeagle; Onion Valley; **Riverside Co**.: Anza; 18 mi. w Blythe; Keen Camp; Sage; White Water; **Sacramento Co**.: Sacramento; **San Bernardino Co**.: San Bernardino Mountains, Dollar Lake Trail; 2 mi. w Phelan; 4 mi. n Cajon Junction; Phelan; **San Diego Co**.: Julian; near Buckman Spring; 2 mi. n Warner Springs; Borrego Valley; Pine Valley; Del Mar; **Sacramento Co**.: Lone Mt.; Lobos Creek; **San Luis Obispo Co**.: 5 mi. w Nipomo; **Santa Barbara Co**.: Santa Cruz Island; **Shasta Co**.: 3 mi. sw Old Station; Bridge Camp; **Sierra Co**.: Gold Lake; Sardine Lakes; Sierra Valley; Smith Mill 15 mi. se Sierraville; **Siskiyou Co**.: Macdoel; **Trinity Co**.: Butter Creek 12 mi. se Junction City; 10 mi. w Coffee Creek Ranger Station.; Mountain Meadow Ranch Coffee Crk; **Tuolumne Co**.: Dodge Ridge; **Yolo Co**.: Rumsey; 4 mi. w Davis; **Yuba Co**.: 18 mi. ne Marysville.

Hosts. Kurczewski (1967) reported *fletcheri* females ovipositing in the cells of *Tachysphex similis* Rohwer (Crabronidae).

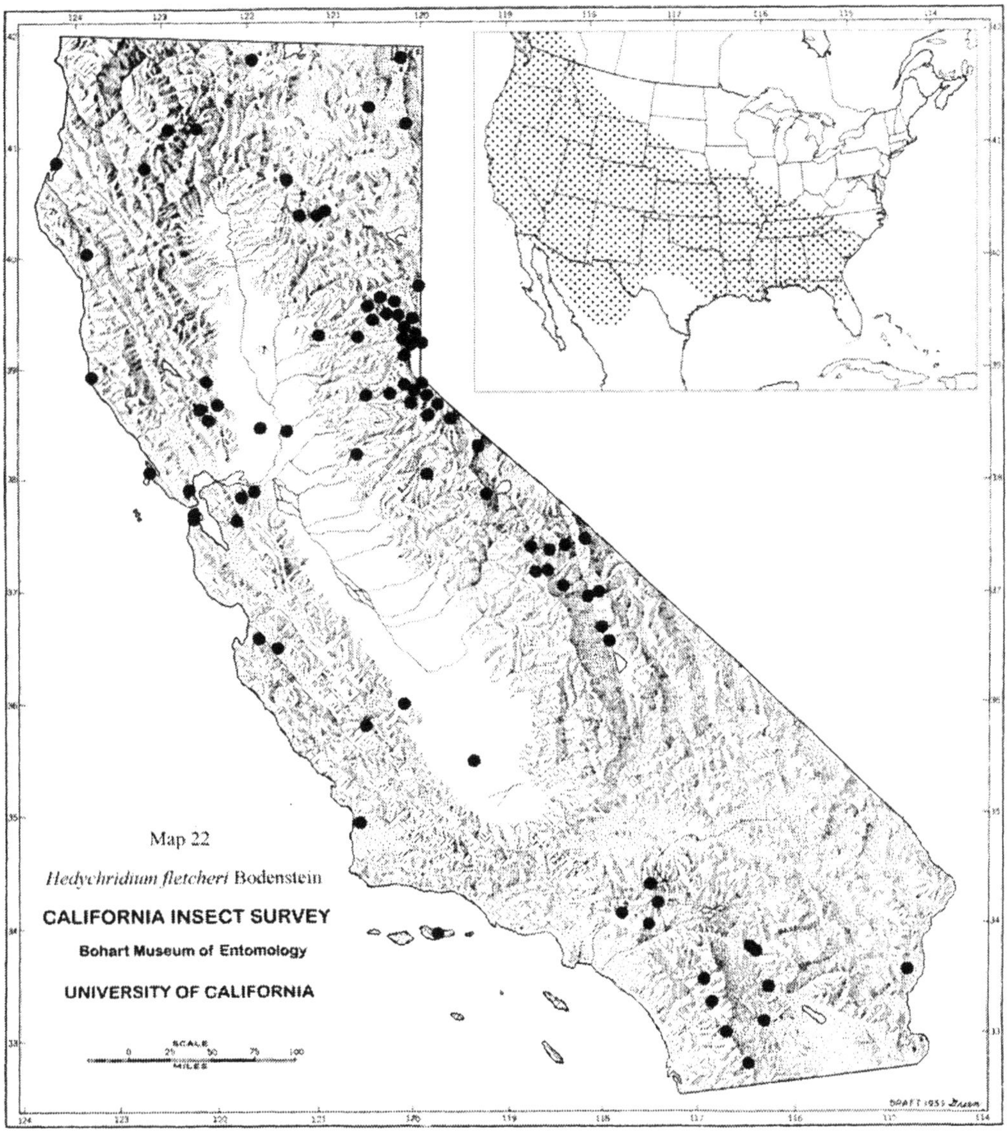

Map 22. California distribution of *Hedychridium fletcheri*. Inset: overall distribution.

Discussion. *H. fletcheri* is a widespread, highly variable species. It is the most abundant *Hedychridium* species in North America. Large specimens have flagellomere I 1.7-1.9x as long as broad and less than 1.5x as long as the pedicel, the brow is well-developed, overhanging an area of fine dense punctation, and the face is unusually long, with the subantennal distance subequal to the pedicel length. In smaller specimens these features are less pronounced. This variation may be due to a wide host range but there is little host data one way or another.

***Hedychridium frontis* Kimsey**

Map 23

Hedychridium frontis Kimsey, in Bohart & Kimsey 1978:612. Holotype ♂; USA: California, Riverside Co., 18 mi. w Blythe (UCDC).

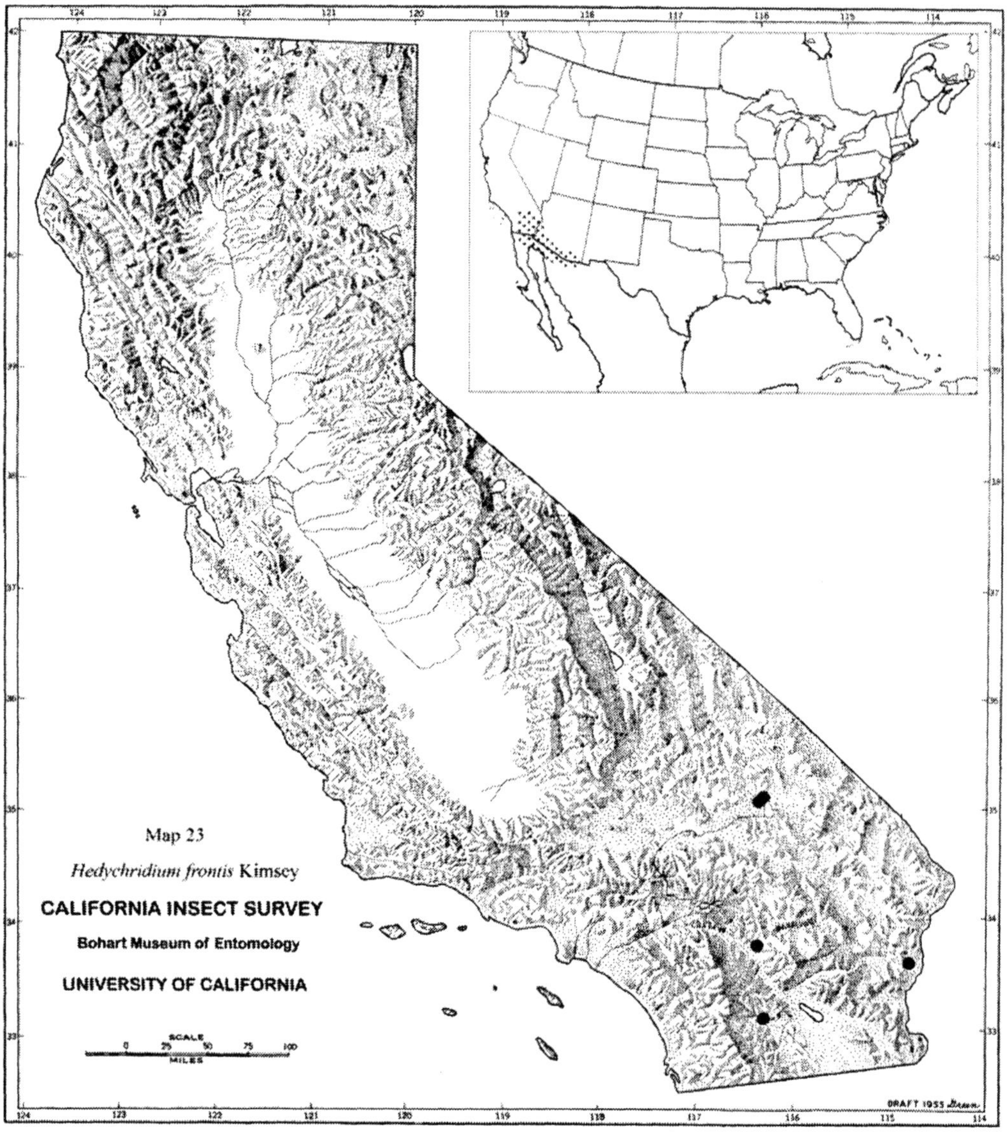

Map 23. California distribution of *Hedychridium frontis*. Inset: overall distribution.

Geographic Range (Map 23). Lower Sonoran Zone of southern California, Arizona, and Sonora, Mexico; 35 specimens were seen, including the type series.

California Records. Collected April and September; **Riverside Co.**: 18 mi. w Blythe; Thousand Palms; **San Bernardino Co.**: Cronise Wash; 15 mi. w Baker; **San Diego Co.**: Borrego Valley.

Hosts. Unknown.

Discussion. This is a tiny species (averaging 2 mm in length), which is greenish in males and coppery to purple in females. Diagnostic features include flagellomere I short, 1.5x as long as broad and subequal to the pedicel length, and medial vein straight. *H. frontis* is probably most similar to *antennatum*, which does not occur in California. Additional characteristics include sternum III greenish in males and brown in females, tegula with greenish highlights, the brow V-shaped and Rs stub as long as the medial vein.

Hedychridium frugale Bohart

Map 24

Hedychridium frugale Bohart, in Bohart & Kimsey 1978:613. Holotype ♂; USA: California, Imperial Co., Glamis (UCDC).

Geographic Range (Map 24). Upper and Lower Sonoran Zones of southern California, Arizona, Idaho, Nevada, Oregon, and in southern Alberta, Canada; 86 specimens were examined, including the type series.

California Records. Collected March through October; **Imperial Co.**: Palo Verde 6 mi. n Glamis; Algodones Dunes, 7 mi. se Glamis; **Inyo Co.**: Deep Springs; **Kern Co.**: 19 mi. w Shafter; **Mono Co.**: Tom's Place; **Orange Co.**: Santiago Canyon; **Riverside Co.**: Thousand Palms; 18 mi. w Blythe; **Sacramento Co.**: Sacramento; **San Bernardino Co.**: Baker; 27 mi. n Vidal; Needles; **San Diego Co.**: Borrego Valley; **Santa Barbara Co.**: Goleta; **Ventura Co.**: Lockwood Creek; Foster Park; **Yolo Co.**: Davis.

Hosts. Unknown.

Discussion. Similar to *cornutum* as discussed under that species, both have the body covered with fine dense microreticulation. *H. frugale* can be distinguished from *cornutum* by the longer flagellomere I (2.0-2.5x as long as broad), malar space longer than 0.9 MOD, flagellum slender, and widely spaced punctures on the head and thorax. Most individuals are coppery to red, although a few may be primarily blue to green.

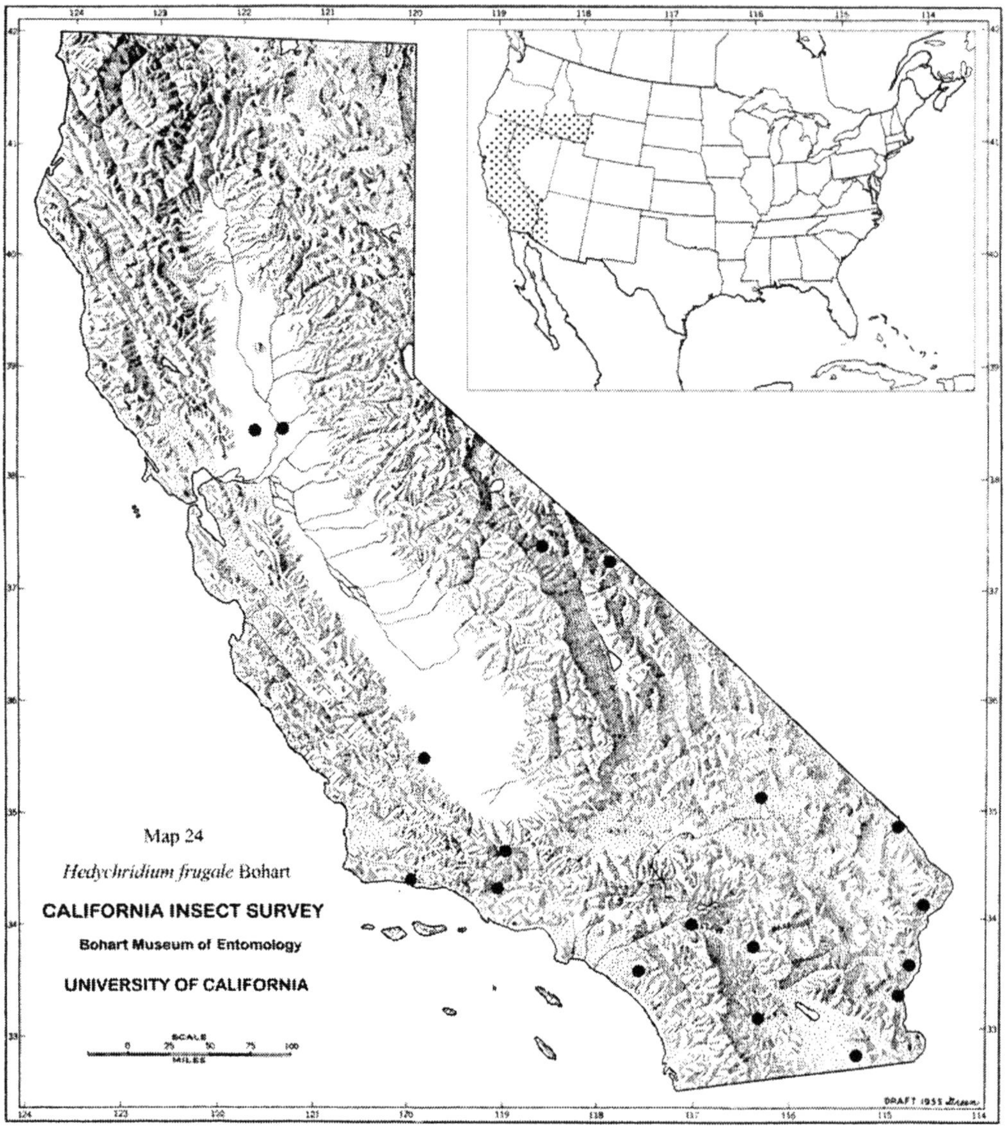

Map 24. California distribution of *Hedychridium frugale*. Inset: overall distribution.

Hedychridium gemmatum Kimsey

Map 25

Hedychridium gemmatum Kimsey, in Bohart & Kimsey 1978:614. Holotype ♂; USA: California, Riverside Co., 18 mi. w Blythe (UCDC).

Geographic Range (Map 25). This species occurs in the far south of California, Arizona, New Mexico and Texas; 12 specimens were seen, including the type series

California Records. Collected in October; **Riverside Co**.: 18 mi. w Blythe; **San Diego Co**.: Borrego Valley.

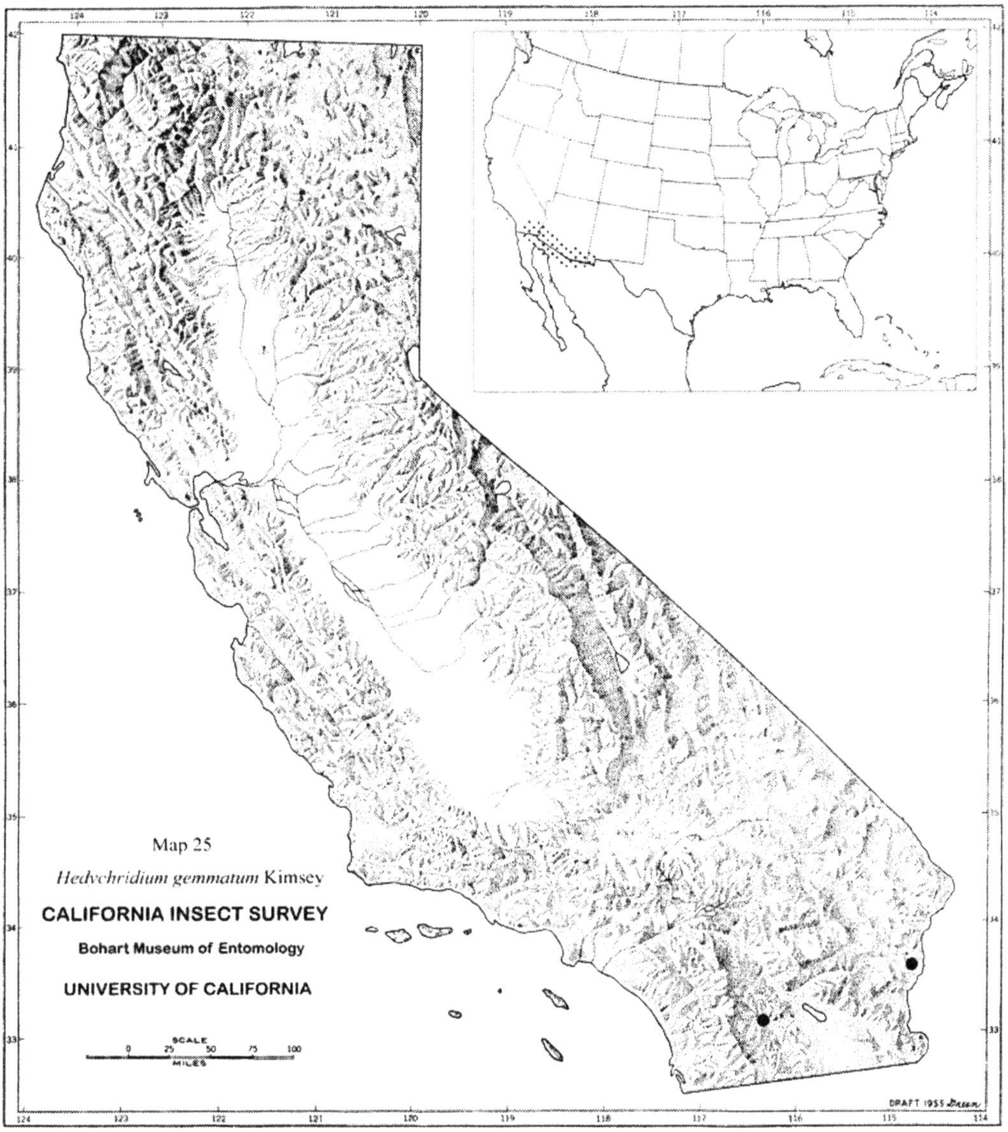

Map 25. California distribution of *Hedychridium gemmatum*. Inset: overall distribution.

Hosts. Unknown.

Discussion. This is a tiny species (about 2 mm long) related to *frontis* and *antennatum*, based on the straight medial vein and flagellomere I 1.5x as long as the pedicel. Distinguishing features include the brown tegula and flagellum, medial vein longer than Rs stub, flagellomere V as long as wide, and brow smoothly rounded, traversed by coarse punctation.

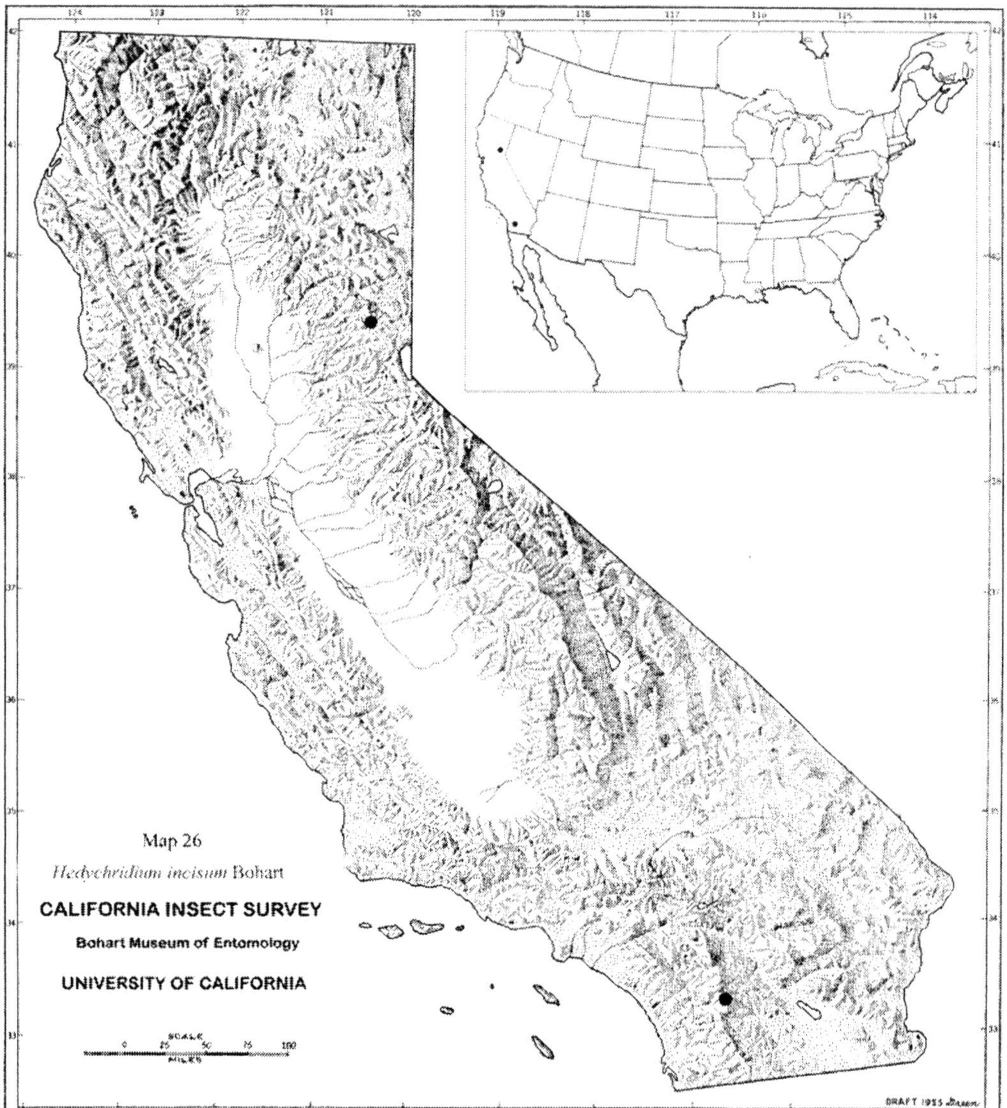

Map 26. California distribution of *Hedychridium incisum*. Inset: overall distribution.

Hedychridium incisum Bohart

Map 26

Hedychridium incisum Bohart, in Bohart & Kimsey 1978:615. Holotype ♂; USA: California, Sierra Co., Sierra Valley (UCDC).

Geographic Range (Map 26). This species is a California endemic found in the Upper Sonoran Zone of eastern and southern California; 5 specimens were studied.

California Records. Collected in October; **San Diego Co**.: 9 mi. s Warner Springs; **Sierra Co**.: Sierra Valley.

Hosts. Unknown.

Discussion. The short pedicel (less than half the length of flagellomere I), arched medial vein, cylindrical flagellum, and undifferentiated punctation between brow and lower frons indicate a relationship between *crassum, purum* and *incisum*. *H. incisum* can be distinguished by the unmodified tergum II, flagellomere I 2.3x as long as broad and 2.2x as long as the pedicel, wings infuscate and tergum III posteromedially indented. The modification of tergum III is the most distinctive feature of this species. This California endemic is rarely collected.

Hedychridium leucostigma Bohart

Map 27

Hedychridium leucostigma Bohart, in Bohart & Kimsey 1978:616. Holotype ♂; USA: California, San Diego Co., Borrego Valley (UCDC).

Geographic Range (Map 27). Southern California deserts, Arizona and in southern British Columbia; 22 specimens were seen.

California Records. Collected March through May, October; **Imperial Co**.: Palo Verde; 6.2 mi. n Glamis; **Inyo Co**.: Owens Lake; **Riverside Co**.: 18 mi. w Blythe; **San Diego Co**.: Borrego Valley.

Hosts. Unknown.

Discussion. These tiny pale wasps are structurally closest to *argenteum*, based on the straight medial vein, brown tegula, reddish flagellum and the dense silvery pubescence on the scapal basin. *H. leucostigma* differs from *argenteum* in having sparse pubescence on the body, yellow wing venation, a shorter, stouter flagellum, and flagellomere I shorter or as long as the pedicel length.

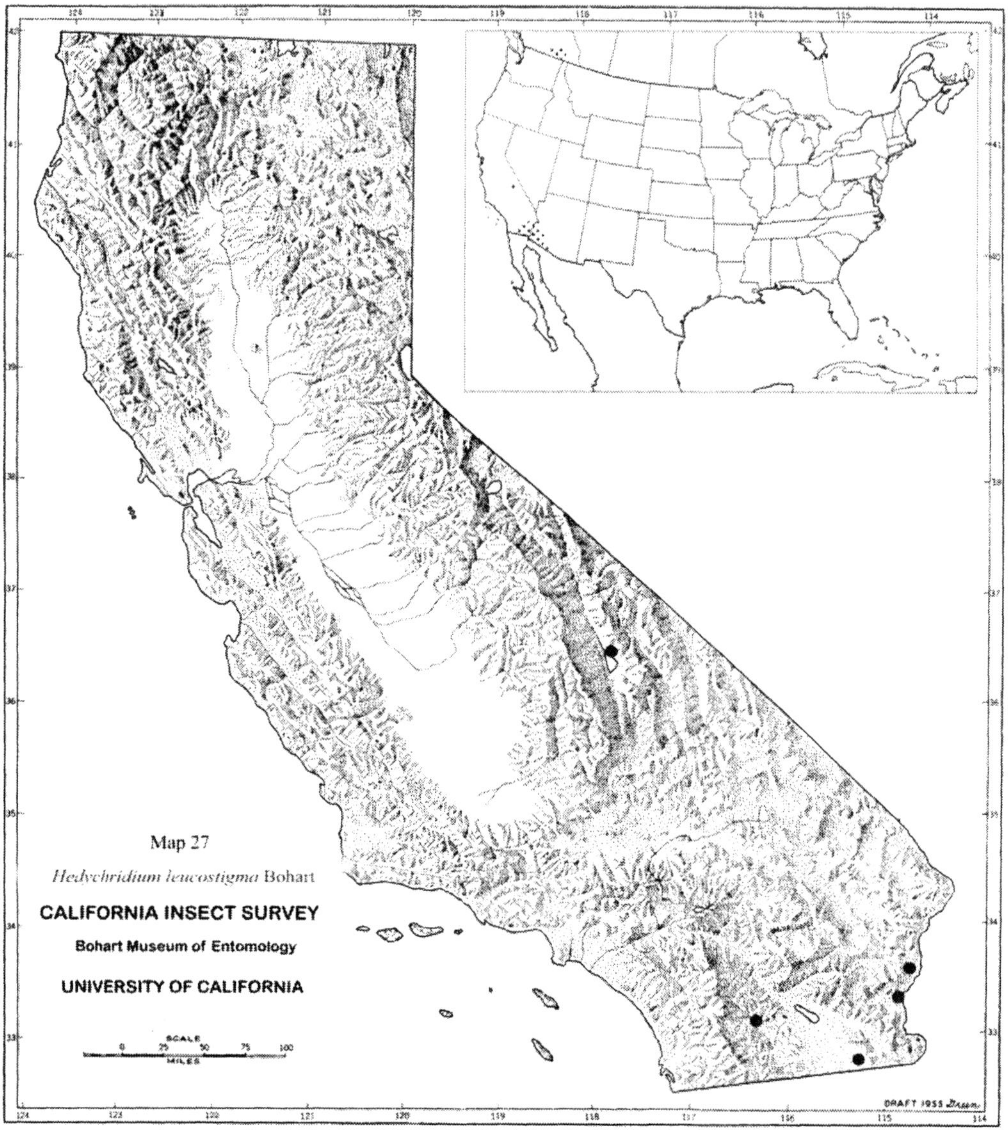

Map 27. California distribution of *Hedychridium leucostigma*. Inset: overall distribution.

Hedychridium lividum Bohart

Map 28

Hedychridium lividum Bohart, in Bohart & Kimsey 1978:617. Holotype ♂; USA: California: Riverside Co., Thousand Palms (UCDC).

Geographic Range (Map 28). Southern California deserts, in the Lower Sonoran Zone; 32 specimens were studied.

California Records. Collected March through October; **Imperial Co.**: Glamis; **Riverside Co.**: Thousand Palms; 19 mi. w Blythe; **San Diego Co.**: Borrego Valley.

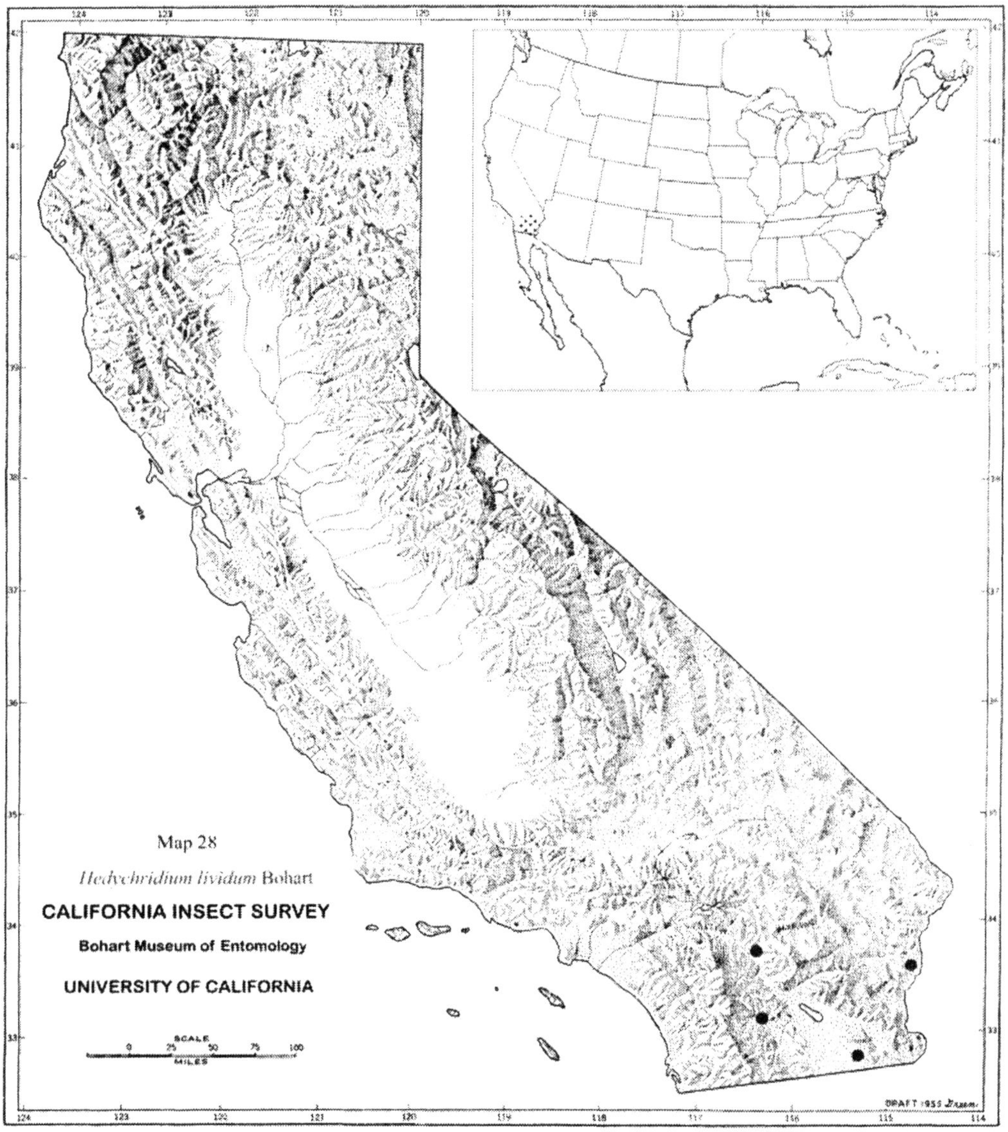

Map 28. California distribution of *Hedychridium lividum*. Inset: overall distribution.

Hosts. Unknown.

Discussion. Diagnostic features of this species are the short flagellomere I (1.5-1.7x as long as broad), malar space 0.3-0.6 MOD long, medial vein strongly curved; brow indistinct; frons sparsely pubescent, tergum II with large basomedial black area, and sternum III and tegula brown.

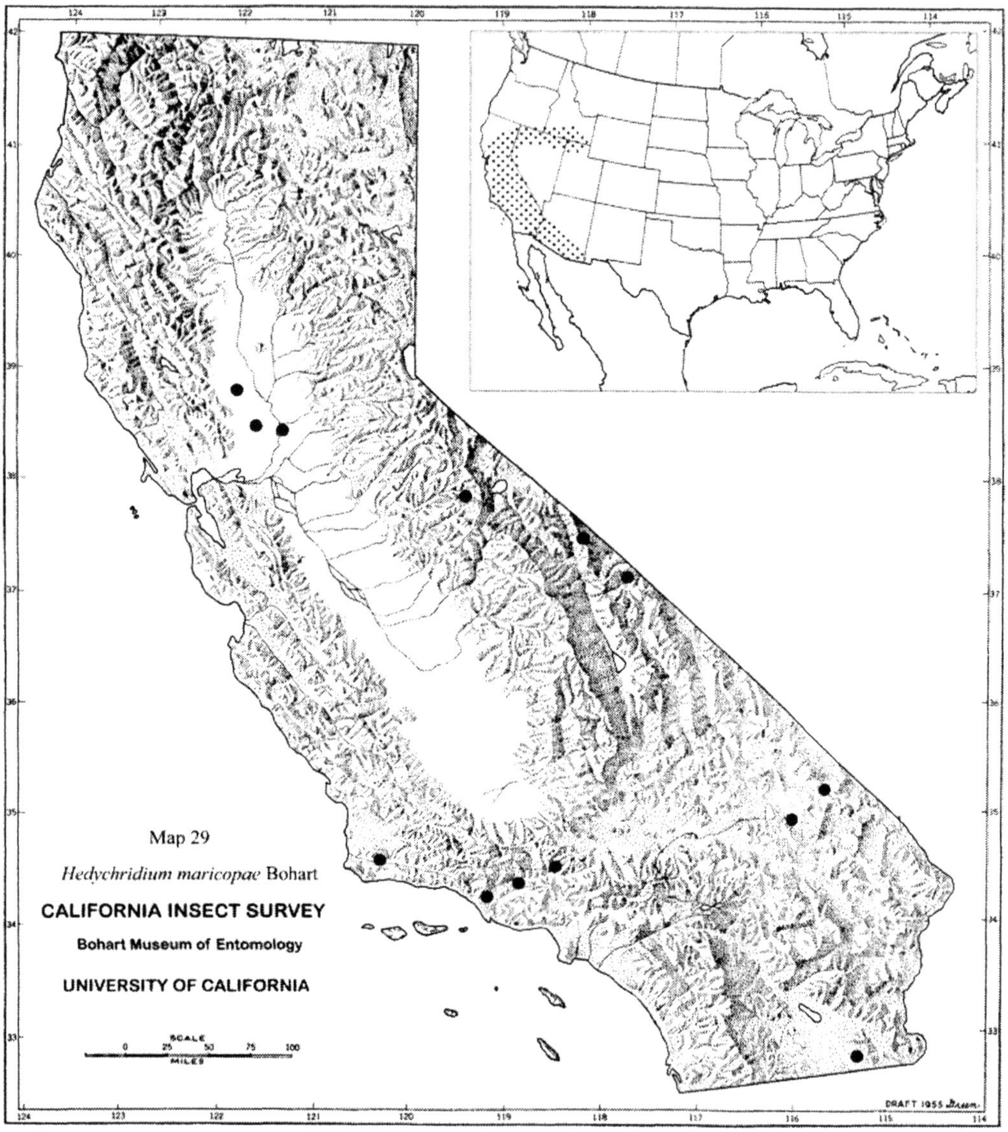

Map 29. California distribution of *Hedychridium maricopae*. Inset: overall distribution.

Hedychridium maricopae Bohart

Map 29

Hedychridium maricopae Bohart, in Bohart & Kimsey 1978:618. Holotype ♂; USA: Arizona, Maricopa Co., Wickenburg (USU).

Geographic Range (Map 29). *H. maricopae* occurs in the Upper and Lower Sonoran Zones of California, Nevada, Idaho, and Arizona; 38 specimens were studied.

California Records. Collected March through September; **Colusa Co**.: Arbuckle; **Imperial Co**.: Glamis; **Inyo Co**.: Deep Springs; **Mariposa Co**.: Yosemite National Park; **Mono Co**.: White Mts.; **Sacramento Co**.: Sacramento; **Santa Barbara Co**.: Santa Ynez; **San Bernardino Co**.: 7 mi. sw Kelso; Ivanpah; **Ventura Co**.: Foster Park; **Yolo Co**.: Yolo.

Hosts. Unknown.

Discussion. Similar to *amabile*, both species have a long flagellomere I (2.5x as long as broad and 1.3x as long as the pedicel), medial vein nearly straight, tegula brown, lower frons densely silvery and male sternum III blue. The short malar space, 0.7-0.8 MOD long will distinguish *maricopae* from *amabile*.

Hedychridium menkei Bohart

Map 30

Hedychridium menkei Bohart, in Bohart & Kimsey 1978:619. Holotype ♂; USA: California, Tuolumne Co., Leland Meadow (UCDC).

Geographic Range (Map 30). Primarily found in the western USA and Canada (Alberta, British Columbia and Ontario), in the Upper Sonoran to Canadian Zones; also recorded from Florida and Nebraska; 86 specimens were seen.

California Records. Collected June through October; **Alpine Co**.: Lake Winnemucca; Carson Pass; Hope Valley; **El Dorado Co**.: South Lake Tahoe; Echo Lake; Grass Lake; **Lassen Co**.: Hallelujah Junction; **Modoc Co**.: Cedar Pass campground; **Mono Co**.: White Mountains; Mill Creek Canyon, 7 mi. se Hwy 395; **Nevada Co**.: Truckee; Sagehen Creek; **Sacramento Co**.: Carmichael; **San Bernardino Co**.: Big Bear Lake; **San Francisco Co**.: Lands End; Ingleside; **Santa Barbara Co**.: Goleta; **Sierra Co**.: Yuba Pass; Sierraville; **Siskiyou Co**.: McBride Springs; **Trinity Co**.: 10 mi. n Coffee Creek Rangers Station; **Tuolumne Co**.: Leland Meadow; Soda Springs Yosemite National Park; **Yolo Co**.: Davis.

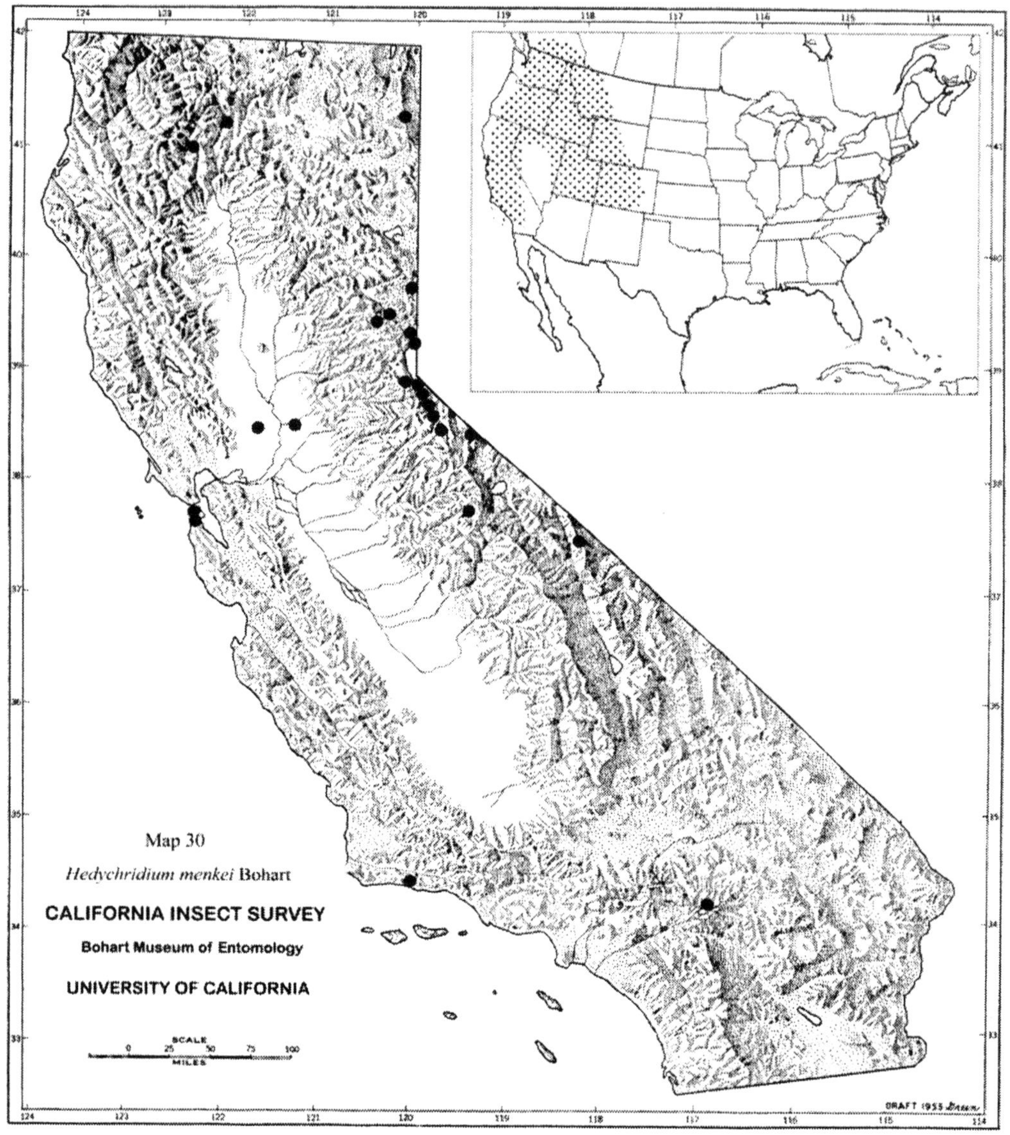

Map 30. California distribution of *Hedychridium menkei*. Inset: overall distribution.

Hosts. Unknown.

Discussion. The long flagellomere I, short malar space, curved medial vein and sparse pubescence on the frons indicates a relationship with *bilobatum* and *dimidiatum*. Distinguishing features include the cross-ridging of the lower face nearly reaching the eye margin, tegula and sternum III brown, and tergum II with a large indistinctly edged black area.

Hedychridium milleri Kimsey

Hedychridium milleri Kimsey, in Bohart & Kimsey 1978:620. Holotype ♂; USA: California, Lake Co., Borax Lake (UCDC).

Geographic Range. This species is endemic to central California; 51 specimens were seen, only from the type series.

California Records. Collected in June, **Lake Co**.: Borax Lake.

Hosts. Unknown.

Discussion. In this species the brow is indistinct, the scapal basin is flattened and sparsely setose and the tegula is brown. *H. milleri* is similar to *paulum* based on the straight medial vein, slightly infuscate wings, bluish sterna II-III, flagellomere I 1.8-2.0x as long as broad and about as long as the pedicel, short malar space (0.6 MOD long) and subantennal distance 0.6 MOD long.

This species has only been collected in the type locality, where it must be locally abundant, based on the large number of specimens in the type series. However, no one has returned to this site since the original series was collected.

Hedychridium mirabile Kimsey

Map 31

Hedychridium mirabile Kimsey, in Bohart & Kimsey 1978:620. Holotype ♂; USA: California, Yolo Co., Rumsey (UCDC).

Geographic Range (Map 31). Western USA; mostly in the Upper Sonoran and Transition Zones; 253 specimens.

California Records. Collected April through October; **Alameda Co**.: Corral Hollow; **Imperial Co**.: Glamis; **Inyo Co**.: Eureka Valley; Big Pine Creek, 2 mi. e Lone Pine; 2 mi. s Lone Pine; Little Lake; 13 mi. s Death Valley Junction; **Kern Co**.: Kernville; **Lake Co**.: Lakeport; **Los Angeles Co**.: Glendale; **Mono Co**.: Bridgeport; **Monterey Co**.: 1 mi. s Soledad; Arroyo Seco; **Orange Co**.: Irvine Park; **Riverside Co**.: 8 mi. w Blythe; 20 mi. w Blythe; **Sacramento Co**.: Grand Island; Sacramento; **San Bernardino Co**.: 3 mi. s Kramer Junction; Kelso Dunes; **San Diego Co**.: Borrego Valley; 2 mi. n Warner Springs; **San Joaquin Co**.: Corral Hollow 8 mi. sw Tracy; **Santa Clara Co**.: Isabel Creek; **Siskiyou Co**.: Shasta City; **Solano Co**.: Fairfield; **Yolo Co**.: Rumsey; 2 mi. nw Rumsey; Davis; Elkhorn Ferry; **Yuba Co**.: 18 m ne Marysville.

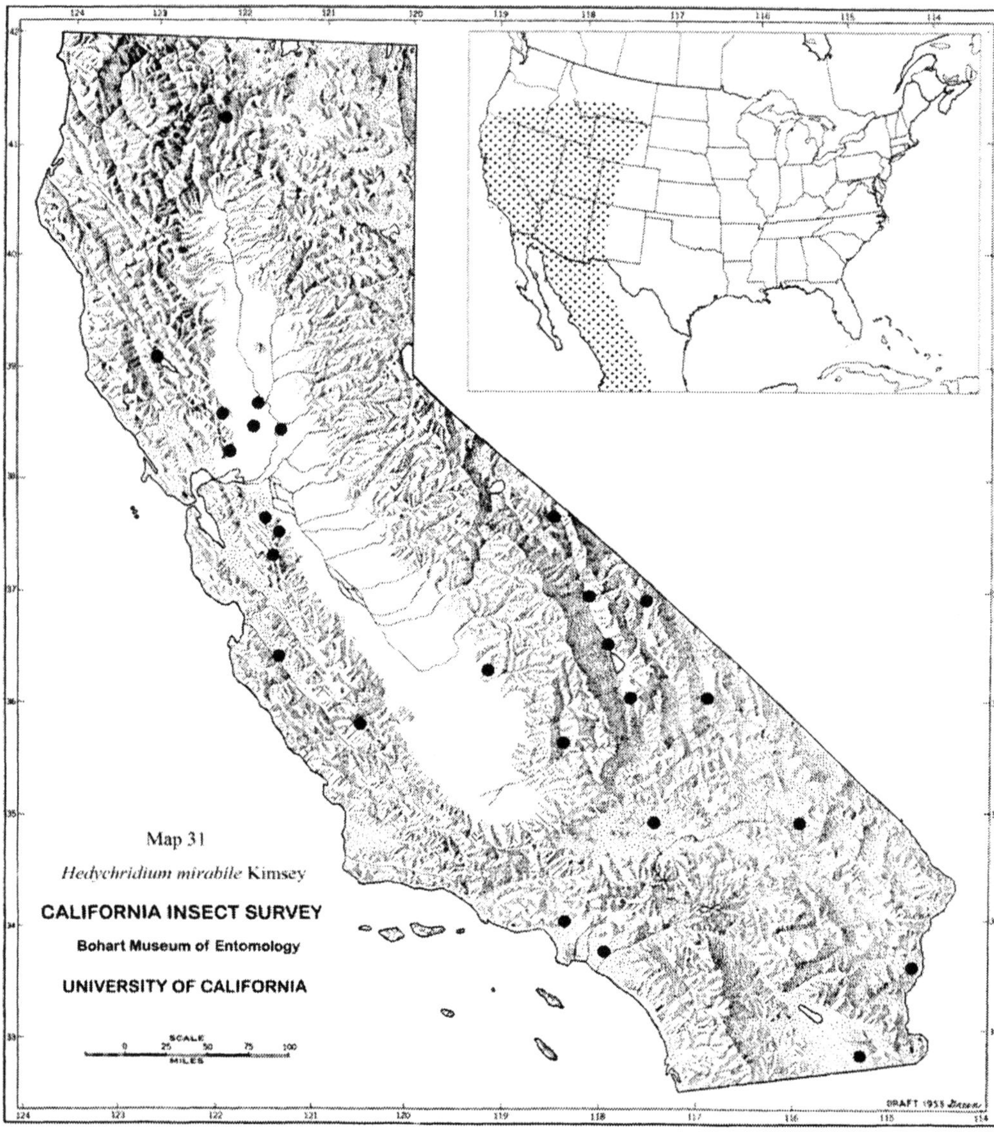

Map 31. California distribution of *Hedychridium mirabile*. Inset: overall distribution.

Hosts. Unknown.

Discussion. The prominent brow, densely silvered scapal basin, flagellomere I less than twice as long as broad and subequal to the pedicel length, and the curved medial vein suggest a relationship with *fletcheri*. Distinguishing features include the Rs stub shorter than the medial vein, sternum II brown, flagellomere V longer than wide, tegula brown, and subantennal distance 0.5-0.7 MOD long. As in *amabile*, some individuals are all blue or green, others have bright coppery to gold markings. Both color types occur in some localities.

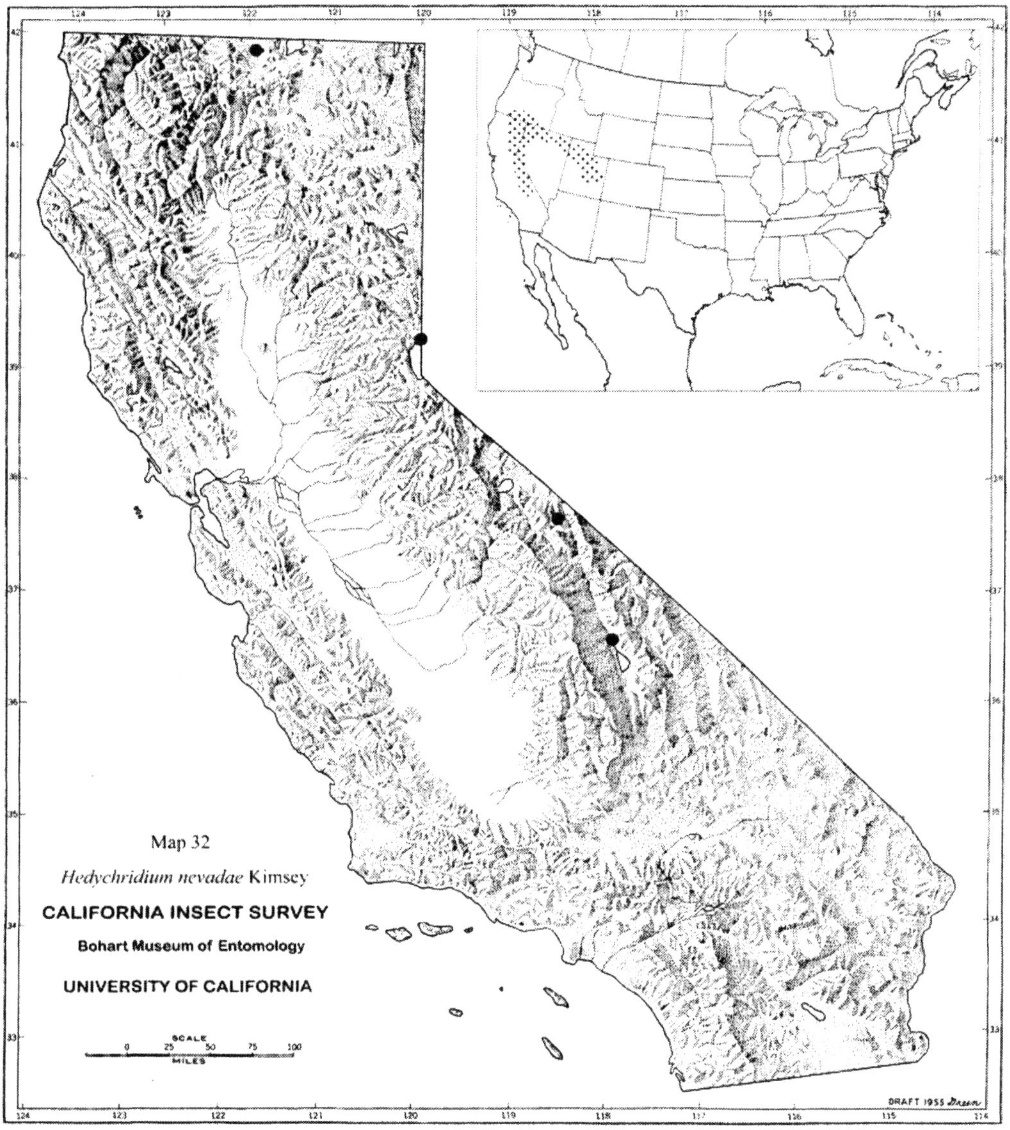

Map 32. California distribution of *Hedychridium nevadae*. Inset: overall distribution.

Hedychridium nevadae Kimsey

Map 32

Hedychridium nevadae Kimsey, in Bohart & Kimsey 1978:621. Holotype ♂; USA: California, Inyo Co.: Lone Pine (UCDC).

Geographic Range (Map 32). Western USA, mostly in Transition Zones; 13 specimens were seen.

California Records. Collected August through October; **Inyo Co**.: Lone Pine; **Mono Co**.: Benton; **Siskiyou Co**.: Macdoel.

Hosts. Unknown.

Discussion. *H. nevadae* is one of the largest *Hedychridium* species, averaging 6-8 mm long. It has a number of distinctive and unusual traits, including the long, slender and indistinctly toothed mandible and pronotum flat to the apex of the collar. Additional diagnostic features include the subantennal distance more than 2 MOD long, scapal basin with sparse dark setae, malar space about 1 MOD long, flagellomere I twice as long as broad, and body dark green to purple.

Hedychridium olene Kimsey

Fig. 22, Map 33

Hedychridium olene Kimsey, in Bohart & Kimsey 1978:622. Holotype ♂; USA: California, Nevada Co., Boca (UCDC).

Geographic Range (Map 33). Western USA, Mexico (Baja California) and Canada (Ontario, Saskatchewan), mostly in Transition and Canadian Zones; 23 specimens were studied.

California Records. Collected April through October; **Humboldt Co**.: Arcata; **Inyo Co**.: Antelope Springs; **Nevada Co**.: Boca; Sagehen Creek; **Riverside Co**.: Cottonwood Springs; 18 m w Blythe; **San Bernardino Co**.: New York Mts.; **San Diego Co**.: Borrego Valley; **San Luis Obispo Co**.: Camatta Canyon; **Sierra Co**.: Sierraville; Sierra Valley; **Tuolumne Co**.: Leland Meadow; Dodge Ridge; Strawberry; **Yolo Co**.: Davis.

Hosts. Unknown.

Discussion. The most distinctive feature of *olene* is the elbowed forefemur. Although this modification is sometimes seen in palearctic and afrotropical species of *Hedychridium*, *Holopyga* and *Xerochrum, olene* is the only Western Hemisphere species of *Hedychridium* with an elbowed forefemur. Additional diagnostic features include flagellomere I more than twice as long as broad, sternum II brown, medial vein curved, malar space about 0.5 MOD long, and the lower frons sparsely pubescent.

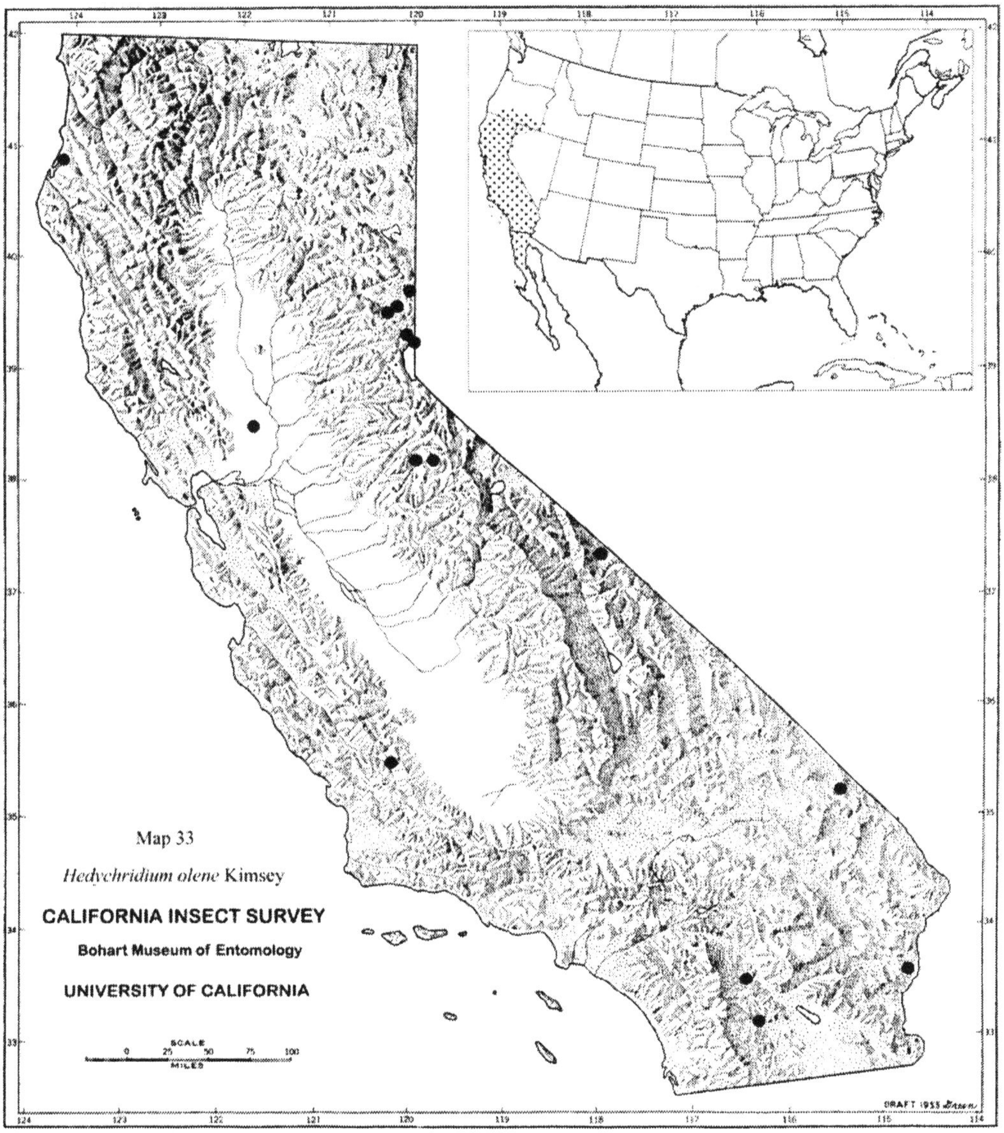

Map 33. California distribution of *Hedychridium olene*. Inset: overall distribution.

Hedychridium paulum Bohart

Map 34

Hedychridium paulum Bohart, in Bohart & Kimsey 1978:623. Holotype ♂; USA: California, Nevada Co., Sagehen Creek (UCDC).

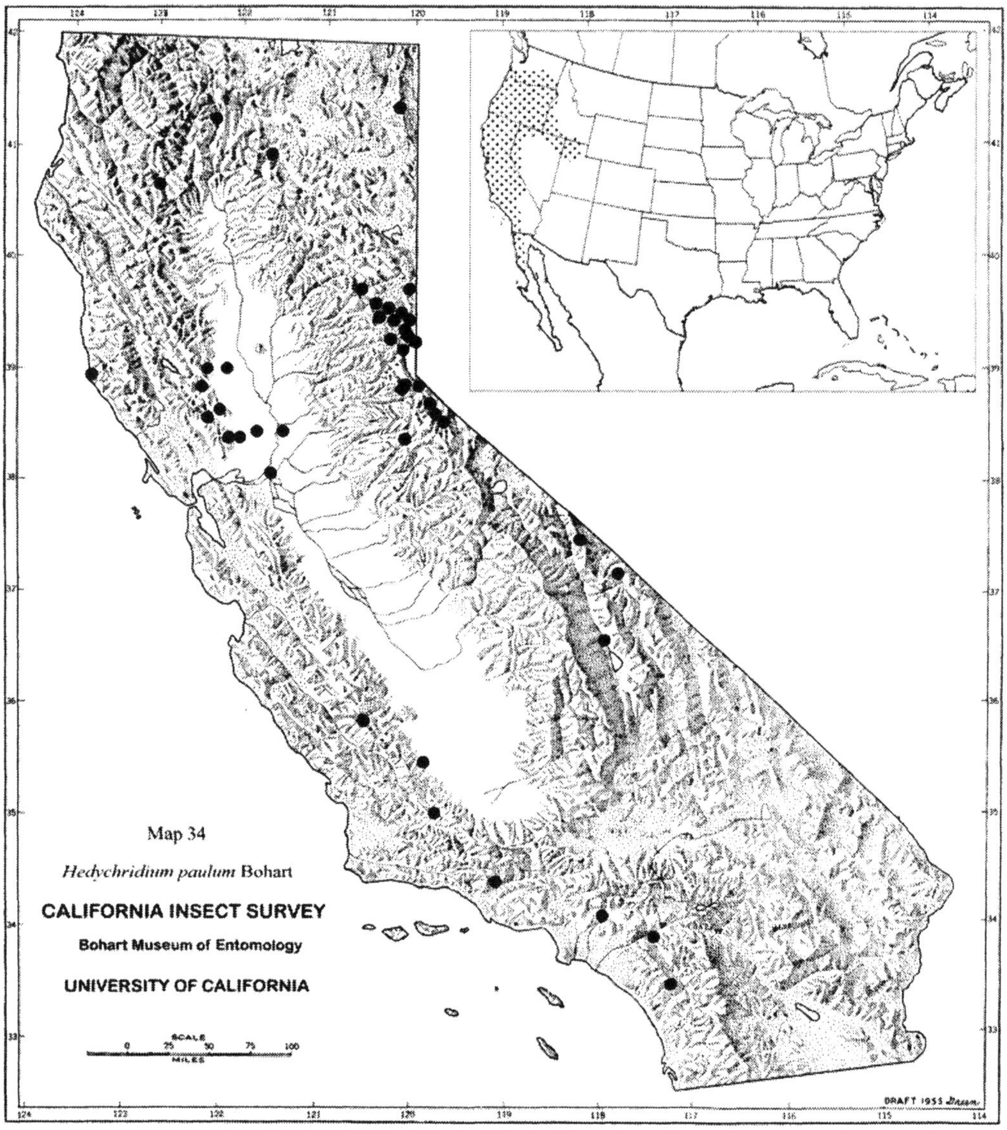

Map 34. California distribution of *Hedychridium paulum*. Inset: overall distribution.

Geographic Range (Map 34). Western USA and Baja California, Mexico; mostly in the Canadian Zone; 283 specimens were examined.

California Records. Collected April through October; **Alpine Co.**: Carson Pass; Hope Valley; **Amador Co.**: Pioneer; **Colusa Co.**: 21 mi. sw Williams; Stonyford; **El Dorado Co.**: 1 mi. s Meyers; Fred's Place; Grass Lake; **Inyo Co.**: Deep Springs Lake; 3 mi. w Lone Pine; **Kern Co.**: 19 mi. w Shafter; **Lassen Co.**: Hallelujah Junction; **Lake Co.**: north fork Cache Creek; **Los Angeles Co.**: Tanbark Flat;

Mendocino Co.: 4 mi. e Point Area; **Modoc Co.**: 5.5 mi. e Cedarville; **Mono Co.**: White Mountains; **Monterey Co.**: Arroyo Seco; **Napa Co.**: 4 mi. nw Berryessa; **Nevada Co.**; Sagehen Creek; Boca; Russell Valley; **Placer Co.**: Carnelian Bay Lake Tahoe; **Plumas Co.**: Johnsville; **Riverside Co.**: 1 mi nw Mira Loma; 3 mi. nw Temecula Canyon; **Sacramento Co.**: Sacramento; Grand Island near Isleton; **San Luis Obispo Co.**: 3 mi. s Oceano; **Shasta Co.**: Hat Creek; **Sierra Co.**: Sierra Valley; Sierraville; Sattley; Independence Lake; Yuba Pass; Weber Lake; **Siskiyou Co.**: Mt. Shasta City; **Trinity Co.**: Junction City; **Ventura Co.**: Foster Park; **Yolo Co.**: Davis; Rumsey; Elkhorn Ferry; 8 mi. w Winters; Putah Canyon.

Hosts. Unknown.

Discussion. *H. paulum* is abundant in the western states. Individuals range in color from blue green to intensely coppery red. Diagnostic features include flagellomere I 1.5x as long as broad, malar space 0.5-0.7 MOD long, medial vein almost straight, brow and lower frons closely and evenly punctate; subantennal distance less than 1 MOD long, and tegula and sterna II-III greenish.

Hedychridium politum Bohart

Map 35

Hedychridium politum Bohart, in Bohart & Kimsey 1978:624. Holotype ♂; USA: California, Nevada Co., Sagehen Creek (UCDC).

Geographic Range (Map 35). Western USA, mostly in the Transition and Canadian Zones; 223 specimens were studied.

California Records. Collected April through September; **Alpine Co.**: Caples Lake; Ebbetts Pass; Hope Valley; Carson Pass; **Contra Costa Co.**: Danville; **El Dorado Co.**: Fallen Leaf Lake; 2 mi. e Ice House; Grass Lake; **Fresno Co.**: Huntington; **Inyo Co.**: Eureka Valley; **Kern Co.**: Mt. Pinos; **Mariposa Co.**: Yosemite National Park; **Mono Co.**: Mammoth Lake; Sheep Mt. White Mountains; **Nevada Co.**: Sagehen Creek; Jackson Lake; Fuller Lake; Hobart Mills; **Placer Co.**: Carnelian Bay, Lake Tahoe; **Plumas Co.**: 14 mi. w Johnsville; **Santa Barbara Co.**: Santa Cruz Island; **Shasta Co.**: Snow Mountain Rd.; **Sierra Co.**: 10 mi. s Sierraville; Sierra Buttes; Weber Lake; Independence Lake; **Siskiyou Co.**: Panther Meadows, Mt. Shasta; McBride Spring; **Trinity Co.**: Mountain Meadow Ranch Headquarters Coffee Creek; Big Flat Coffee Creek; **Tulare Co.**: Giant Trees; **Tuolumne Co.**: Strawberry; Leland Meadow; Boundary Hill Research Area Yosemite National Park.

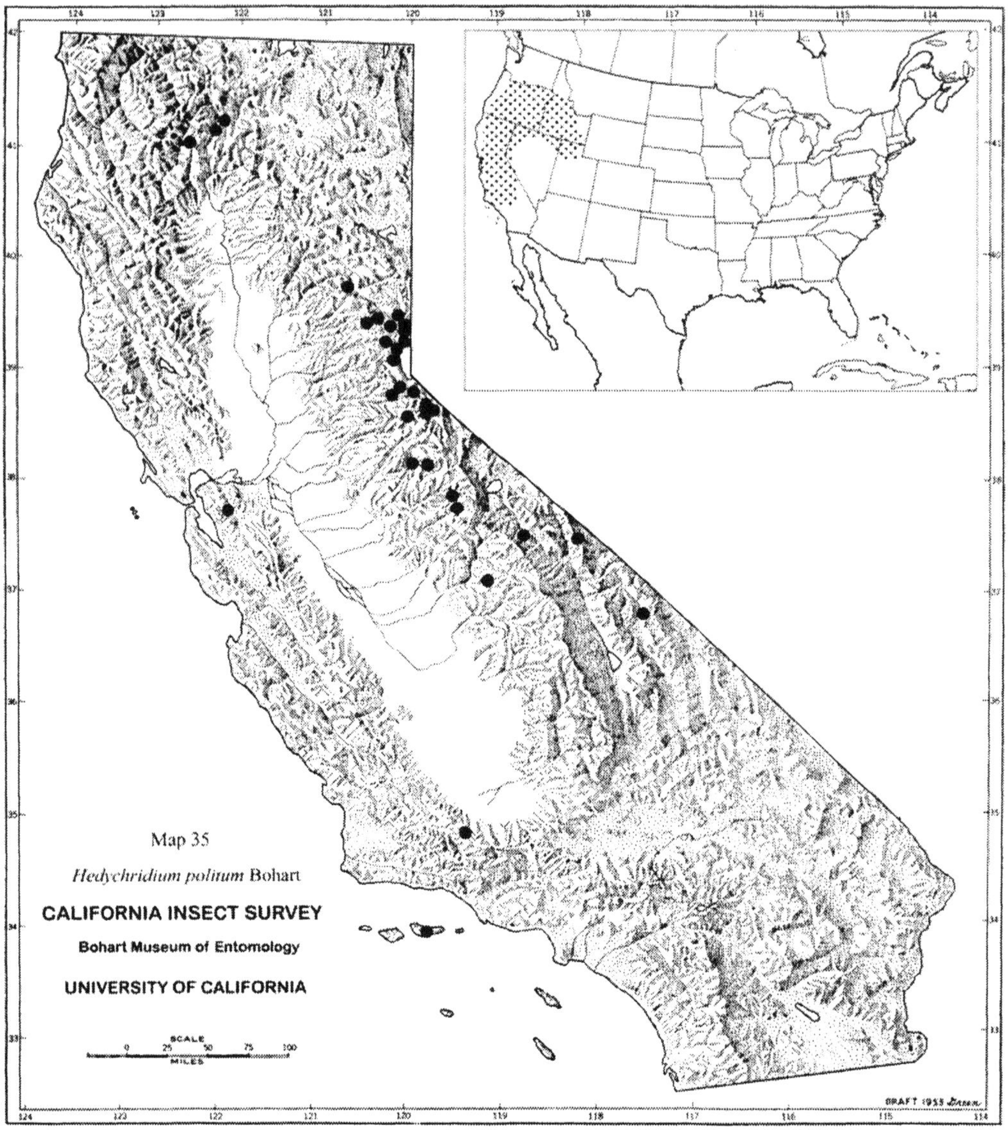

Map 35. California distribution of *Hedychridium politum*. Inset: overall distribution.

Hosts. Unknown.

Discussion. This is a greenish blue species that shares a number of characteristics with *dimidiatum*, including flagellomere I more than twice as long as broad, malar space about 0.5 MOD long, lower frons sparsely setose and medial vein curved. In addition, the tegula is green, tergum III is broadly rounded apically, the scapal basin has a weak medial zone of cross-ridging, and the male sternum III blue.

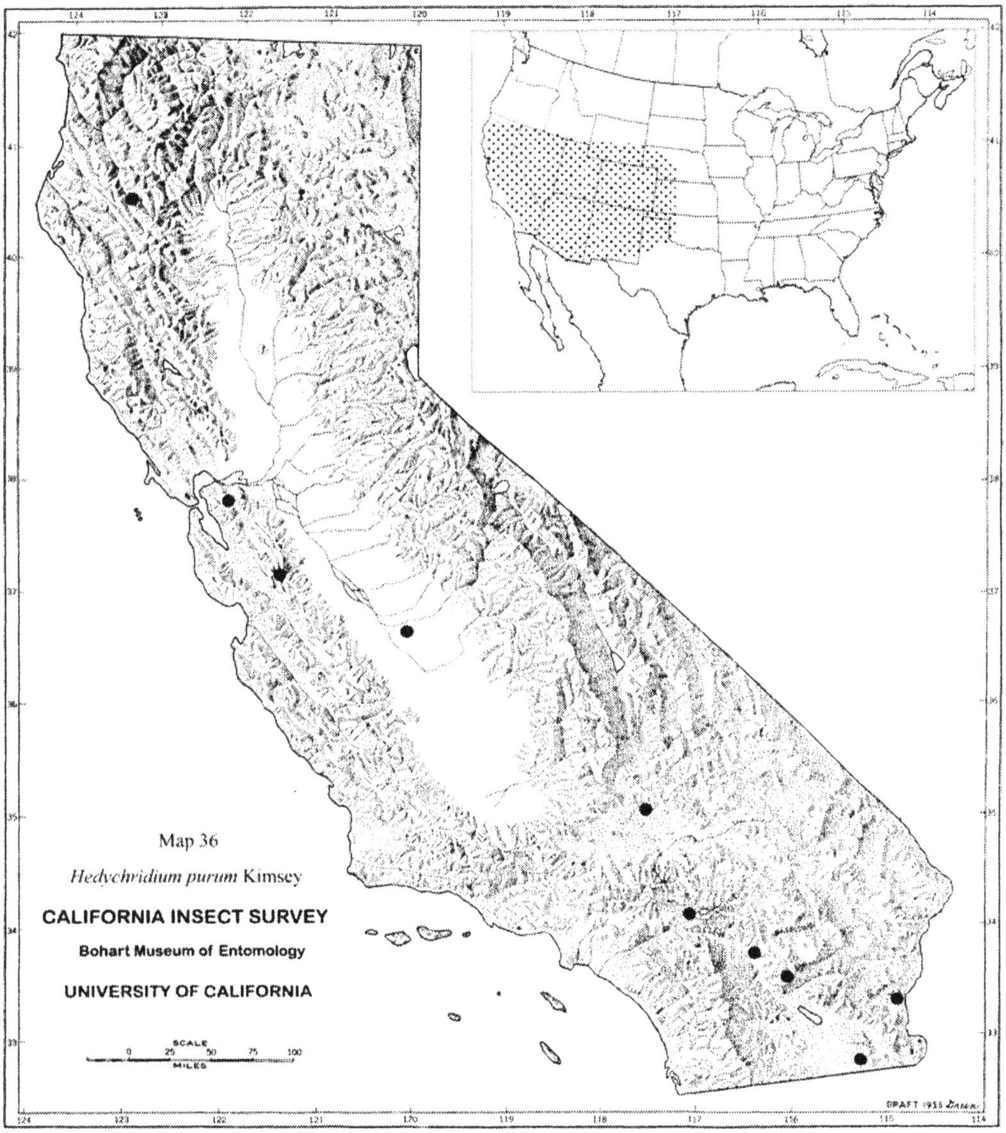

Map 36. California distribution of *Hedychridium purum*. Inset: overall distribution.

Hedychridium purum Kimsey

Map 36

Hedychridium purum Kimsey, in Bohart & Kimsey 1978:625. Holotype ♂; USA: Utah, Box Elder Co., Kelton (UCDC).

Geographic Range (Map 36). Western USA and northwestern Mexico (Baja California Norte), Upper Sonoran Zones; 10 specimens were studied.

California Records. Collected April through August; **Contra Costa Co**.: Danville; **Fresno Co**.: Panoche Rd.; **Imperial Co**.: 3.5 mi nw Glamis; Glamis; Palo Verde 6 mi. n Glamis; **Riverside Co**.: Thousand Palms; Box Canyon; **San Bernardino Co**.: 4 mi. s Kramer Junction; Colton Hills; **Santa Clara Co**.: San Antonio Valley; **Trinity Co**.: Butter Creek.

Hosts. Unknown.

Discussion. The most distinctive feature of *purum* is the whitish color of the wing membrane. In addition, Rs is short and as long as the stigmal apex, tegula is blue to green, and the female flagellum has some red or yellow. *H. purum* shares a number of features with *crassum* as discussed under that species.

Hedychridium rasile Bohart

Map 37

Hedychridium rasile Bohart, in Bohart & Kimsey 1978:626. Holotype ♂; USA: California, Nevada Co., Sagehen Creek (UCDC).

Geographic Range (Map 37). Canadian Zone of California and Nevada; 56 specimens were seen.

California Records. Collected June through August; **Alpine Co**.: 1 mi. s Monitor Pass; Winnemucca; Highland Lakes; Carson Pass; **Humboldt Co**.: Hoopa; **Modoc Co**.: 10 mi. sw Eagleville Patterson Meadow; **Nevada Co**.: Sagehen Creek; Boca; Russell Valley; **Placer Co**.: Carnelian Bay and Incline Beach, Lake Tahoe; **Sierra Co**.: Independence Lake.

Hosts. Unknown.

Discussion. A small greenish blue species, *rasile* resembles *amabile* based on flagellomere I more than twice as long as broad, the nearly straight medial vein, and the brown tegula. Diagnostic features of *rasile* are the extensively cross-ridged mesopleuron, weakly pubescent lower frons, malar space 0.5 MOD long, male sternum III blue, and flagellomere I less than twice the pedicel length.

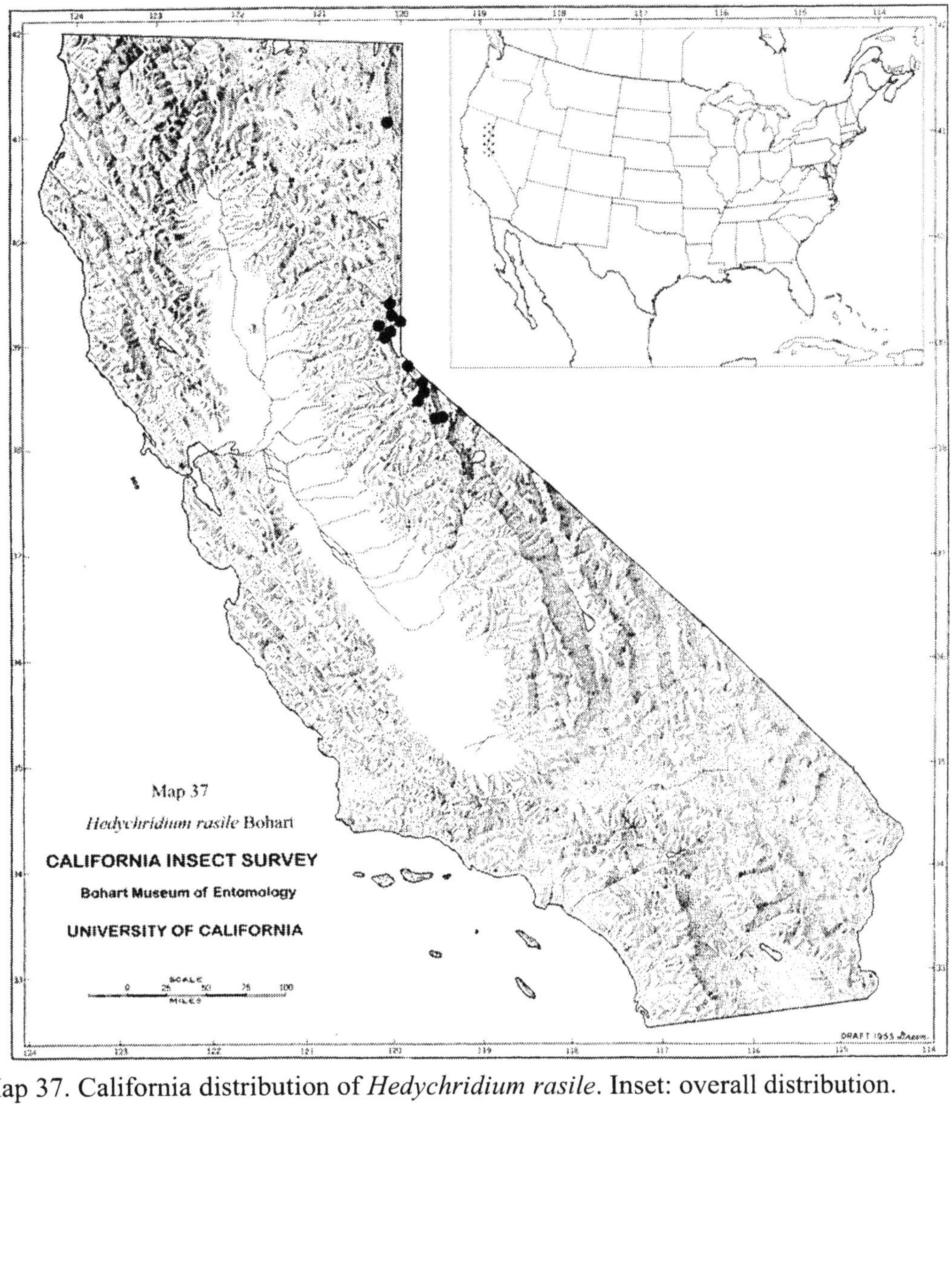

Map 37. California distribution of *Hedychridium rasile*. Inset: overall distribution.

Hedychridium solierellae Bohart & Brumley

Map 38

Hedychridium solierellae Bohart & Brumley 1967:234. Holotype ♂; USA: California, Colusa Co., Arbuckle (UCDC).

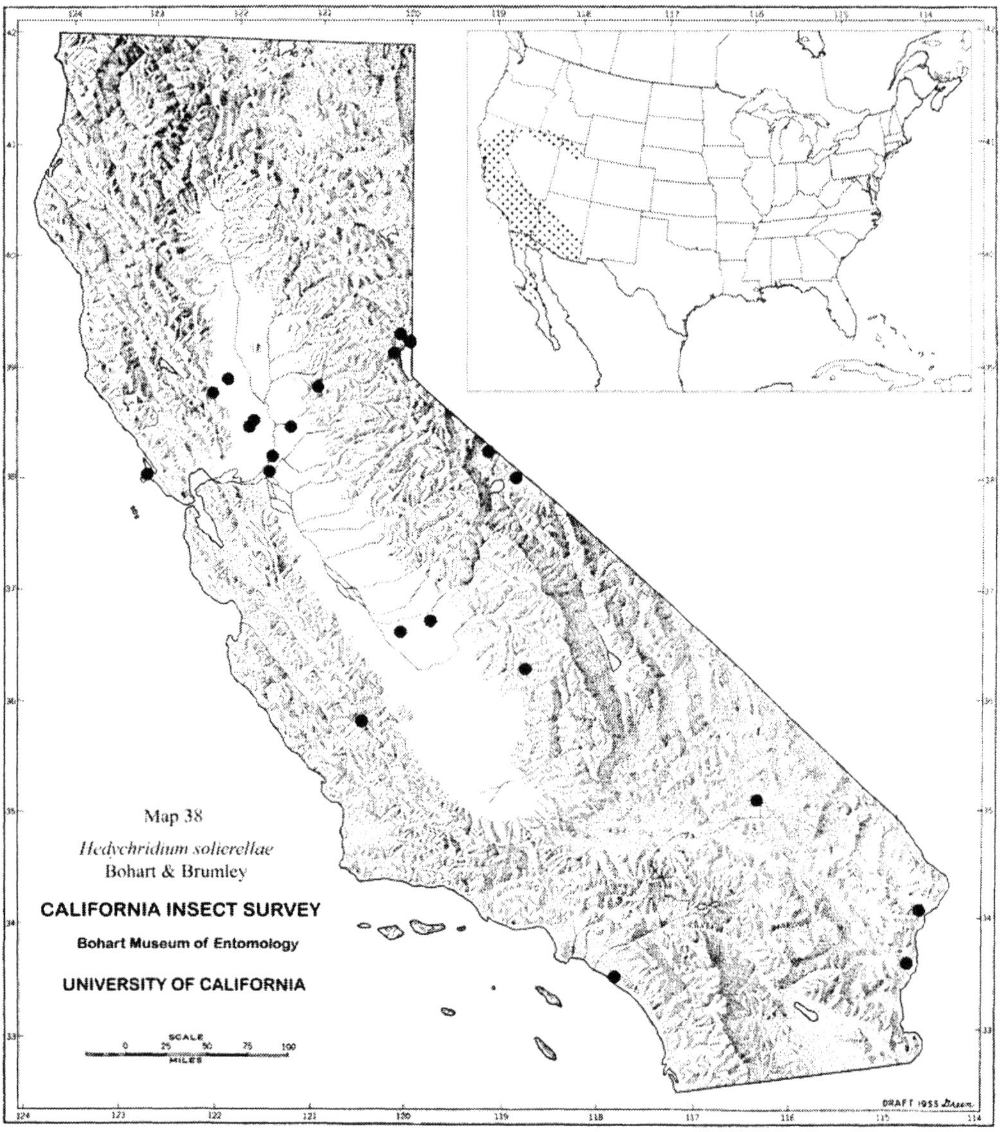

Map 38. California distribution of *Hedychridium solierellae*. Inset: overall distribution.

Geographic Range (Map 38). Western USA, southern Canada (British Columbia), and Baja California, Mexico, all west of 100th meridian; 367 specimens were studied.

California Records. Collected March through October; **Colusa Co**.: Arbuckle; **Fresno Co**.: Fresno; Panoche Rd.; **Marin Co**.: Point Reyes; **Mono Co**.: Mill Creek Canyon, 7 mi. s Hwy 395; **Monterey Co**.: Arroyo Seco; **Nevada Co**.: Boca; Sagehen Creek; **Orange Co**.: Laguna Beach; **Placer Co**.: 4 mi. s Rocklin; Carnelian Bay, Lake Tahoe; **Riverside Co**.: 18 mi. w Blythe; **Sacramento Co**.: Grand Island; Sacramento River levee; Sacramento; Carmichael; **San Bernardino Co**.: 5 mi. n Vidal; 15 mi. w Baker; **San Diego Co**.: Borrego Valley; **Tulare Co**.: Ash Mountain; **Yolo Co**.: Davis; 5 mi. e Woodland; 5 mi. e Guinda.

Hosts. This species has been reared from the nests of the crabronid wasps *Solierella peckhami* (Ashmead) and *S. plenoculoides* (W. Fox) (Carrillo and Caltagirone 1970, Parker and Bohart 1968).

Discussion. This is another small greenish blue species related to *amabile*, with a straight medial vein, flagellomere 2.5x as long as broad, and a brown tegula. *H. solierellae* can be separated from *amabile* by the 1 MOD long malar space, tergum II with a large dark basomedial spot and nearly contiguous punctation, and the male sternum III brown.

Genus *Hedychrum* Latreille

Generic Diagnosis. Face with sparse erect setae; scapal basin deeply concave, with transverse cross-ridging; malar space less than 1 MOD long; base of oral fossa with sharp tooth; pronotum with anteromedial pit; notauli sulciform; mesopleuron rounded, with indistinct omaulus and short scrobal sulcus; metanotum rounded; propodeum without medial enclosure and with medial carina; mid and hindtibia with pit or depression on inner surface; hindfemur enlarged, anterior surface brown and microreticulate in males; tarsal claws with subparallel tooth, appearing apically bifid; forewing medial vein slightly curved, arising at cu-a, stigma slender and apically acute; tergum III usually with lateral tooth and subapically swollen; female sternum III with subbasal sulcus extending toward midline and often with apicomedial tubercle; volsella divided into digitus and cuspis.

Hosts. Few hosts have been reported for *Hedychrum,* but all are crabronids in the subfamily Philanthinae. Byers (1978) reared an undetermined species of *Hedychrum* from *Cerceris halone* Banks.

Distribution. This genus is found worldwide, with the greatest diversity in the Palearctic Region (table 1). Only three species of *Hedychrum* occur in California.

Discussion. The genus *Hedychrum* has a number of unique characteristics that make it easy to recognize, and unlike most other chrysidines the sexes are easy to distinguish. Both sexes have a pit on the inner surface of the hindtibia and bifid tarsal claws. Females have a subbasal sulcus on sternum III and associated medial tooth or lobe. The anterior surface of the female hindfemur is smooth and shiny green, whereas in males the hindfemur is reticulate and dull brown. The American species were treated by French (in Bohart and Kimsey1982).

Key to the California Species of *Hedychrum*

1 Sterna II-III brown, although II or III may have small medial spot of green or blue ....*boharti* French

- Sterna II-III extensively green to blue...2

2 Hindtibial and tergum III apical setae black, mostly 1.5 MOD long; tergum I smooth subbasally ecarinate; female sternum III apical tooth not bilobed (Fig. 20) ....*nigropilosum* Mocsáry

- Hindtibial setae white to light brown, tergum III apical setae white to black, both tergum III apical setae and hindtibial setae mostly 1 MOD long; tergum I basally with three or more longitudinal carinae; female sternum III apical tooth bilobed ....*parvum* Aaron

Hedychrum boharti French

Fig. 23, Map 39

Hedychrum boharti French, in Bohart & Kimsey 1982:76. Holotype ♂; USA: California, Mono Co., 11 mi. n Bridgeport (UCDC).

Geographic Range (Map 39). Upper Sonoran Zone in the western USA and Baja California, Mexico; 76 specimens were seen.

California Records. Collected May through October; **Alpine Co**.: Bear Valley; **Colusa Co**.: 10 mi. sw Stonyford; **Contra Costa Co**.: Antioch; **El Dorado Co**.: 8 mi. ne Ice House Rd.; **Inyo Co**.: Big Pine; 5 mi. e Big Pine; Independence; **Lassen Co**.: Hallelujah Junction; **Los Angeles Co**.: Tanbark Flat; Junction City; Elizabeth Canyon; La Crescenta; **Modoc Co**.: 2 mi. e Canby; **Mono Co**.: 11 mi. n Bridgeport; Montgomery City; 3 mi. e Benton; **Monterey Co**.: Arroyo Seco; **Napa Co**.: Monticello Dam; **Nevada Co**.: Boca; **Orange Co**.: Irvine Park; **Plumas Co**.: Blairsden; Taylorsville; **Riverside Co**.: Temecula Canyon; 9 mi. w Beaumont; San Timoteo Canyon; Saboda Springs; **San Benito Co**.: 24 mi. sw Idria; **San Diego Co**.: East El Cajon; San Diego; **San Joaquin Co**.: Tracy; **San Luis Obispo Co**.: 5 mi. ne Santa Margarita; Creston; **Santa Barbara Co**.: Yucaipa 3 mi. se Santa Barbara; Figorea Mts.; Bluff Camp San Rafael Mts.; Goleta; Santa Cruz Island; 4 mi. e Los Prietos; Santa Ynez Mts.; **Santa Cruz Co**.: 4 mi. e Boulder Creek; **Siskiyou Co**.: Mt. Shasta City; **Sonoma Co**.: Petaluma; **Stanislaus Co**.: Del Puerto Canyon; **Trinity**

Co.: Junction City; **Ventura Co.**: Foster Park; Sespe Canyon; **Yolo Co.**: Putah Canyon; Woodland; Davis; 5 mi. e Woodland.

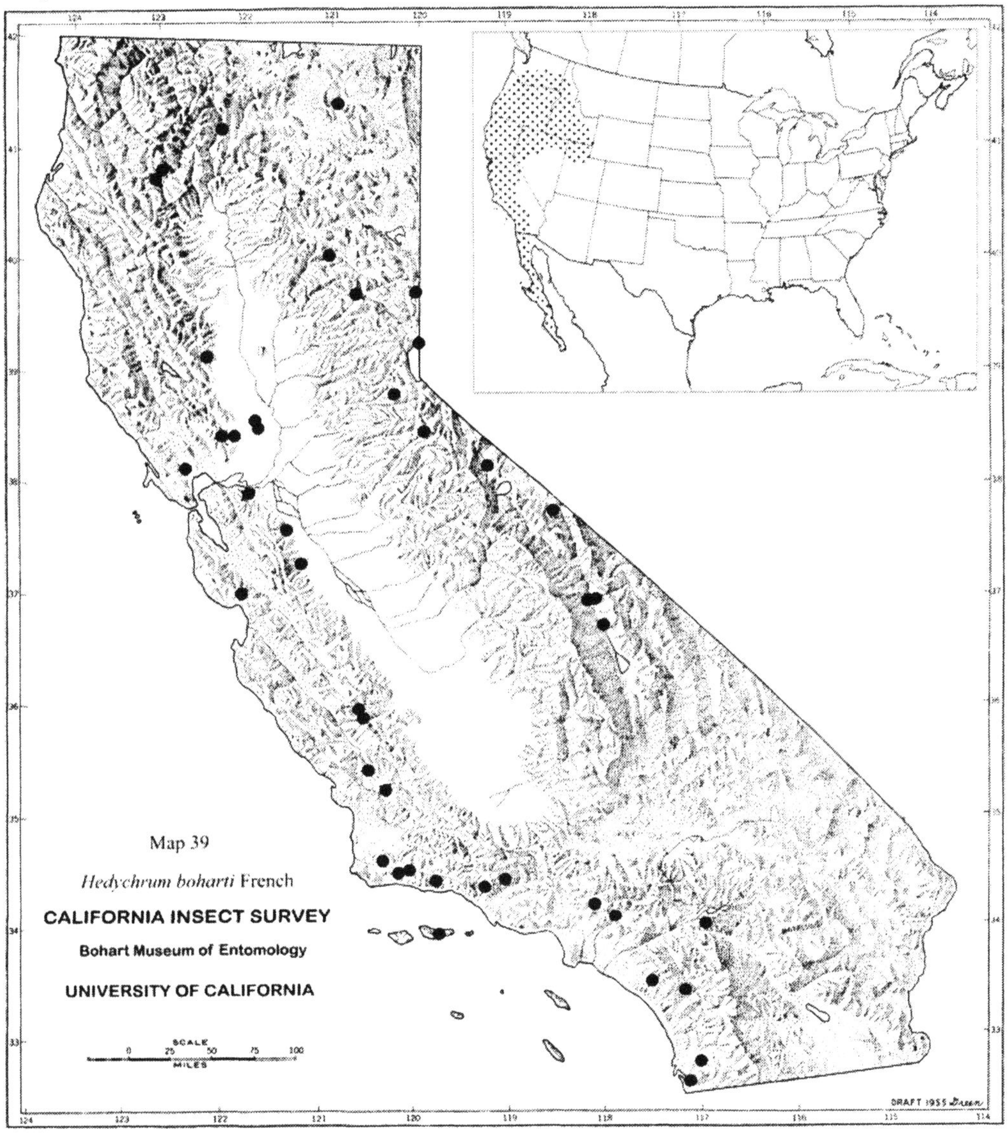

Map 39. California distribution of *Hedychrum boharti*. Inset: overall distribution.

Hosts. Unknown.

Discussion. *Hedychrum boharti* is a distinctive species, the forefemur has a strong subbasal angle, the hindtibial pit extends nearly the entire length of the tibia, and the sterna are all brown. This species most closely resembles the non-Californian species *spiloventer* French.

Hedychrum nigropilosum Mocsáry

Fig. 20, Map 40

Hedychrum nigropilosum Mocsáry 1889:162. Lectotype ♂ (designated by French 1982); USA: California, "Mariposa" (RNH).

Geographic Range (Map 40). Western USA, rarely in western Canada (Alberta, British Columbia) and northern Mexico (Distrito Federal, Hidalgo); 763 specimens were studied.

California Records. Collected June through August; **Alpine Co**.: Red Lake; **Amador Co**.: 6 mi. w Tragedy Springs; **El Dorado Co**.: 6 and 8 mi. n Ice House Rd.; Uncle Tom's Cabin; **Humboldt Co**.: 1 mi. nw Weott; **Inyo Co**.: Mammoth; **Lake Co**.: Blue Lakes; **Lassen Co**.: Hallelujah Junction; Bridge Creek Camp; **Los Angeles Co**.: Claremont; **Mono Co**.: Benton; Hot Creek; Devils Gate; Leavitt Meadow; **Monterey Co**.: Arroyo Seco; **Nevada Co**.: Sagehen Creek; Boca; Hobart Mills; **Placer Co**.: Emigrant Gap; **Plumas Co**.: Greenville; **Santa Barbara Co**.: Santa Ynez Mts.; 4 mi. e Los Prietos; **Santa Cruz Co**.: Santa Cruz; **Sierra Co**.: Independence Lake; Weber Lake; Sierraville; 15 mi. se Sierraville; Kyburz Flat; Yuba Pass; **Trinity Co**.: 10 mi. n Carrville; 10 mi. n Coffee Creek Ranger Station; **Tuolumne Co**.: Dodge Ridge; Strawberry; Dardanelles; **Ventura Co**.: Sespe Canyon.

Hosts. This species was recorded entering the nests of *Eucerceris flavocincta* Cresson (Crabronidae) by Bohart and Powell (1956).

Discussion. The commonest species of *Hedychrum* in California are *nigropilosum* and *parvum* and these species can be easily confused. The long black pubescence of *nigropilosum* will distinguish it from *parvum*, which has short whitish setae. This difference is particularly distinctive on the hindtibia. Small individuals of *nigropilosum* may have brownish setae on the body but the hindtibia will still have black setae. In addition, the anterior face of metasomal tergum I is smooth in *nigropilosum* but has three carinae in *parvum*. Finally, the apicomedial tooth on the female tergum III is simple in *nigropilosum* (Fig. 20) and bilobate in *parvum*.

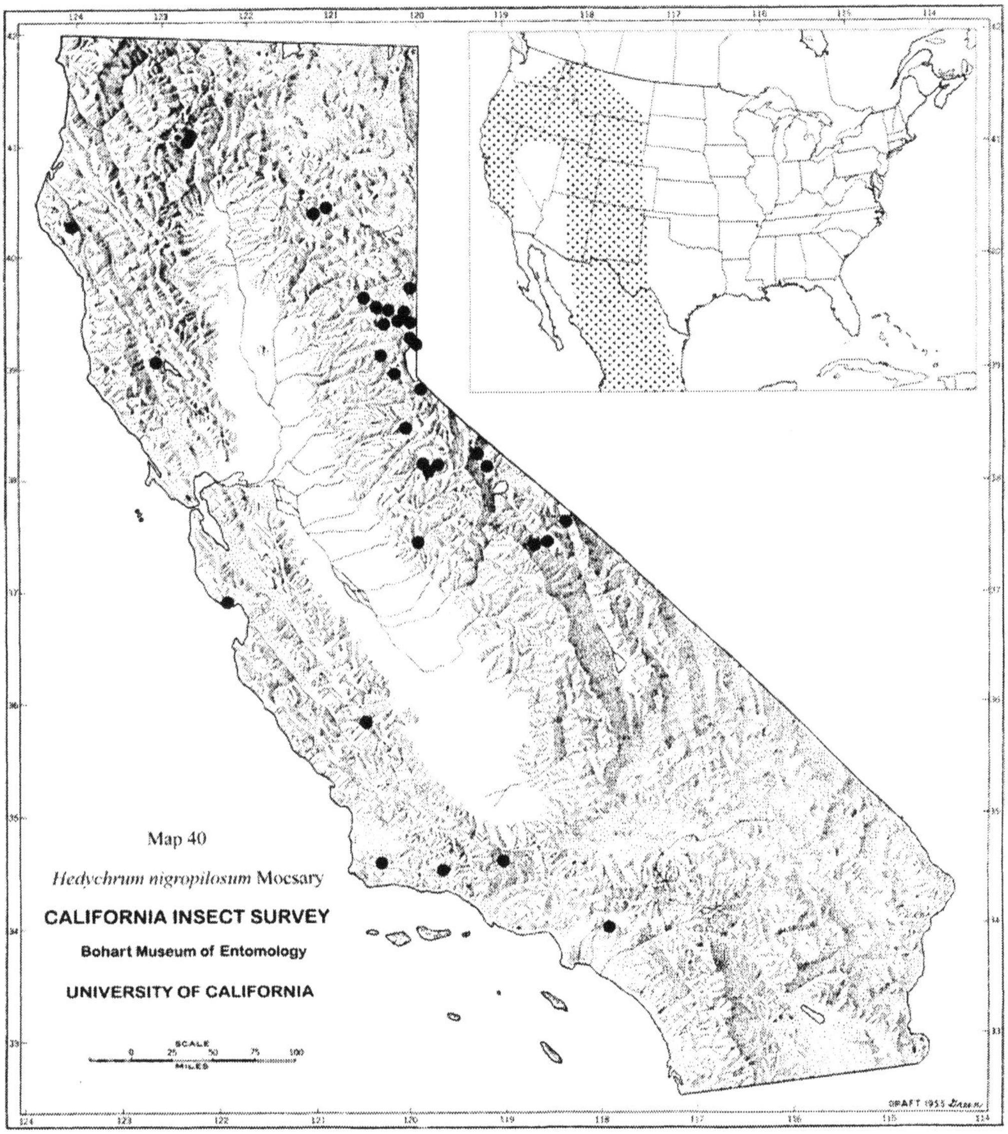

Map 40. California distribution *Hedychrum nigropilosum*. Inset: overall distribution.

Hedychrum parvum Aaron

Map 41

Hedychrum violaceum variety *parvum* Aaron 1885:223. Lectotype ♂ (designated by Cresson 1928); USA: "Montana" (ANSP).

Hedychrum polygoni Rohwer 1909:87. Holotype ♀; USA: Colorado, Boulder (USNM).

Geographic Range (Map 41). Widespread in the USA from New York to Florida and west to California, but apparently not in Texas or the Carolinas, also found in Canada (Alberta, British Columbia); 1132 specimens have been examined.

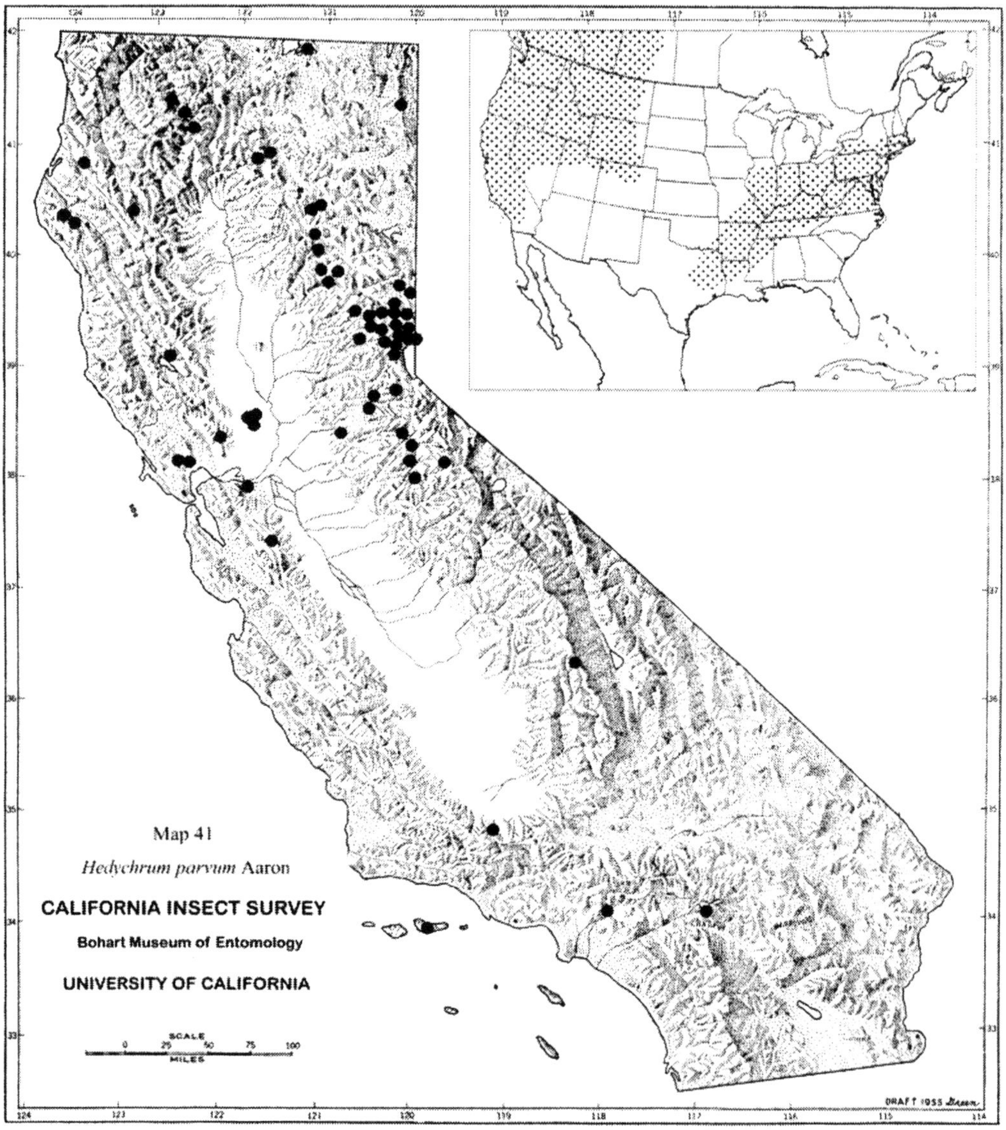

Map 41. California distribution of *Hedychrum parvum*. Inset: overall distribution.

California Records. Collected May through September; **Alameda Co.**: Cedar Mt.; **Amador Co.**: 6 mi. w Tragedy Spring; Ione; **Contra Costa Co.**: Antioch; **El Dorado**

Co.: Pollock Pines; 6-8 mi. e Ice House Rd.; Pine Hill w Rescue; **Humboldt Co.**: 1 mi. nw Weott; Kneeland; 2 mi. s Redwood Valley; **Kern Co.**: Mill Potrero; **Lake Co.**: Lower Lake; **Lassen Co.**: Bridge Creek Camp; Summit Camp; Hallelujah Junction; **Los Angeles Co.**: Tanbark Flat; **Modoc Co.**: Cedarville; 5 mi. ne Clear Lake; **Napa Co.**: Mt. Veeder; **Nevada Co.**: Sagehen Creek; Hobart Mills; Boca; Russell Valley; Fuller Lake; Jackson Lake; **Placer Co.**: Carnelian Bay, Lake Tahoe; Emigrant Gap; Nyack; Brockway Summit; **Plumas Co.**: 4 mi. w Quincy; Vinton; Onion Valley; Chester; Lake Almanor; Butterfly Valley; **San Bernardino Co.**: Green Canyon se Sugarloaf; Dollar L Trail; **Santa Barbara Co.**: Santa Cruz Island; **Shasta Co.**: Hatt Creek; Moose Camp; **Sierra Co.**: Kyburz Flat; Independence Lake; Yuba Pass; Sierraville; Sierra Valley; 15 mi. se Sierraville; Weber Lake; **Siskiyou Co.**: Etna; **Sonoma Co.**: Petaluma; Boyes Hot Spring; **Trinity Co.**: 10 mi. w Coffee Creek Ranger Station; Scott Mountain; **Tulare Co.**: Mineral King; **Tuolumne Co.**: Bumble Bee; Leland Meadow; Strawberry; Dodge Ridge; **Yolo Co.**: 2 and 5 mi. e Woodland; Davis; Woodland.

Hosts. Evans (1970) reared this species from *Eucerceris fulvipes* Cresson (Crabronidae).

Discussion. Although *parvum* is the commonest *Hedychrum* species in the Nearctic Region, *nigropilosum* is the commonest species collected in California. The two can be separated by the characteristics discussed under *nigropilosum*.

Genus *Holopyga* Dahlbom

Generic Diagnosis. Facial setae sparse, erect; scapal basin deep, transversely cross-ridged; vertex with interocellar sulcus; malar space less than 1 MOD long; genal carina present; pronotum with anteromedial pit; mesopleuron strongly angulate, with well-developed verticaulus, omaulus, signum and scrobal carina; notauli sulciform; scutellar wing fossa without anterior lobe; forefemur usually angulate subbasally with ventral carina; hindtarsal claw with 2-5 subsidiary teeth; forewing medial vein strongly arched, arising at or slightly before cu-a, costa dilated medially, stigma short and broad, R1 clearly indicated; tergum III apical margin evenly rounded or somewhat indented medially, slightly swollen subapically; volsella with discrete digitus and cuspis.

Discussion. *Holopyga* species occur worldwide, only six species occur in North America and four in California. There are a few characteristics unique to this genus, including the wing bend and the strongly projecting mesopleuron. Most diagnostic features of the group are found in other genera. In *Holopyga* the face has a deeply concave smooth scapal basin, the genal margins are angulate, the tarsal claws have two or more subsidiary, subparallel teeth, the medial vein is strongly angled, the wing is bent slightly at stigmal base, and the mesopleuron strongly projects as an acute angle anterolaterally. Bohart and Kimsey (1982) revised the North American species.

Key to the California Species of *Holopyga*

1 Flagellomere I 2.5-3.5x as long as broad, body length usually more than 6 mm, subantennal distance 1 MOD long or less.................................*ventralis* Say

- Flagellomere I less than 2.3x as long as broad, body length usually much less than 6 mm, subantennal distance often more than 1 MOD long.................................2

2 Forewing Rs stub not reaching as far as apex of stigma, medial vein obtusely arched; female hindtibia with large shallow depression on inner surface ....*mimeca* Kimsey

- Forewing Rs stub reaching as far as or beyond stigmal apex, medial vein sharply arched, often forming nearly right angle; female hindtibia without depression on inner surface...3

3 Tergum III punctures irregular, deep, sharp-edged and nearly contiguous, area before apical rim swollen...*rudis* Kimsey

- Tergum III punctures regular, usually not deep, with rounded edges and usually not contiguous, area before apical rim not swollen in females and only slightly swollen in males...*horus* Aaron

Holopyga horus Aaron

Map 42

Holopyga horus Aaron 1885:220. Lectotype ♀ (designated by Cresson 1928); USA: Montana (ANSP).

Hedychrum continuum Aaron 1885:224. Lectotype ♂ (designated by Cresson 1928); USA: "W. T." (ANSP).

Geographic Range (Map 42). USA (CA, MT, NM, SD) and Mexico (Baja California Sur, Puebla, Sinaloa), generally west of 100th meridian; 453 specimens were studied.

California Records. Collected March through September; **Alpine Co.**: Hope Valley; Ebbetts Pass; Red Lake; Lake Winnemucca; 1.5 mi. ne Red Lake; **Amador Co.**: 6 mi. w Tragedy Spring; **Contra Costa Co.**: Antioch; Pacheco Pass; Mt. Diablo; **El Dorado Co.**: 1 mi. sw Meyers; 4 mi. s Meyers; Grass Lake; South Lake Tahoe

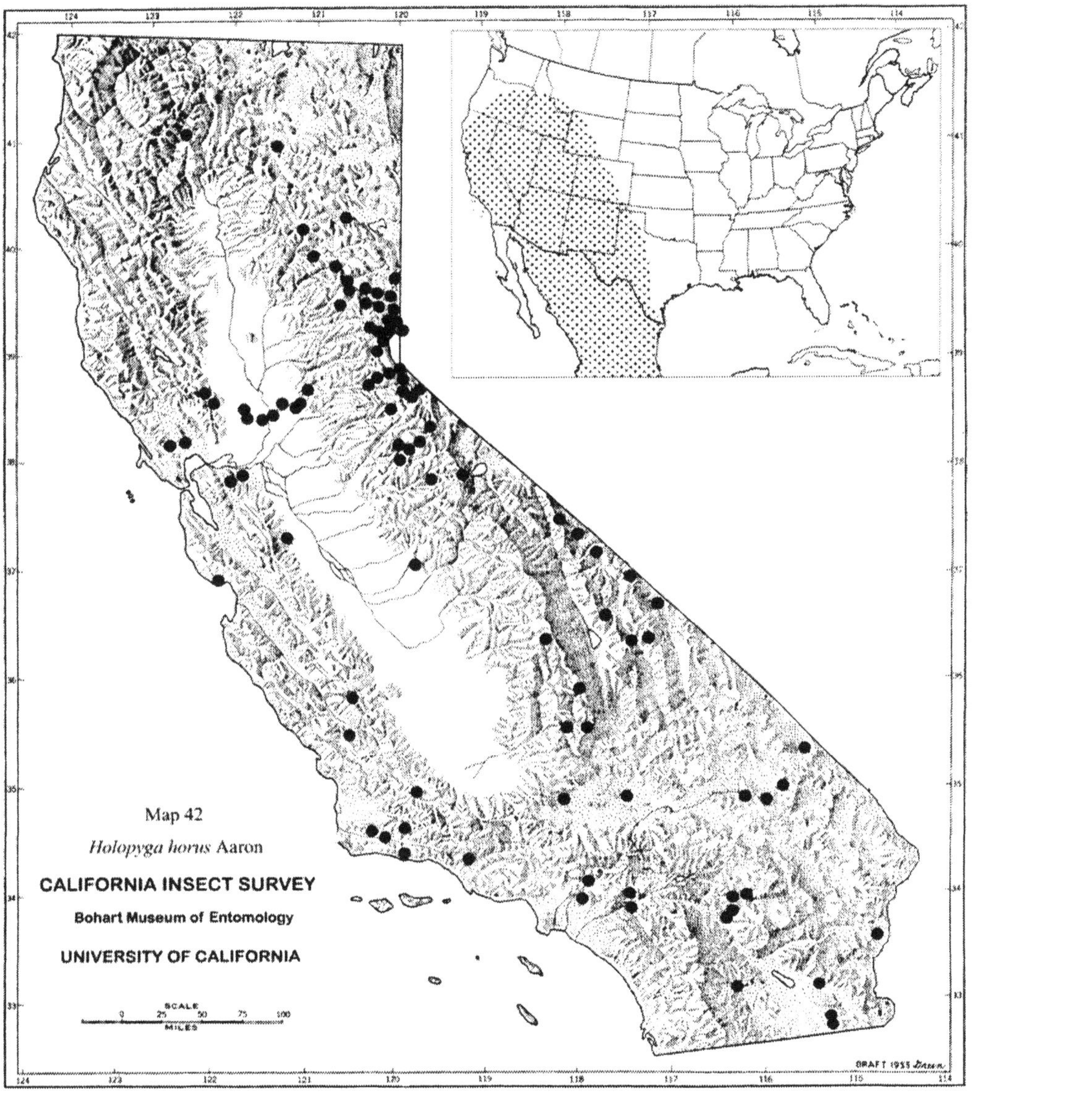

Map 42. California distribution of *Holopyga horus*. Inset: overall distribution.

1.5 mi. w Camp Richardson; 6-8 mi. n Ice House Rd.; **Imperial Co**.: 3-4 mi. n Glamis; Glamis; 1 mi. nw Niland; **Inyo Co**.: 3 mi. w Big Pine; Little Lake; Deep Springs; 25 mi. e Lone Pine; Eureka Valley dunes; Antelope Spring; Big Pine; Argus Mts.; Westgard Pass; Darwin Falls; Inyo Mts.; Panamint Spring; **Kern Co**.: Rosamond; Sand Canyon 3 mi. w Brown; 16 mi. s Inyokern; **Lassen Co**.: Hallelujah Junction; 2 mi. n Susanville; **Los Angeles Co**.: Claremont; 11 mi. ne San Gabriel Canyon Rd.; **Madera Co**.: s fork Merced River; **Mono Co**.: Sonora Pass; White Mts.; Dead Man's Creek; **Monterey Co**.: Arroyo Seco; **Napa Co**.: Samuel Springs (beneath Lake Berryessa); **Nevada Co.**: Boca; Prosser Dam; Sagehen Creek; Hobart Mills; Truckee; Jackson Lake; 10 mi. ne Sagehen Creek; Russell Valley; **Placer Co.**: Carnelian Bay Lake Tahoe; Boca; Brockway Summit; Folsom Lake; **Plumas Co.**: 4 mi. w Quincy; Graeagle; Buck's Lake; Eagle Lake; Lake Almanor; **Riverside Co.**: Riverside; Thousand Palms; 5 mi. e Chiriasc Summit; 18 mi. w Blythe; Andreas Canyon; 9 mi. w Beaumont; Coyote Wash; Pinto Basin, Joshua Tree National Monument; **Sacramento Co.**: Citrus Heights; Folsom; Sacramento; Fair Oaks; American River; **San Bernardino** Co.: 6 mi. s Ivanpah; 6 mi. nw Adelanto; Atolia; 1 mi. s Adelanto; Kelso; 12 mi. s Baker; Cronise Wash; **San Diego Co.**: Borrego Valley; **San Luis Obispo Co.**: 4 mi. sw Creston; California City; **San Mateo Co.**: 4 mi. e Portola; **Santa Barbara Co.**: Santa Cruz Island; Santa Ynez Mts.; Bluff Camp San Rafael Mts.; Aliso Canyon 6 mi. sw New Cuyama; Goleta; 2 mi. e Solvang; **Santa Clara Co.**: San Antonio Valley; **Santa Cruz Co.**: Santa Cruz; **Shasta Co.**: 6 mi. e McArthur; **Sierra Co.**: Sierraville; Sierra Valley; Gold Lake; 5 mi. e Weber Lake; Weber Lake; Independence Lake; Yuba Pass; Sattley; Sardine Lakes.; Kyburz Flat; **Siskiyou Co.**: Montague; **Sonoma Co.**: Glen Ellen; Petaluma; **Stanislaus Co.**: Del Puerto Canyon; **Trinity Co.**: Carrville; **Tulare Co.**: Mineral King; Ash Mountain; **Tuolumne Co.**: 11 mi. n Strawberry; Strawberry; Kennedy Meadow; Dardanelles; Dodge Ridge; Leland Meadow; Tuolumne Meadows; **Ventura Co.**: Sespe Canyon; **Yolo Co.**: Davis; Rumsey; West Sacramento; 7 mi. e Woodland; Putah Canyon.

Hosts. Unknown.

Discussion. *Holopyga horus* is the commonest species of *Holopyga* in the western states. It is also highly variable, with punctation ranging from large and nearly contiguous, to small and two puncture diameters apart. Size and color also vary, with individuals ranging from 3-6 mm in length, and from brassy green to purple in body color. Diagnostic characteristics include the hindtibial claw with one subapical and one subbasal tooth and male flagellomere I usually less than twice as long as broad.

Holopyga mimeca Kimsey

Map 43

Holopyga mimeca Kimsey, in Bohart & Kimsey 1982:26. Holotype ♂; USA: California, Riverside Co., Desert Center (CNC).

Geographic Range (Map 43). Desert areas of California, Nevada and Arizona; 5 specimens were seen.

California Records. Collected April through May; **Riverside Co.**: Desert Center; **San Diego Co.**: Borrego.

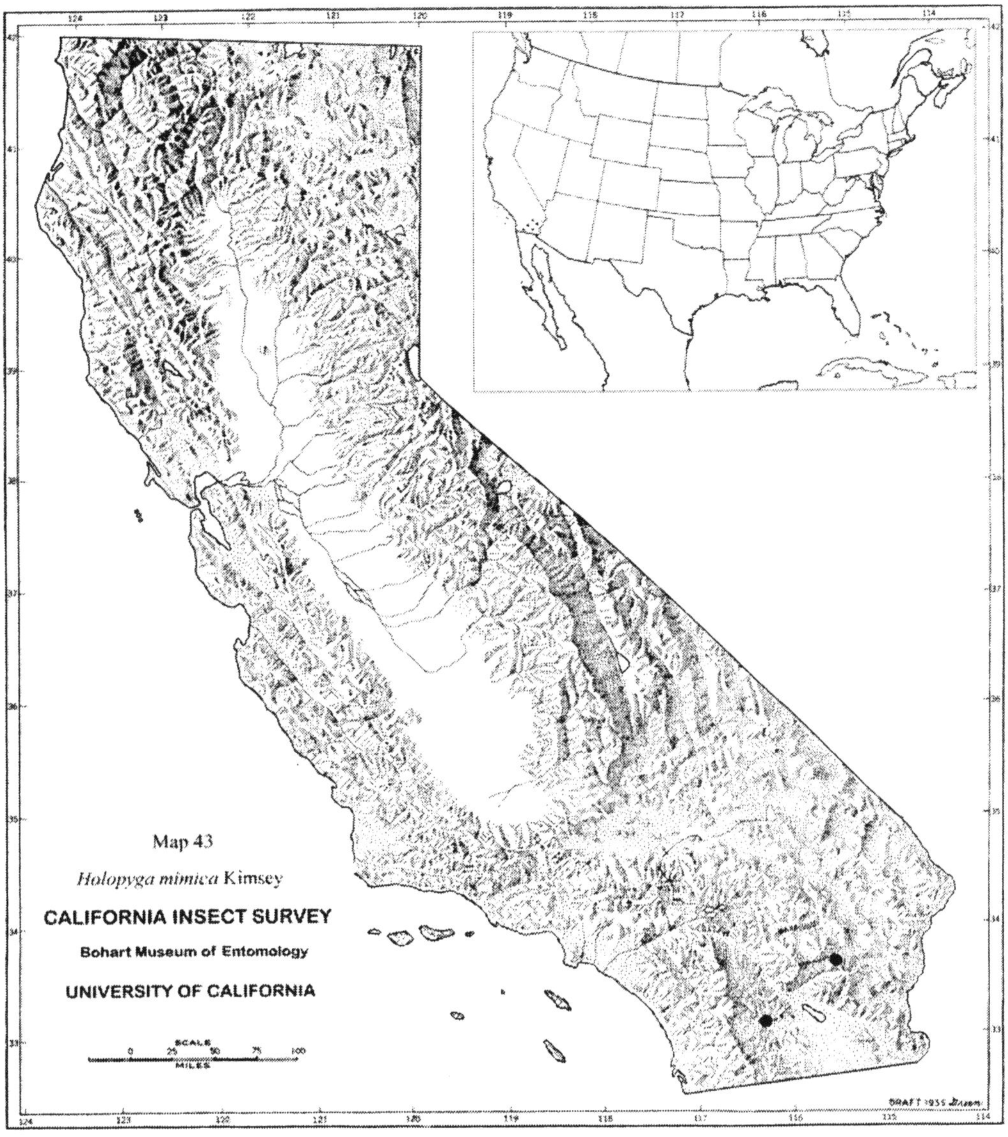

Map 43. California distribution of *Holopyga mimica*. Inset: overall distribution.

Hosts. Unknown.

Discussion. *H. mimeca* can be immediately distinguished from other *Holopyga* species by the unusually short Rs stub, short flagellomere I, medial vein obtusely arched, and scapal basin with a large shallow depression on the inner surface.

Map 44. California distribution of *Holopyga rudis*. Inset: overall distribution.

Holopyga rudis **Kimsey**

Map 44

Holopyga rudis Kimsey, in Bohart & Kimsey 1982:26. Holotype ♂; USA: Arizona, Cochise Co., Portal (UCDC).

Geographic Range (Map 44). Desert areas of northern Mexico (Chihuahua, Durango, Puebla, Zacatecas, Jalisco, Sonora, Baja California Sur), and southern California, Arizona, Texas and New Mexico; 16 specimens were seen.

California Records. Collected July through September; **San Diego Co**.: Pine Valley; **Santa Barbara Co**.: Goleta; **Tuolumne Co**.: Strawberry.

Hosts. Unknown.

Discussion. The distinctive deep punctures, especially on tergum III will immediately distinguish *rudis* from other *Holopyga*. Otherwise this species resembles *horus*, except that tergum III is swollen before the apical rim and is not apicomedially indented. This species is rarely collected in California.

***Holopyga ventralis* (Say)**

Fig. 3, Map 45

Hedychrum ventralis Say 1824:330. Type sex unknown; USA: Pennsylvania (destroyed).

Hedychrum aspera Brullé 1846:52. Holotype ♀; "N. Amer." (MNHN).

Holopyga dohrni Dahlbom 1854:48. Holotype ♂; USA: Locality restricted to N.Y. (Bohart & Kimsey 1982) (MZLU).

Holopyga compacta Cresson 1865b:304. Holotype ♀; USA: Colorado (ANSP).

Geographic Range (Map 45). Widespread in the USA, Canada (Alberta), and Mexico (Baja California Sur and Norte, Veracruz, Morelia, Guerrero, Chihuahua, Puebla, Jalisco, Michoacan, Zacatecas, Coahuila, Oaxaca, San Luis Potosi), southward into El Salvador and Cuba; 715 specimens were studied.

California Records. Collected May through October; **Alameda Co**.: Berkeley; **Alpine Co**.: Hope Valley; Carson Pass; **Amador Co**.: 6 mi. w Tragedy Spring; Plymouth; **El Dorado Co**.: 6-8 mi. n Ice House Rd.; Echo Lake; Silver Lake; Grass Lake; Meyers; **Humboldt Co**.: 5 mi. e Kneeland; **Imperial Co**.: Palo Verde; **Inyo**

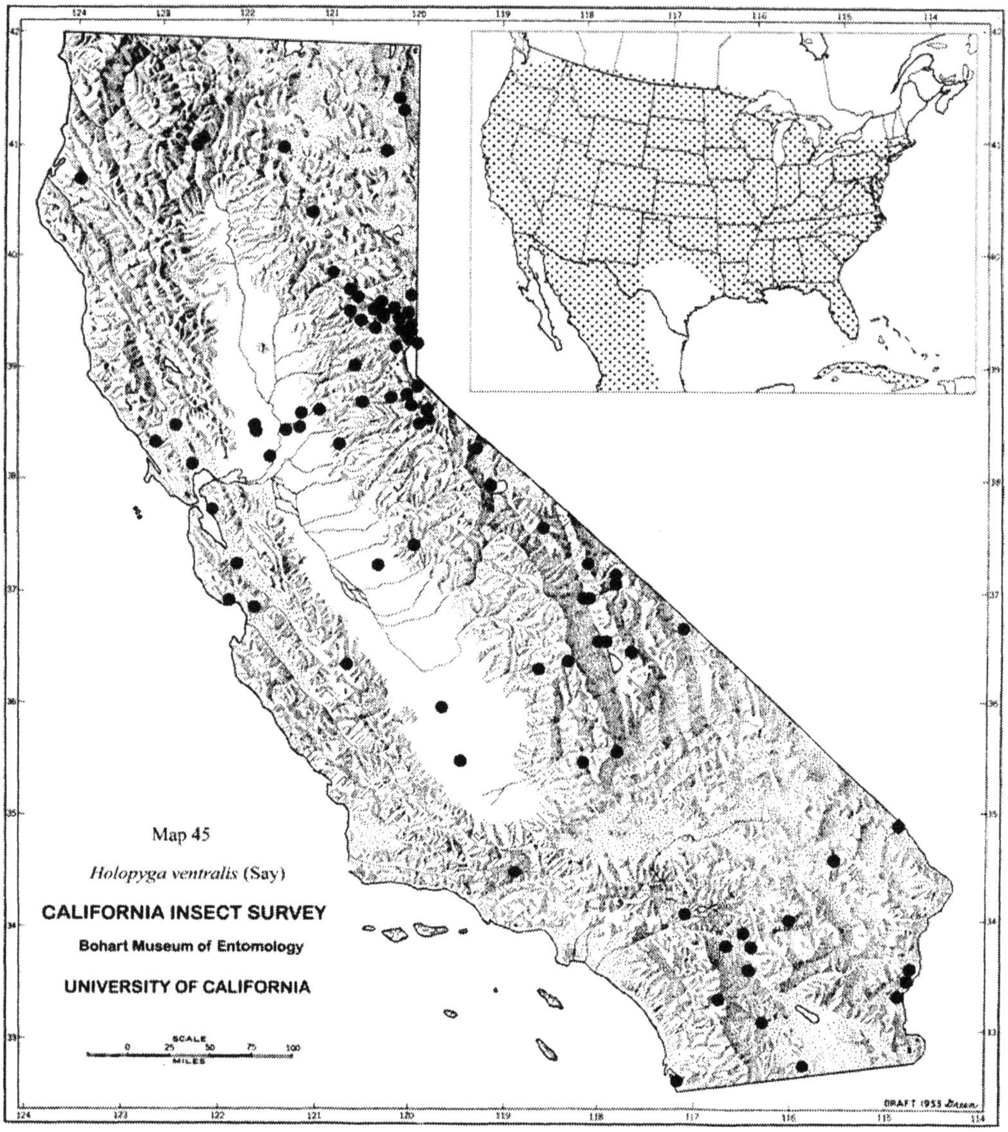

Map 45. California distribution of *Holopyga ventralis*. Inset: overall distribution.

Co.: Big Pine Creek; Alabama Hills; 12 mi. e Keeler; 3 mi. e Big Pine; Deep Springs Lake; Antelope Spring; Westgard Pass; Eureka Valley dunes; Lone Pine; Balleret; Brown; 7 mi. n Parchers Camp; **Kern Co.**: Short Canyon; Wasco; **Kings Co.**: Corcoran; **Lassen Co.**: Bridge Creek Camp; Madeline Plains; Hallelujah Junction; Sugarville; **Mariposa Co.**: Mariposa; **Merced Co.**: Merced; **Modoc Co.**: 7 mi. e Lake City; Cedar Pass; **Mono Co.**: 11 mi. n Bridgeport; Benton Station. Tom's Place; Mono Lake; **Nevada Co.**: Russell Valley; Prosser Dam; Hobart Mills; Jackson Lake; 10 mi. ne Sagehen Creek; Sagehen Creek; Boca; 3 mi. n Boca; **Placer Co.**: Folsom;

Emigrant Gap; **Plumas Co.**: 4 mi. w Quincy; Johnsville; Graeagle; **Riverside Co.**: Thousand Palms; Coyote Wash; 18 mi. w Blythe; 10 mi. e White Water; Hopkins Well; Andreas Canyon Palm Springs; **Sacramento Co.**: Sacramento; American River; Fair Oaks; **San Bernardino Co.**: Needles; Wildwood Canyon; 29 Palms; 6 mi. nnw Granite Pass; **San Benito Co.**: 6 mi. s Idria; **San Diego Co.**: Chula Vista; 2 mi. n Warner Springs; Borrego Valley; **Santa Clara Co.**: Santa Clara; **Santa Cruz Co.**: 9 mi. ne Soquel; Santa Cruz; **Shasta Co.**: Hat Creek; **Sierra Co.**: Independence Lake; Gold Lake; Yuba Pass; 15 mi. se Sierraville; Sierraville; Sattley; Sierra Buttes; Sardine Lakes; **Sonoma Co.**: Occidental; Tolay; 4 mi. w Plantation; **Trinity Co.**: Carrville; Coffee Creek Ranger Station; **Tulare Co.**: Mineral King; Sequoia National Park Ash Mt.; **Tuolumne Co.**: Leland Meadow; Kennedy Meadow; Strawberry; **Ventura Co.**: Sespe Canyon; **Yolo Co.**: Davis; 5 mi. e Woodland; Clarksburg.

Hosts. *Bicyrtes quadrifasciata* (Say) and *B. fodiens* (Handlirsch) have been reported as hosts of *ventralis* (Evans 1966).

Discussion. The multidentate tarsal claws, with 4-5 subsidiary teeth, large body size (6-8 mm long), long flagellomere I (more than three times as long as broad), brown sterna, and arched medial vein will distinguish *ventralis* from other *Holopyga* species.

Genus *Microchridium* Bohart

Generic Diagnosis. Face flat, without indication of scapal basin, microreticulate; malar space long, about half eye height; body nonmetallic, impunctate, sculpturing somewhat striatiform; flagellum I length subequal to breadth; mesopleuron gently rounded, nearly bisected by scrobal sulcus, without omaulus or scrobal carina; notauli deeply sulciform; scutellar wing fossa without anterior lobe; scutellum with longitudinal groove; hindtibia with small subapical pit on inner surface; tarsal claw with one subparallel medial tooth; forewing venation restricted to basal third of wing, cu-a vaguely indicated, medial vein straight arising at origin of cu-a, stigma short, blunt and as long as Rs; tergum III flattened and one-fourth as long as II; volsella elongate and undivided.

Distribution. This monotypic genus is endemic to southwestern North America.

Hosts. *Microchridium minutum* was reared from a species of *Ammoplanellus* (Crabronidae) (Halsted 1987).

Discussion. No other chrysidid in North America resembles *Microchridium*. This monotypic genus is immediately recognizable by its tiny size, non-metallic black color, reticulate body sculpture and reduced wing venation.

Microchridium minutum Bohart

Figs. 11, 25, 34, Map 46

Microchridium minutum Bohart 1980:132. Holotype ♂; USA: California, San Bernardino Co., Cronise Wash (UCDC).

Map 46. California distribution of *Microchridium minutum*. Inset: overall distribution.

Geographic Range (Map 46). Western USA, Nebraska, and Baja California, Mexico; 29 specimens were studied.

California Records. Collected April through August; **Imperial Co**.: Fish Creek; 10 mi. w Palo Verde; 3 mi. n Glamis; **Inyo Co**.: Westgard Pass; Eureka Valley; **Riverside Co**.: 18 mi. w Blythe; San Timoteo Canyon; Deep Canyon; **San Bernardino Co**.: Cronise Wash; Mountain Pass; **San Diego Co**.: Borrego Valley; **Santa Barbara Co**.: 45 mi. ne Santa Barbara.

Discussion. Features that distinguish the genus *Microchridium* are also diagnostic for the species. This tiny chrysidid is probably more common than the number of museum specimens would indicate. Its dark, non-metallic color and tiny size make it nearly impossible to collect intentionally and even when collected by sweeping specimens often end up sorted into other groups of Hymenoptera.

Genus *Minymischa* Kimsey

Generic Diagnosis. Scapal basin shallow or flat, with few cross ridges; malar space less than 1 MOD long; pronotum gently rounded, without anterior declivity, carinae or pits; mesopleuron rounded without carinae; notauli sulciform; scutellar wing fossa without anterior lobe; propodeal posterior surface without carinae or enclosures; female fore and midtarsi expanded and flattened (Fig. 21); tarsal claws with one submedial, subparallel tooth; male hindtarsal claw apically bifid; forewing medial vein nearly straight, arising beyond cu-a; Rs one-fifth as long as medial vein or shorter; volsella undivided and less than 5x as long as broad, with stout apical setae.

Hosts. Unknown.

Discussion. Endemic to western North America, all three species of *Minymischa* occur in California. *Minymischa* is most similar to *Pseudolopyga*, but can be distinguished by the shorter Rs stub and oddly flattened female foretarsomere. The genus and new species were described by Kimsey (in Bohart and Kimsey1982).

Key to the California Species of *Minymischa*

1 Body brightly bicolored, terga bright yellowish green to brassy, thorax dark green to blue; flagellomere II longer than broad........................*deserticola* Kimsey

\- Body concolorous, terga and thorax black, dark blue or purple; flagellomere II about as long as broad 2

2 Flagellomere X length subequal to breadth; forewing without brown stained vein remnants beyond medial cell.......................................*arenicola* Kimsey

- Flagellomere X longer than wide; forewing with brown-stained vein remnants beyond medial cell and extending Rs vein............................*ventura* Kimsey

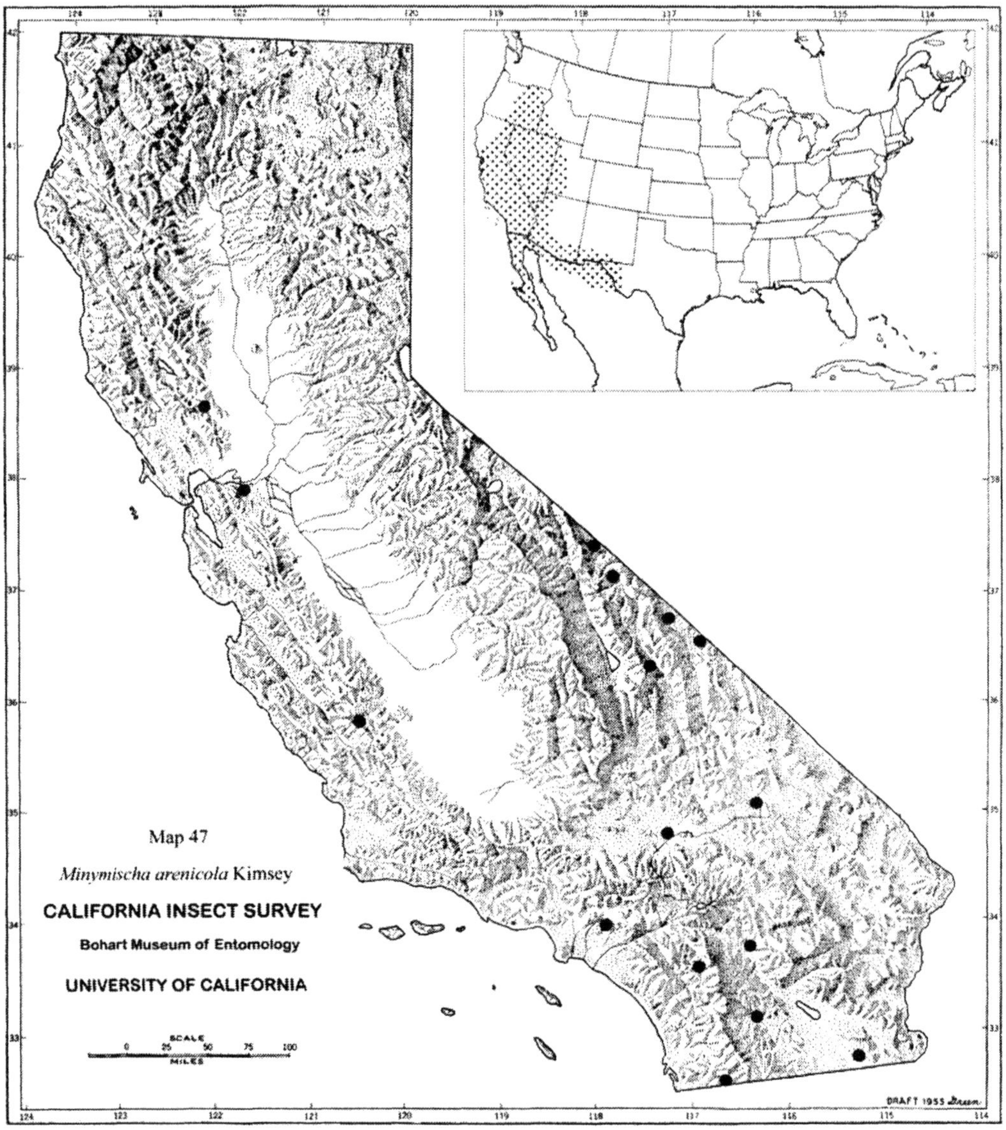

Map 47. California distribution of *Minymischa arenicola*. Inset: overall distribution.

***Minymischa arenicola* Kimsey**

Figs. 21, 26, 29, Map 47

Minymischa arenicola Kimsey 1982:353. Holotype ♂; USA: California, San Bernardino Co., Barstow (UCDC).

Geographic Range (Map 47). Widespread in the USA west of the 100th meridian and in northern Mexico (Chihuahua, Baja California); 66 specimens were studied.

California Records. Collected March through May; **Contra Costa Co**.: Antioch; **Imperial Co**.: Glamis; **Inyo Co**.: 7 mi. s Oasis; Deep Springs; Eureka Valley dunes; Skidoo Junction Death Valley; Darwin Falls; **Los Angeles Co**.: Claremont; **Monterey Co**.: Arroyo Seco; **Napa Co**.: Monticello Dam; **Riverside Co**.: Hemet; Thousand Palms; **San Bernardino Co**.: Cronise Wash; 5 mi. n Barstow; **San Diego Co**.: 2 mi. n Borrego Springs; Campo.

Discussion. *Minymischa arenicola* most closely resembles species of *Pseudolopyga* in size and color. It can be distinguished from other species of *Minymischa* by the dark color, and flagellomeres II and X as broad as long or broader.

Minymischa deserticola Kimsey

Map 48

Minymischa deserticola Kimsey 1982:354. Holotype ♂; USA: California, San Diego Co., Borrego Springs (UCDC).

Geographic Range (Map 48). Southern desert areas of California, Arizona and Nevada; 18 specimens were studied.

California Records. Collected March through May; **Imperial Co**.: Fish Creek; **Inyo Co**.: Eureka Valley dunes; **Riverside Co**.: Thousand Palms; 18 mi. w Blythe; **San Bernardino Co**.: Kramer Junction; **San Diego Co**.: Borrego Springs; Split Mt.; north end Borrego Clark Lake.

Discussion. The brightly colored metasoma makes *deserticola* the most distinctive species of *Minymischa*. In addition, flagellomeres I and II are subequal in length and the wings are weakly stained.

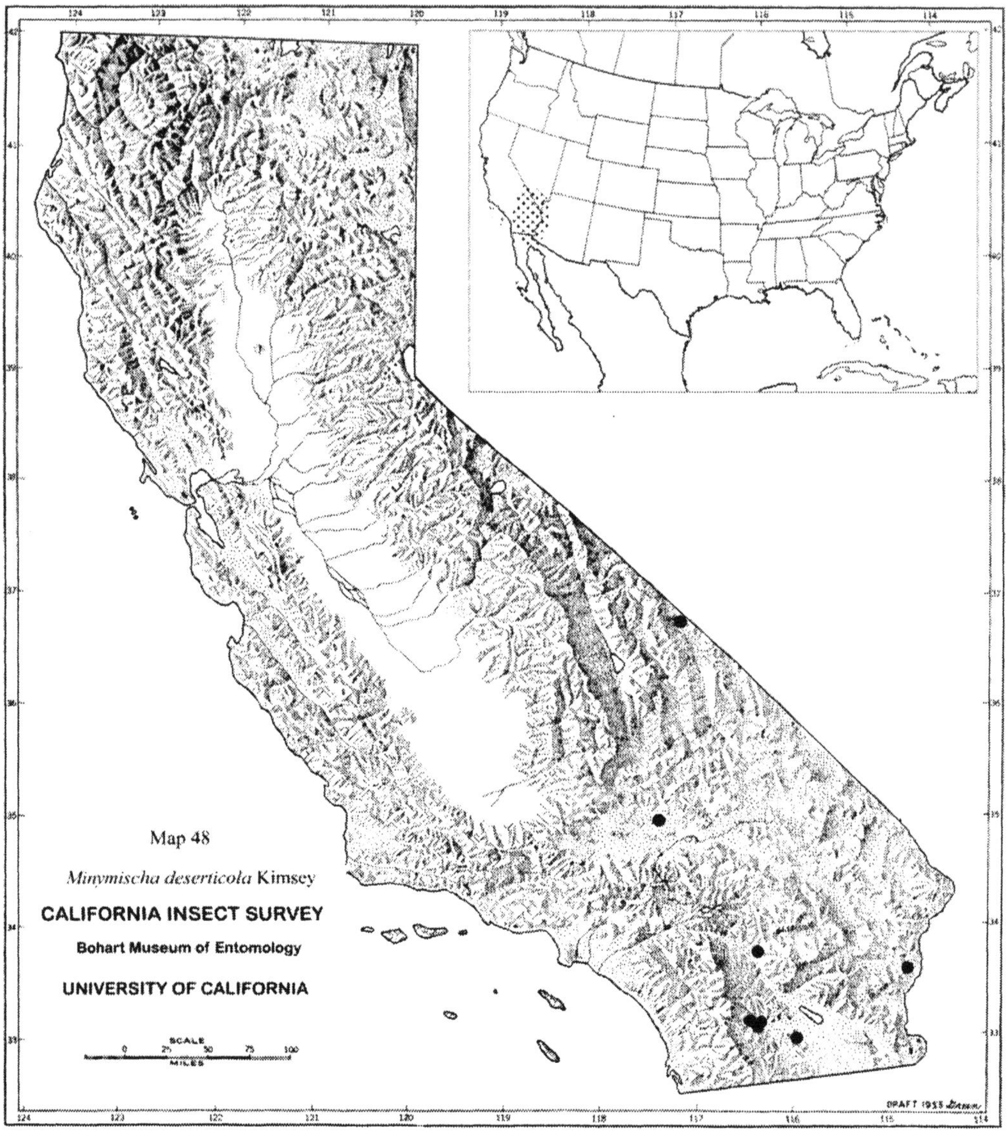

Map 48. California distribution of *Minymishca deserticola*. Inset: overall distribution.

Minymischa ventura **Kimsey**

Map 49

Minymischa ventura Kimsey 1982:354. Holotype ♂; USA: California, Ventura Co., Quatal Canyon (UCDC).

Geographic Range (Map 49). Upper Sonoran Zone of Oregon and California; 3 specimens were studied.

California Records. Collected April and May; **Ventura Co.**: Quatal Canyon.

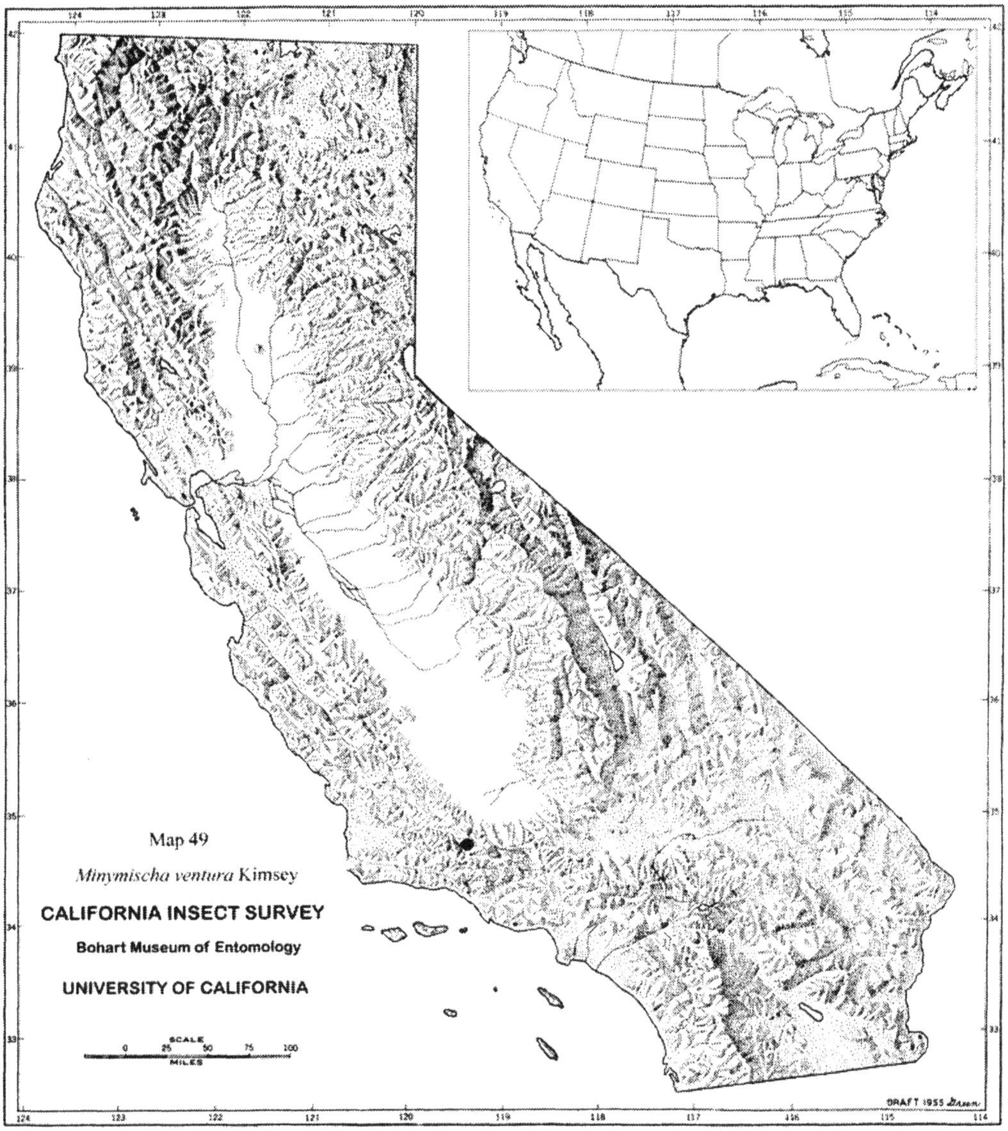

Map 49. California distribution of *Minymischa ventura*. Inset: overall distribution.

Discussion. This rarely collected California endemic can be recognized by the stained vein remnants on the forewing, flagellomere II as long as broad and flagellomere X longer than broad, and the dark body color.

Genus *Omalus* Panzer

Generic Diagnosis. Scapal basin usually deep, smooth, asetose; malar space less than 1 MOD long, bisected horizontally by genal carina; head cupped posteriorly, with carinate postocular edge; pronotum deeply impressed laterally; vertex, pronotum medially and scutum impunctate; mesopleuron with scrobal sulcus horizontal, extending from lateroventral margin of pronotum to scrobe, with single carina dorsally, transpleural carina reaching apex of propodeal angle; scutellum with two flattened areas along anterior margin; metanotum evenly rounded; forewing medial cell asetose, medial vein strongly arched and arising before cu-a, stigma short, broad, apically rounded; forefemur ventrally carinate, often subapically broadened, tarsal claws with 2-3 subsidiary teeth; metasomal terga I-III strongly convex, tergum III apical margin often sinuate laterally, occasionally transparent, usually without apicomedial notch.

Hosts. *Omalus* species are nest parasites of crabronid wasps in the subfamily Pemphredoninae.

Discussion. The most distinctive features of *Omalus* are the diffuse or obsolescent scutal punctation and the rounded metanotum. This is a small genus with four North American species (only three occur in California), and several species in South America and Eurasia. *Omalus* was revised in Bohart & Kimsey (1982).

Key to the California Species of *Omalus*

1 Tergum III apical margin with medial notch well-developed and about 1 MOD deep or deeper, notauli deeply impressed and entire...............*aeneus* (Fabricius)

\- Tergum III apical margin, with shallow medial dent less than 1 MOD deep, or entire; notauli incomplete or indicated only by fine line..........................2

2 Notauli complete from anterior to posterior margin ending in posterior pit, but indicated by fine line; tergum III with yellowish transparent rim about 1 MOD long..*iridescens* (Norton)

\- Notauli obsolescent, represented only by posterior pits (Fig. 13); tergum III with narrow brown rim, much less than 1 MOD long.............*glomeratus* (Buysson)

Omalus aeneus (Fabricius)

Figs. 15, 33, Map 50

Chrysis aeneus Fabricius 1787:284. Holotype ♀; Germany: "Halae Saxonum" (ZMUC).

Omalus laeviventris Cresson 1865b:303. Lectotype ♀ (designated by Cresson 1916); USA: "Colorado Territory" (ANSP).

Elampus cyanescens Provancher 1881:303. Lectotype ♀ (designated by Bohart and Kimsey 1982); Canada (LU).

Omalus diversus Aaron 1885:213. Holotype ♀; USA: California (ANSP).

Geographic Range (Map 50). Widespread in the Nearctic and Palearctic Regions; 822 specimens were studied.

California Records. Collected April through October; **Alameda Co**.: Oakland; Fremont; **Alpine Co**.: Woodfords; **Amador Co**.: 6 mi. w Tragedy Spring; **Calaveras Co**.: Big Meadow; **Contra Costa Co**.: Contra Costa Canyon; **El Dorado Co**.: 8 mi. n Ice House Rd.; Echo Lake; 3 mi. e Uncle Tom's Cabin; Emigrant Gap; **Inyo Co**.: Big Pine Creek; Darwin Falls; **Lassen Co**.: Susan River Camp; **Los Angeles Co**.: Tanbark Flat; **Marin Co**.: Mill Valley; **Modoc Co**.: 24 mi. sw Alturas; **Mono Co**.: Mill Creek Canyon; Mammoth Lake; Dead Man's Creek; **Napa Co**.: Pope Valley; Mt. Veeder; **Nevada Co**.: Sagehen Creek; Boca; Jackson Lake; Fuller Lake; **Placer Co**.: Carnelian Bay, Lake Tahoe; Dutch Flat; 5 mi. w Foresthill; **Plumas Co**.: Bucks Lake; 4 mi. w Storrie; Blairsden; Antelope Valley; Johnsville; Graeagle; **San Bernardino Co**.: San Bernardino Mts.; **San Diego Co**.: Mt. Laguna; **Santa Barbara Co**.: Los Padres National Forest Upper Oso Campground; **Santa Clara Co**.: Stanford University; **Santa Cruz Co**.: Felton Spring; **Shasta Co**.: Viola; Moose Camp; **Sierra Co**.: Yuba Pass; Weber Lake; Sierra Valley; **Sonoma Co**.: Kenwood; **Trinity Co**.: Scott Mt.; Eagle Creek; **Tulare Co**.: Mineral King; **Tuolumne Co**.: Mather; Strawberry; Leland Meadow; **Yolo Co**.: Clarksburg.

Hosts. Krombein (1979) reared *aeneus* from *Passaloecus cuspidatus* F. Smith and *Pemphredon* sp. (Crabronidae).

Discussion. Diagnostic features of *aeneus* include the well-developed, deeply impressed notauli, the distinct and usually acute apicomedial notch on tergum III, the brown-tinted wings and well-punctured scutellum. *O. aeneus* are moderate-sized wasps ranging from 3-5 mm long. This is the only truly holarctic chrysidid.

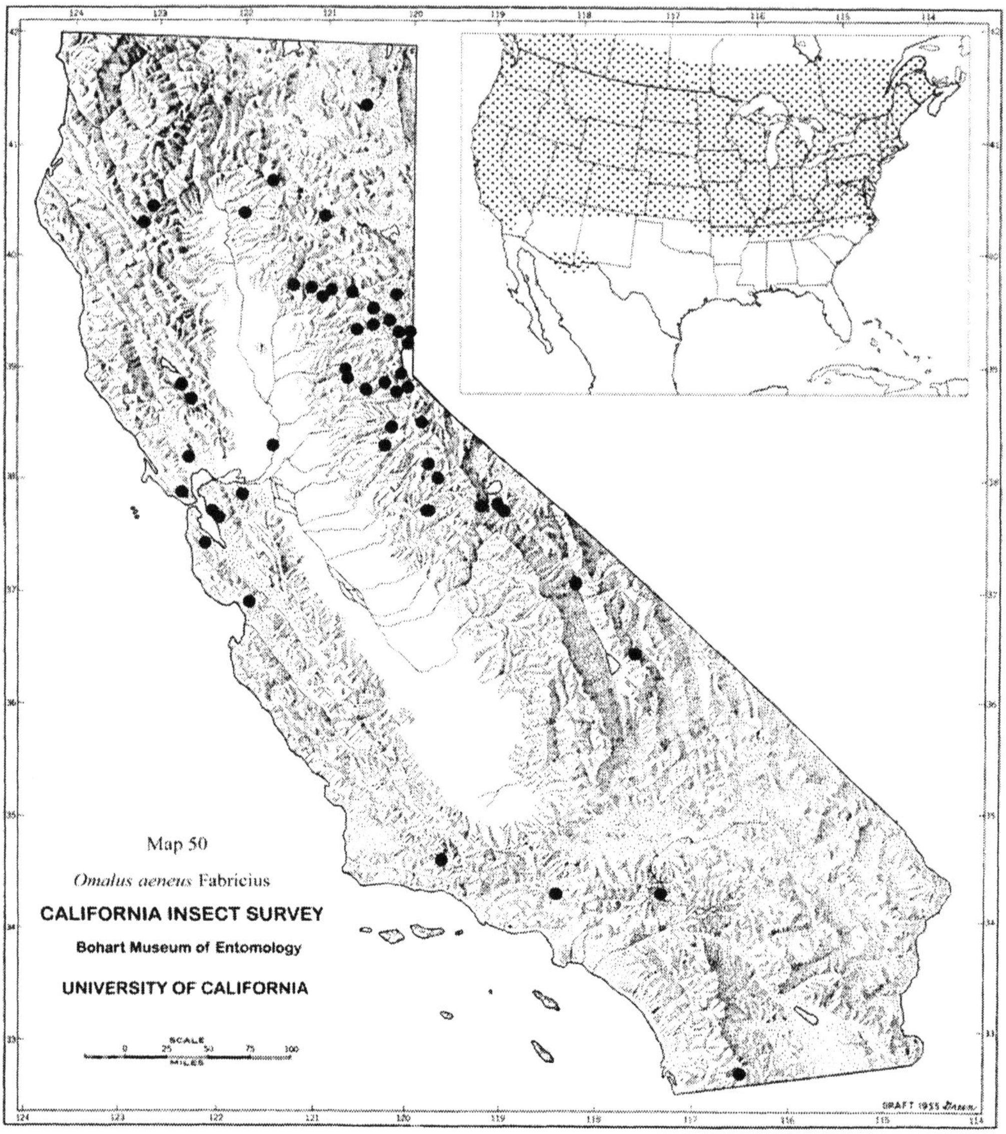

Map 50. California distribution of *Omalus aeneus*. Inset: overall distribution.

Omalus glomeratus (du Buysson)

Fig. 13, Map 51

Ellampus glomeratus du Buysson 1901:98. Holotype ♂; USA: "Washington Territory" (NHMV).

Geographic Range (Map 51). USA west of the 100th meridian, as well as in Florida, Oklahoma, and Canada (British Columbia); in the Lower Sonoran to Canadian Zones; 251 specimens were studied.

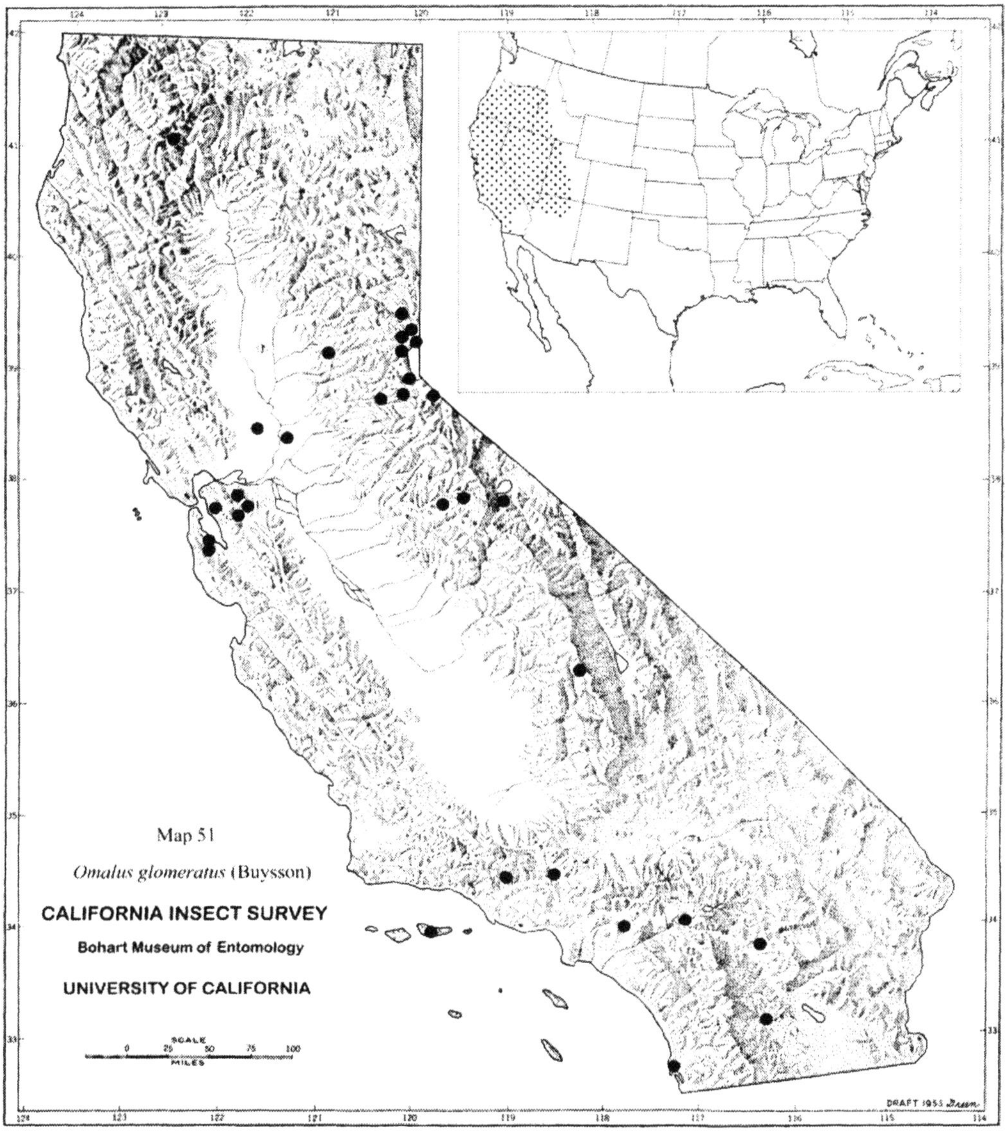

Map 51. California distribution of *Omalus glomeratus*. Inset: overall distribution.

California Records. Collected May through August; **Alameda Co**.: Fish Ranch Rd. Berkeley; **Alpine Co**.: Woodfords; **Contra Costa Co**.: Moraga; Mt. Diablo; Tassajara; **El Dorado Co**.: 6 mi. n Ice House Rd.; **Los Angeles Co**.: mountains near Claremont; Tanbark Flat; **Madera Co**.: Oakhurst; **Mono Co**.: Mill Creek Canyon;

Nevada Co.: Boca; Sagehen Creek; Hobart Mills; Grass Valley; **Placer Co.**: Dollar Point near Tahoe City; Carnelian Bay, Lake Tahoe; Emigrant Gap; **Plumas Co.**: Johnsville; **Riverside Co.**: 4 mi. s Palm Desert; **Sacramento Co.**: Sacramento; **San Bernardino Co.**: San Bernardino; Dollar L Trail San Bernardino Mts.; **San Diego Co.**: La Jolla; Borrego; **San Mateo Co.**: Jasper Ridge; Menlo Park; **Santa Barbara Co.**: Santa Cruz Island; **Sierra Co.**: Independence Lake; **Trinity Co.**: Carrville; **Tulare Co.**: Mineral King; **Tuolumne Co.**: Strawberry; Leland Meadow; **Ventura Co.**: Sespe Canyon; **Yolo Co.**: Davis.

Hosts. *O. glomeratus* has been reared from the nests of the crabronid wasps *Stigmus inordinatus* W. Fox (Parker and Bohart 1966) and *Pemphredon grinnelli* (Rohwer) (Bohart and Kimsey 1982).

Discussion. This species can be immediately distinguished by the absence of notauli, and tergum III with a brown rim and apicomedial notch or dent. Some specimens of *iridescens* appear to have lost the notauli, but on close inspection the notauli can be seen on these specimens as fine lines. *O. glomeratus* range from 2.5-4.0 mm long. Males tend to have tergum III all black; in females tergum III is mostly greenish blue.

Omalus iridescens (Norton)

Map 52

Elampus iridescens Norton 1879:234. Lectotype ♀ (designated by Cresson 1928); USA: "Penn." (ANSP).

Elampus marginatus Provancher 1881:304. Lectotype ♀ (designated by Bohart and Kimsey 1982); Canada (LU). Nec Patton 1879.

Geographic Range (Map 52). Transcontinental in the USA, southern Canada (Ontario, British Columbia) and Mexico (Baja California Sur, Veracruz); in Upper Sonoran to Canadian Zones; 565 specimens were seen.

California Records. Collected May through August; **Alameda Co.**: Berkeley; Bay Farm Island; Oakland hills; **Contra Costa Co.**: Contra Costa Canyon; Orinda Cross Rd.; Moraga; **Inyo Co.**: Westgard Pass; **Mendocino Co.**: Guerneville; **Napa Co.**: n slope Mt. Saint Helena; **Nevada Co.**: Sagehen Creek; **Sacramento Co.**: Sacramento; **Santa Barbara Co.**: Santa Cruz Island; **Santa Clara Co.**: Stanford University; **Santa Cruz Co.**: Soquel; **Sierra Co.**: Yuba Pass; **Tulare Co.**: Mineral King.

Hosts. Krombein (1979) recorded *iridescens* from *Stigmus inordinatus* W. Fox, *S. americanus* Packard, *Psenulus trisulcus* (W. Fox) and *Passaloecus* sp. (Crabronidae).

Discussion. The fine notauli and entire and translucent apical rim of tergum III are diagnostic for *iridescens*.

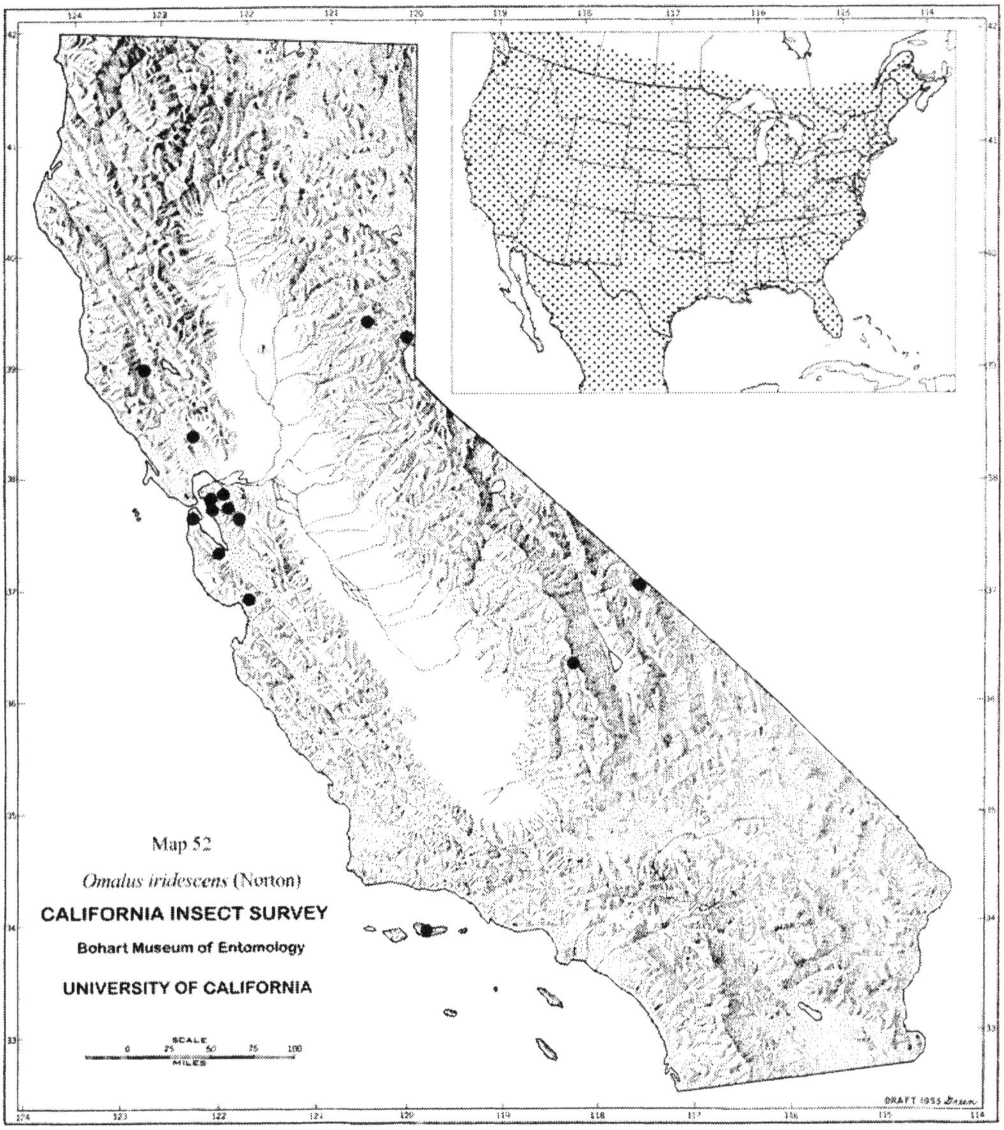

Map 52. California distribution of *Omalus iridescens*. Inset: overall distribution.

Genus *Philoctetes* Abeille de Perrin

Generic Diagnosis. Scapal basin generally smooth, asetose; malar space 1 MOD long or shorter; genal carina generally faint, extending dorsally from mandibular socket; vertex generally densely punctate; pronotum deeply concave laterally, punctate medially; scutum with punctures clumped along notauli, or less commonly more evenly distributed; mesopleuron carinate or rounded, juncture of omaulus, verticaulus and scrobal carina not strongly projecting, signum carina usually present, scrobal sulcus oblique; metanotum usually conical, even mucronate in several species; forewing medial cell asetose, medial vein strongly arched, arising before cu-a, stigma short, broad, apically rounded; forefemur carinate ventrally; tarsal claws with 1-3 subsidiary teeth; terga I-III strongly convex, tergum III lateral margin straight or sinuate, usually deeply notched apicomedially, or bent under, without transparent rim.

Distribution. The majority of *Philoctetes* species are Palearctic, with the highest diversity in Central Asia, one species occurs in Africa and 8 are in the Nearctic Region (all of these are in California).

Hosts. This genus has been reared from crabronid wasps in the subfamily Pemphredoninae.

Discussion. The most distinctive feature of *Philoctetes* species is the clumping of punctures along the notauli on the scutum (Fig. 12). A number species, including *seminudus* and *intermedius*, have the metanotum pointed or projecting posteriorly and somewhat truncate tergum III, which gives them a superficial resemblance to *Elampus* species. The genus was synonymized with *Diplorrhos* Aaron (the junior name) by Kimsey and Bohart (1991).

Key to the California Species of *Philoctetes*

1 Tergum III terminating in narrow apical truncation, apicomedial notch partly membrane filled (Figs. 43); tarsal claw with at least three well-developed subsidiary teeth...*seminudus* (Aaron)

\- Tergum III without discrete apical truncation, apicomedial notch without membrane (as in Fig. 41); tarsal claw with only two well-developed subsidiary teeth (except *granti*)...2

2 Tergum III with sharp or rounded (often nose-like) projection overhanging apical notch (as in Figs. 46, 48)...3

\- Tergum III without projection above apical notch (as in Figs. 45, 47)................4

3 Tergum III with sharp projection, lateral margins of notch protruding as far as overhanging projection in lateral view (Fig. 46).........................*plicatus* (Aaron)

\- Tergum III with blunt and rounded projection above shallow apicomedial emargination, lateral margins of "notch" not protruding as far as overhanging projection in lateral view (Fig. 48)........................*downeyi* (Bohart & Campos)

4 Wing membrane untinted, water clear, terga usually brassy to coppery red ....*telfordi* (Bohart & Campos)

- Wing membrane brown-tinted, terga variously colored..................................5

5 Terga light green, bronzy or brassy, contrasting with thoracic coloration; tergum III obtusely notched apically; metanotum sharply angled in lateral view, slightly hooked apically...*krombeini* (Bohart & Campos)

- Terga blue, dark green or purplish, not contrasting with thoracic color; tergum III notch various; metanotum may be sharply rounded but rarely hooked...............6

6 Metanotum obtusely or acutely pointed in lateral view; scutellum with smooth anteromedial spot usually followed by zone of punctures; tergum III apex in female truncate or subtruncate in lateral view (Fig. 45); body length less than 5 mm...*variatus* (Aaron)

- Metanotum broadly rounded in lateral view; scutellum with smooth medial area extending entire length; tergum III apex in female not truncate in lateral view; body length 5 mm or longer...................................*granti* (Bohart & Campos)

Philoctetes downeyi (Bohart and Campos)

Fig. 48, Map 53

Omalus downeyi Bohart and Campos 1960:246. Holotype ♂; USA: California, El Dorado Co., Camp Richardson (CDFA).

Geographic Range (Map 53). Western states (CA, NV, OR, MT), mostly in the Canadian Zone; 173 specimens were seen.

California Records. Collected June through August; **Alpine Co**.: Lake Winnemucca; Hope Valley; **Amador Co**.: 3 mi. w Jackson; Brockway Summit; 4 mi. n Silver Lake; 1.5 mi. ne Red Lake; **El Dorado Co**.: Camp Richardson; 17 mi. n Truckee; **Lassen Co**.: Susan River Camp; **Mono Co**.: Mill Creek Canyon; **Nevada Co**.: Sagehen Creek; Boca; Fuller Lake; Hobart Mills; **Sierra Co**.: Gold Lake; Independence Lake; Yuba Pass; **Siskiyou Co**.: Mt. Shasta City; **Sonoma Co**.: Cloverdale; **Trinity Co**.: Mountain Meadow Ranch Coffee Crk; **Tuolumne Co**.: Tuolumne Meadow; Dardanelles; Strawberry; **Yolo Co**.: Davis.

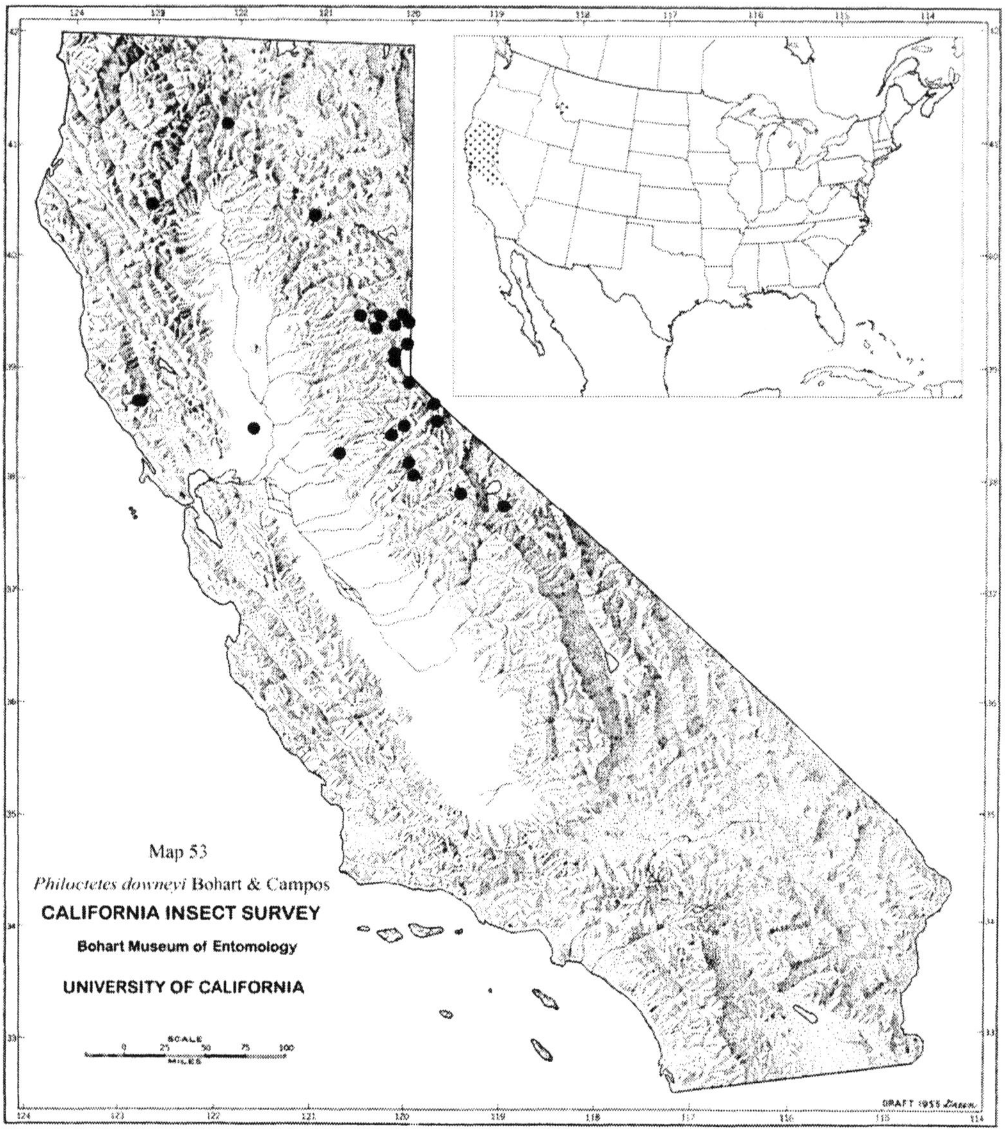

Map 53. California distribution of *Philoctetes downeyi*. Inset: overall distribution.

Hosts. Unknown.

Discussion. Two species of *Philoctetes*, *plicatus* and *downeyi*, have a projection above the apical notch of tergum III. In *downeyi* this projection is rounded, and in *plicatus* it is sharp. Additional diagnostic features of *downeyi* include the black body color, with greenish and bronzy highlights, scutum strongly convex posteriorly, and a few large punctures on the scutellum and metanotum. The metanotum of *downeyi* is produced and sharply angular in lateral view.

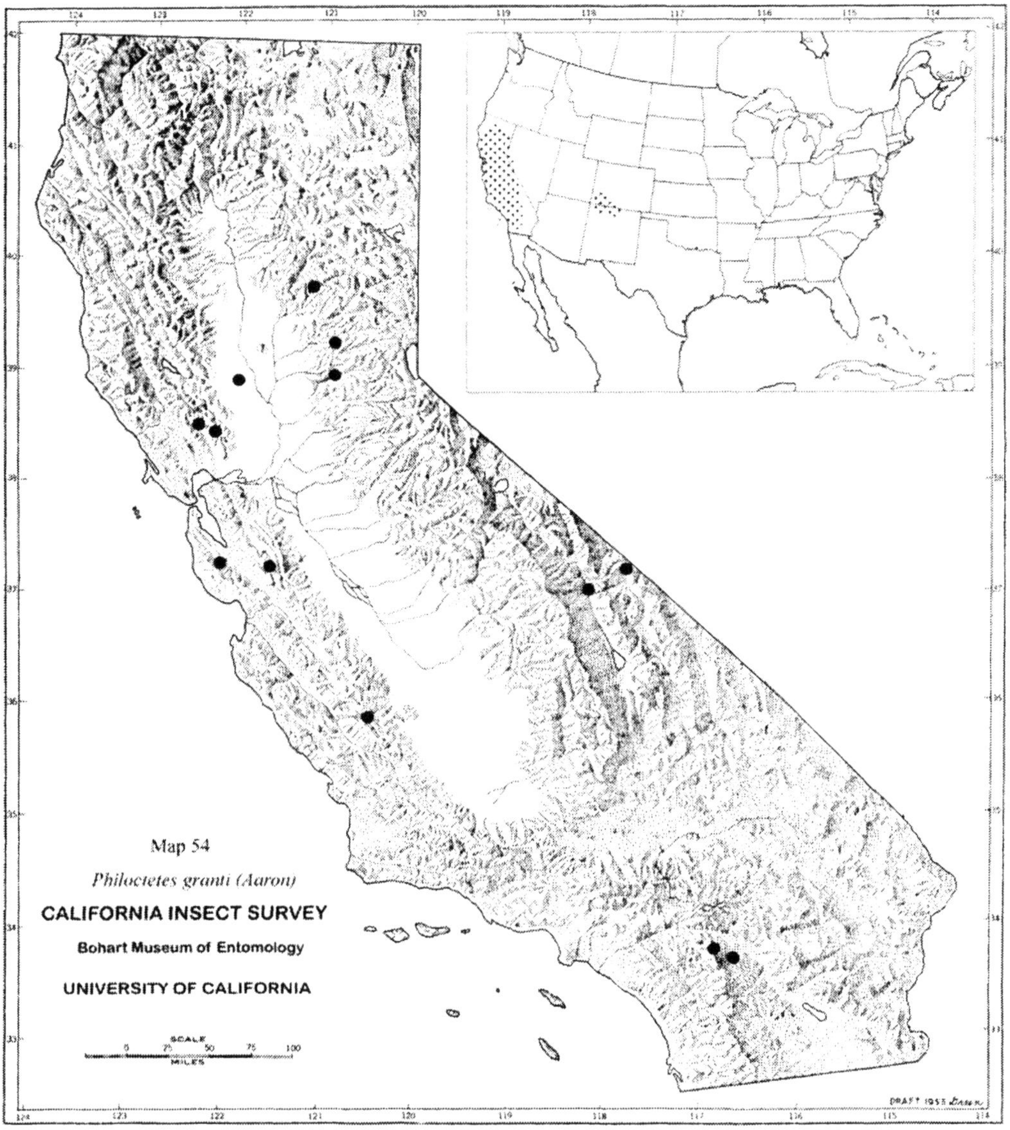

Map 54. California distribution of *Philoctetes granti*. Inset: overall distribution.

Philoctetes granti (Bohart and Campos)

Map 54

Omalus granti Bohart and Campos 1960:247. Holotype ♂; USA: California, Riverside Co., San Jacinto Mts., Idyllwild (CDFA).

Geographic Range (Map 54). Western USA, in the Upper Sonoran and Transition Zones; 30 specimens were studied.

California Records. Collected April through July; **Colusa Co**.: Williams; **Inyo Co**.: Big Pine; Westgard Pass; **Monterey Co**.: Arroyo Seco; **Napa Co**.: Samuel Springs (beneath Lake Berryessa); Pope Valley; **Placer Co**.: Foresthill; **Plumas Co**.: Tobin; **Riverside Co**.: Marion Mt. Camp San Jacinto Mts.; Idyllwild; **Santa Clara Co**.: Uvas Canyon.

Hosts. Unknown.

Discussion. This species can be distinguished by the unnotched tergum III, rounded metanotum, and scutellum with a large smooth area. In addition, the claws have three subsidiary teeth, the male sternum III is metallic colored, and there are relatively few punctures associated with the notauli.

Philoctetes krombeini (Bohart & Campos)

Map 55

Omalus krombeini Bohart & Campos 1960:247. Holotype ♂; USA: Arizona, Tucson (CDFA).

Geographic Range (Map 55). Lower Sonoran deserts of California and Arizona; 62 specimens were studied.

California Records. Collected February through May; **Imperial Co**.: Algodones dunes 7 mi. se Glamis; **Inyo Co**.: Eureka Valley dunes; **Riverside Co**.: Thousand Palms; 16 and 19 mi. w Blythe; **San Diego Co**.: Borrego Valley dunes; Palm Canyon; **San Bernardino Co**.: Kramer Junction.

Hosts. Unknown.

Discussion. This species is close structurally to *variatus*. However, *krombeini* is more brightly colored with the metasomal terga ranging from light green with coppery highlights to bright greenish blue (much darker in *variatus*), wings stained most darkly toward base (evenly stained in *variatus*), and scutal punctures small and dispersed along the notauli (variable, but the punctures are generally larger in *variatus*).

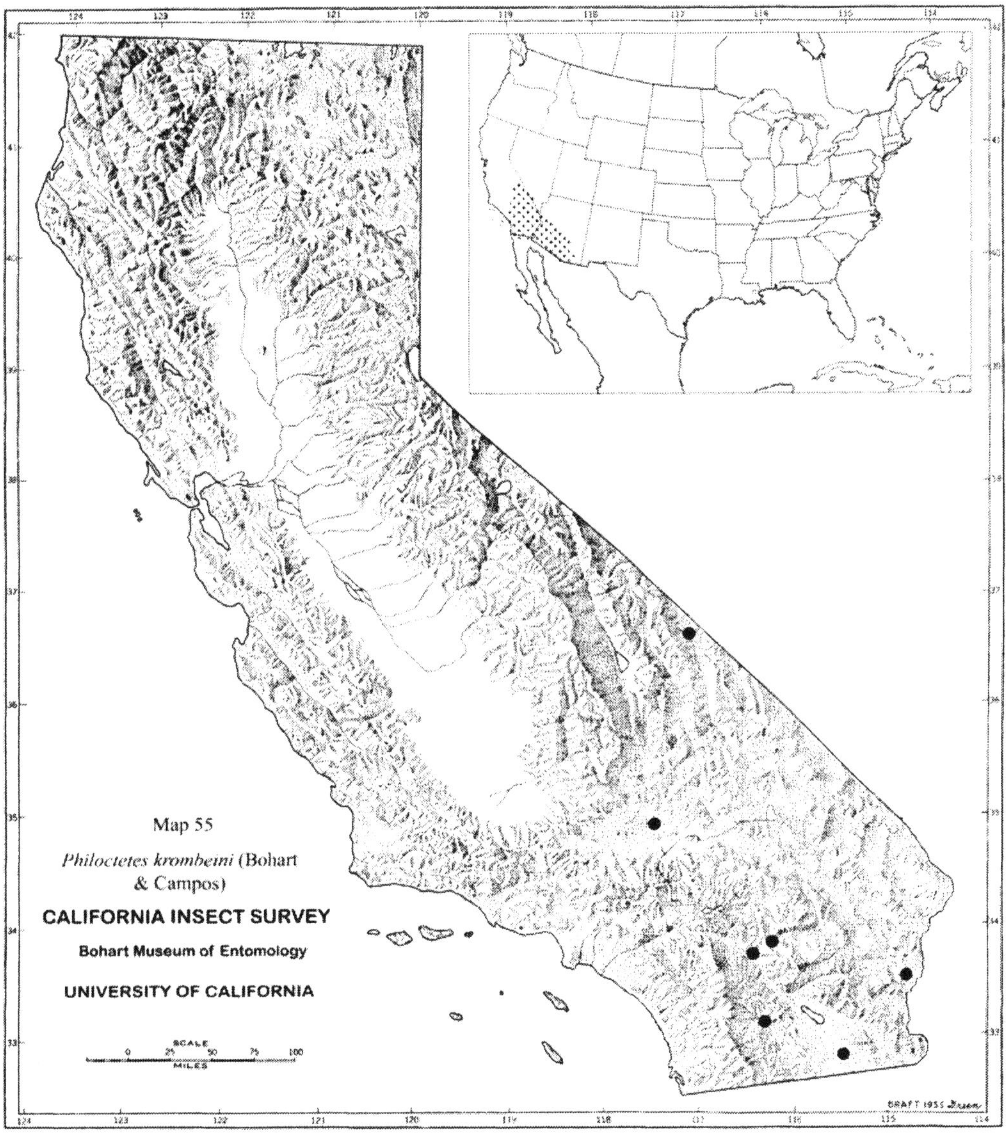

Map 55. California distribution of *Philoctetes krombeini*. Inset: overall distribution.

Philoctetes plicatus (Aaron)

Fig. 46, Map 56

Diplorrhos plicatus Aaron 1885:216. Lectotype ♀ (designated by Cresson 1928); USA: Montana (ANSP).

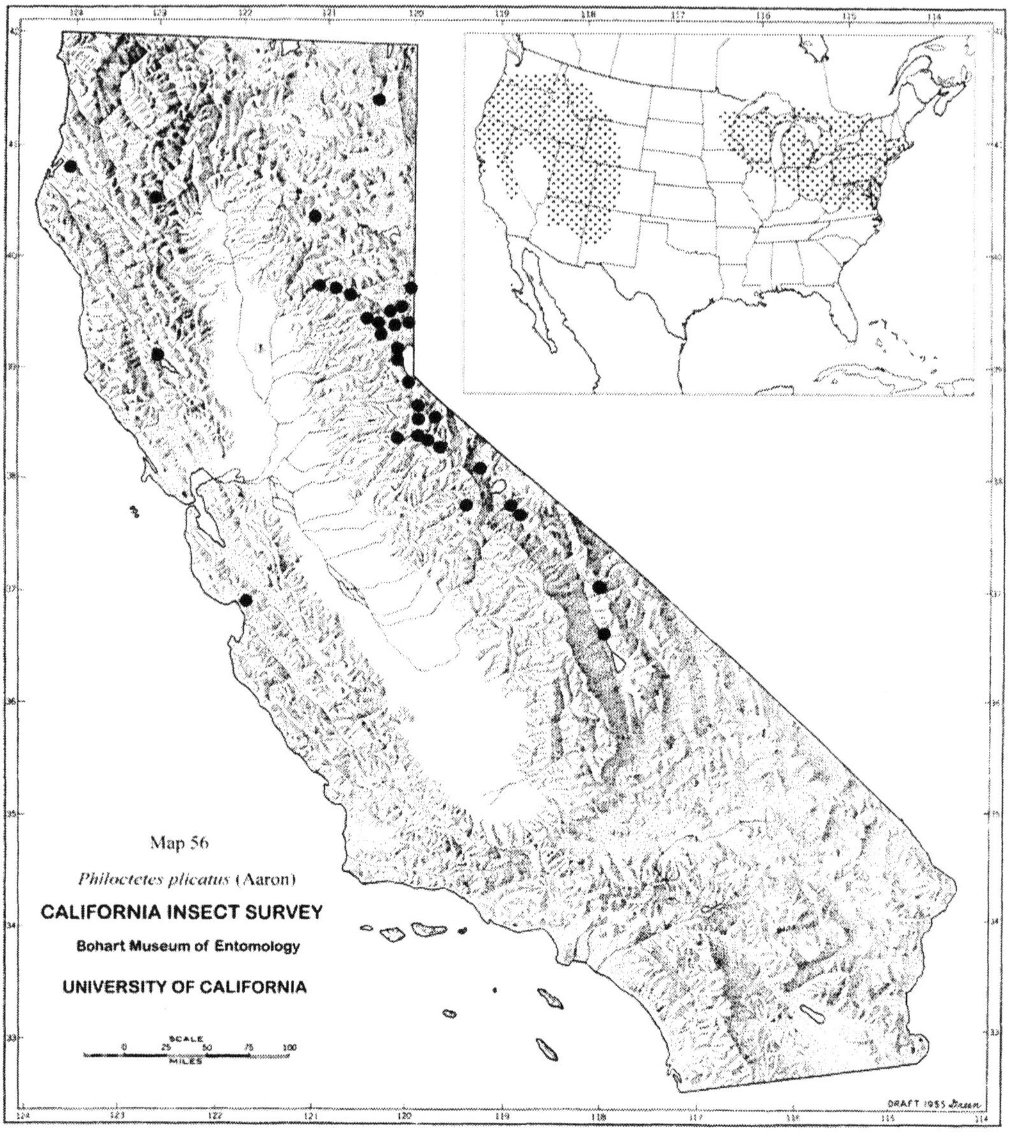

Map 56. California distribution of *Philoctetes plicatus*. Inset: overall distribution.

Geographic Range (Map 56). Mostly west of the 100th meridian, with scattered records from Virginia, New York, Michigan, Alaska, Canada (Yukon Territory, Ontario), and the mountains of Baja California; in the Upper Sonoran to Canadian Zones; 638 specimens were examined.

California Records. Collected April through September; **Alpine Co.**: 1 mi. s Monitor Pass; Carson Pass; Highland Lake; Ebbetts Pass; Lake Winnemucca; Blue Lakes; Hope Valley; Forestdale Meadow; **Amador Co.**: Silver Lake; **El Dorado Co.**:

South Lake Tahoe; **Humboldt Co.**: 1 mi. w Kneeland; **Inyo Co.**: 8 mi. w Independence; **Lake Co.**: Blue Lakes; **Lassen Co.**: Hallelujah Junction; Susan River Camp; **Modoc Co.**: Cedar Pass; **Mono Co.**: Mill Creek Canyon; Blancos Corral; Mammoth Lake; Cottonwood Creek; **Nevada Co.**: Sagehen Creek; Boca; 17 mi. n Truckee; Hobart Mills; Russell Valley; **Placer Co.**: Dollar Point 2 mi. ne Tahoe City; Carnelian Bay, Lake Tahoe; Martis Valley; **Plumas Co.**: Johnsville; Graeagle; 4 mi. w Quincy; **San Bernardino Co.**: Lake Arrowhead; **Santa Cruz Co.**: Soquel; **Sierra Co.**: Sierraville; Yuba Pass; Independence Lake; Sierra Valley; 15 mi. se Sierraville; 9 mi. n Sagehen Creek; **Trinity Co.**: 10 mi. w Coffee Creek Ranger Station; **Tuolumne Co.**: Chipmunk Flat.

Hosts. Unknown.

Discussion. The most distinctive feature of *plicatus* is the sharp projection above the apicomedial notch on tergum III. Males are usually entirely black, females may be extensively blue-green.

Philoctetes seminudus (Aaron)

Figs. 43, 47, Map 57

Notozus seminudus Aaron 1885:218. Lectotype ♂ (designated by Cresson 1928); USA: Washington (ANSP).

Geographic Range (Map 65). Canada (British Columbia, Ontario) and the USA from the Rockies Mountains west to the Pacific Coast, in Transition and Canadian Zones; 375 specimens were seen.

California Records. Collected March through July; **Alameda Co.**: Livermore; **Fresno Co.**: Coalinga; **Monterey Co.**: Arroyo Seco; **Nevada Co.**: Sagehen Creek; **Siskiyou Co.**: Mt. Shasta City; **Trinity Co.**: Carrville; **Tulare Co.**: Ash Mt. Rd., Sequoia National Park.

Hosts. Unknown.

Discussion. This species has a projecting, *Elampus*-like metanotum. Otherwise it is similar to the non-Californian species, *intermedius*. *P. seminudus* are large bodied *Philoctetes* averaging 5-6 mm in length.

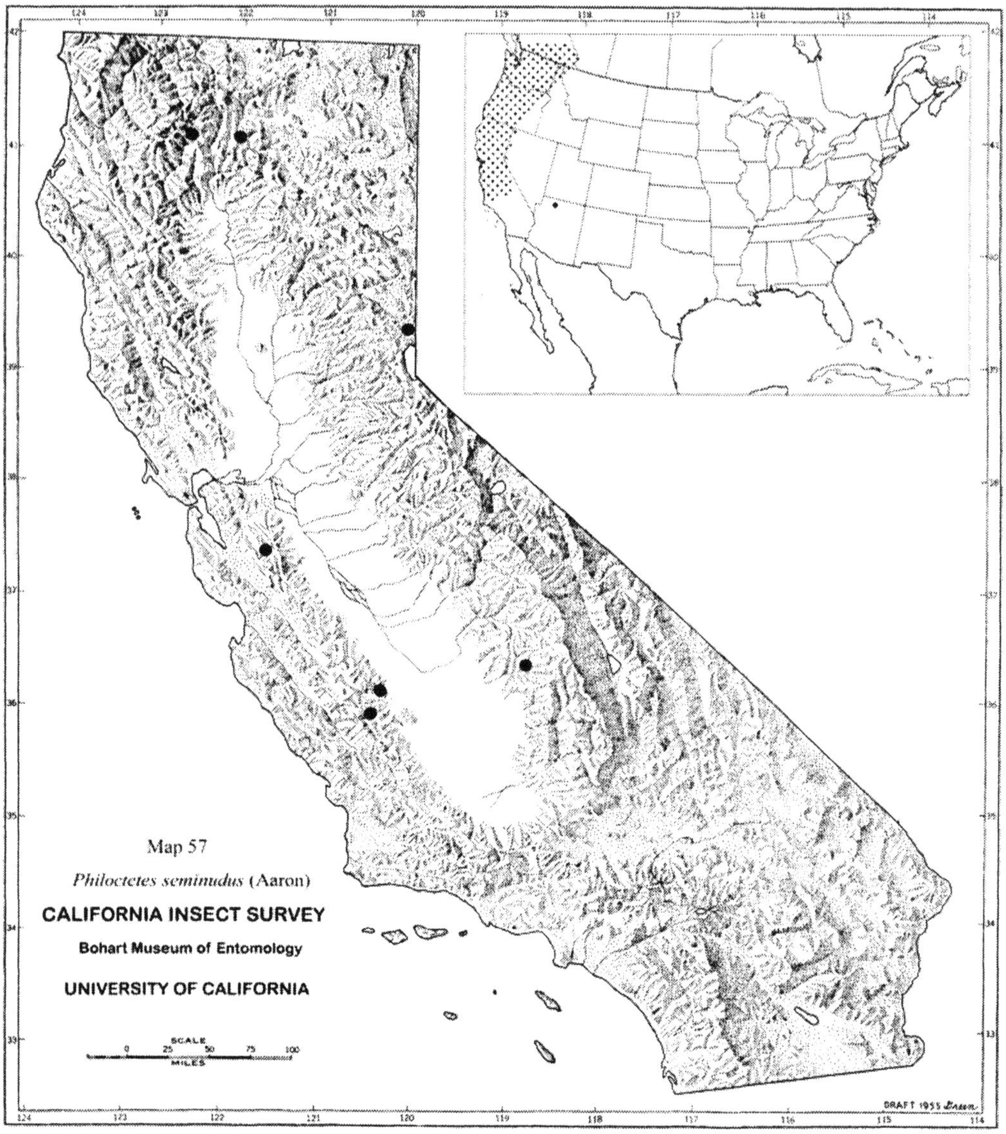

Map 57. California distribution of *Philoctetes seminudus*. Inset: overall distribution.

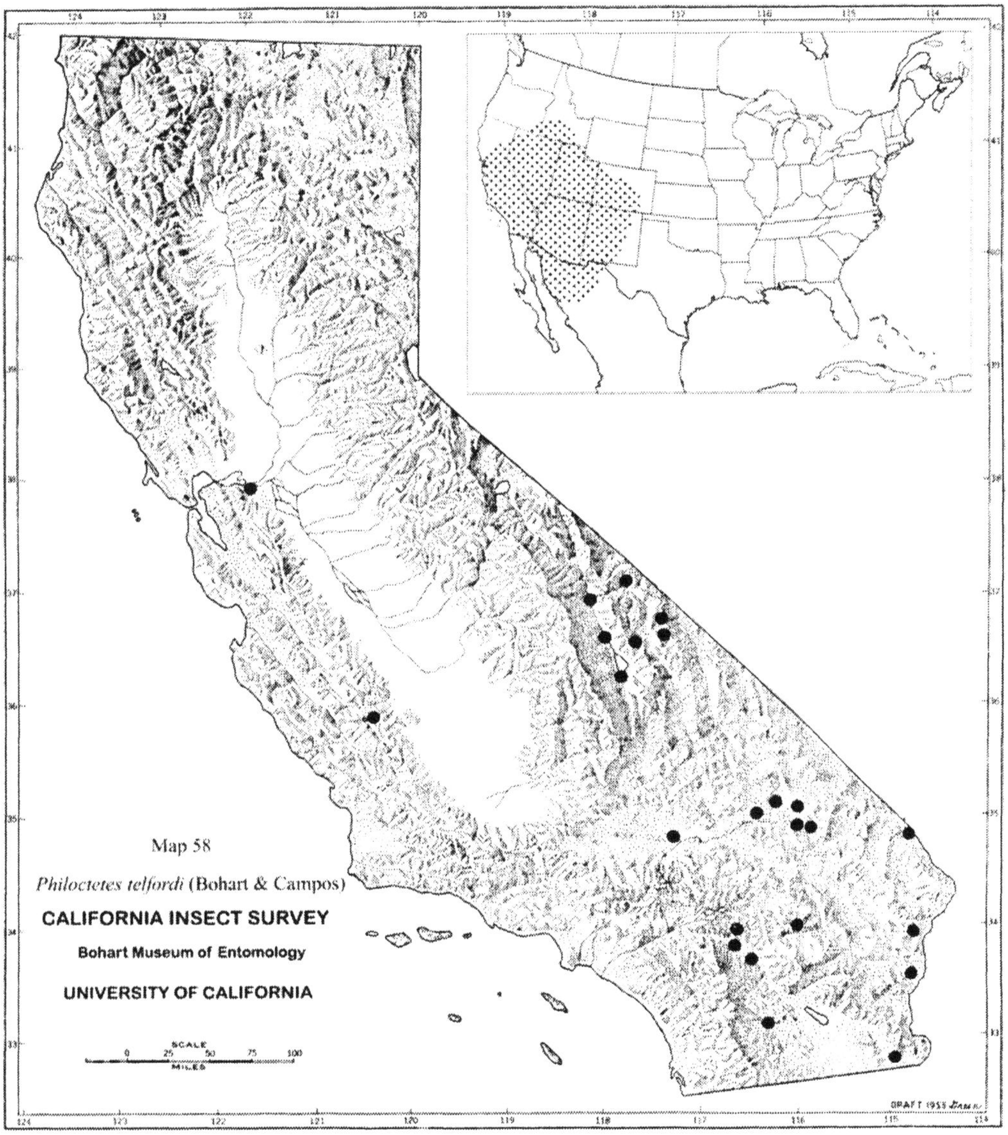

Map 58. California distribution of *Philoctetes telfordi*. Inset: overall distribution.

Philoctetes telfordi (Bohart and Campos)

Map 58

Omalus telfordi Bohart and Campos 1960:249. Holotype ♂; USA: Arizona, Tucson (CDFA).

Geographic Range (Map 58). Lower Sonoran Zone regions from California to Texas, and in northwestern Mexico (Baja California Sur, Sonora); 249 specimens were studied.

California Records. Collected April through August; **Contra Costa Co**.: Antioch; **Imperial Co**.: 6 mi. n Junction Yuma-San Diego Hwy; Palo Verde; **Inyo Co**.: 15 mi. s Big Pine; Darwin Falls; 3-4 mi. e Big Pine; south end Owens Lake; Eureka Valley dunes; 2 mi. e Lone Pine; Surprise Canyon; **Monterey Co**.: Arroyo Seco; **Riverside Co**.: Thousand Palms Canyon; 18 mi. w Blythe; Palm Springs; **San Bernardino Co**.: 10 mi. e Midway; Cronise Wash; Kelso; 27 mi. n Vidal; 15 mi. w Baker; 12 mi. s Baker; Kramer Hills; Needles; Twenty-nine Palms; **San Diego Co**.: Borrego Valley.

Hosts. Unknown.

Discussion. The water clear wings and green to coppery body color will distinguish *telfordi* from other *Philoctetes* species. In addition, this species has a moderately projecting metanotum, obtusely notched tergum III, and small, dispersed punctures along the notauli.

Philoctetes variatus (Aaron)

Figs. 12, 45, Map 59

Elampus variatus Aaron 1885:215. Lectotype ♂ (designated by Cresson 1928); USA: Montana (ANSP).

Elampus cressoni Aaron 1885:215. Lectotype ♀ (designated by Cresson 1928); USA: Colorado (ANSP).

Geographic Range (Map 59). Widespread west of the 100th meridian in the USA, Canada and Mexico (Baja California Sur, Chihuahua), including Alaska and the Northwest Territories; 831 specimens were studied.

California Records. Collected April through August; **Alameda Co**.: Alameda; San Lorenzo; **Alpine Co**.: 2 mi. e Carson Pass; Lake Winnemucca; Carson Pass; Ebbetts Pass; **Butte Co**.: Madrone Lake; **Contra Costa Co**.: Antioch; Mt Diablo; **El Dorado Co**.: El Dorado; **Fresno Co**.: Coalinga Pass; Kings River Canyon; Pioneer Basin; Selma; **Inyo Co**.: Antelope Springs; 3 mi. w Lone Pine; Westgard Pass; Little Lake; Deep Springs; Lone Pine; Independence; Big Pine; 5 mi. e Big Pine; **Kern Co**.: Mill Potrero; Kern Canyon; 3 mi. w Frazier Park; **Kings Co**.: Corcoran; **Lake Co**.: n fork Cache Creek Hwy 20; Borax Lake; **Lassen Co**.: Hallelujah Junction; Susanville; **Los Angeles Co**.: Little Rock; Claremont; Tanbark Flat; **Madera Co**.: Madera; **Marin Co**.: Nicasio; **Merced Co**.: Merced Falls; **Mono Co**.: Mill Creek Canyon; White Mts.; Mammoth Lakes; Paradise Camp; 22-24 mi. n Bishop; **Monterey Co**.: Arroyo

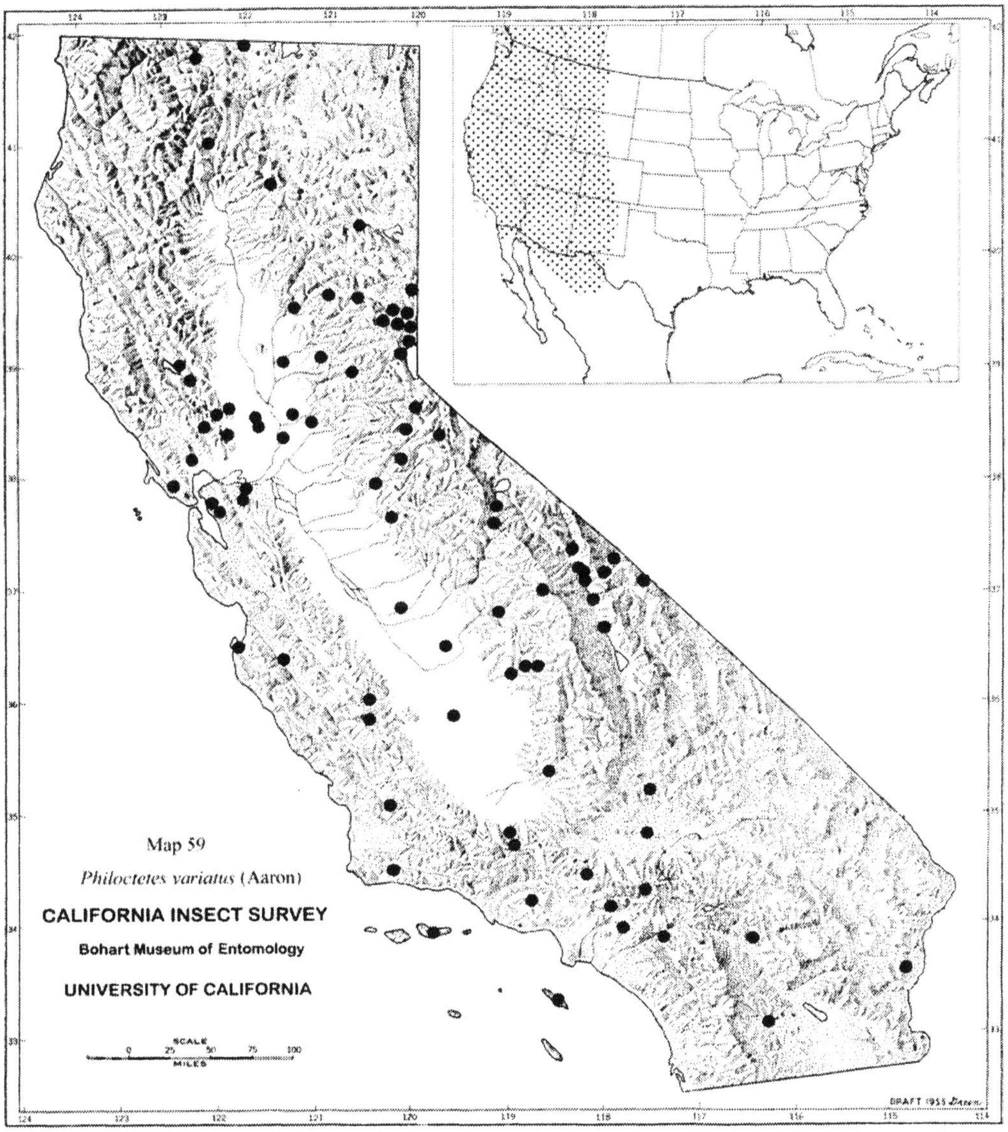

Map 59. California distribution of *Philoctates variatus*. Inset: overall distribution.

Seco; Soledad; Monterey; **Napa Co**.: Pope Valley; Samuel Spring (beneath Lake Berryessa); 4 mi. nw Berryessa; **Nevada Co**.: Sagehen Creek; Boca; Grass Valley; **Placer Co**.: Carnelian Bay Lake Tahoe; **Plumas Co**.: Blairsden; Johnsville; **Riverside Co**.: Thousand Palms; Riverside; 18 mi. w Blythe; **Sacramento Co**.: Sacramento; Carmichael; Auburn; **San Bernardino Co**.: 1 mi. s Adelanto, Kramer Hills, Cajon Pass; **San Diego Co**.: Borrego Valley, Olivehain; **San Luis Obispo Co**.: Pozo; **Santa Barbara Co**.: Santa Ynez Mts., Santa Cruz Island, Santa Catalina Island; **Shasta Co**.: Cassel, Hat Creek; **Sierra Co**.: Sierra Valley, Independence Lake, Sattley, Sierraville; **Siskiyou Co**.: Valentine Caves Lava Bed National

Monument; 13 mi. n Yreka; **Sonoma Co.**: Glen Ellen; **Tuolumne Co.**: Chipmunk Flat; Tuolumne City; **Tulare Co.**: Woodlake; Three Rivers; Ash Mountain; **Ventura Co.**: Sespe Canyon; **Yolo Co.**: Cache Creek Canyon; Davis; 5 mi. e Woodland; Putah Canyon; 5 mi. e Guinda; **Yuba Co.**: 18 mi. ne Marysville.

Hosts. Powell (1964), Parker and Bohart (1966) and Evans (1973) reared this species from nests of the crabronid wasps *Diodontus occidentalis* W. Fox, *Pemphredon grinnelli* (Rohwer), *Stigmus inordinatus* W. Fox, and *Passaloecus cuspidatus* F. Smith.

Discussion. This is a highly variable, widespread species. The metanotum is obtuse to acute in profile but does not overhang the metanotum. Specimens of *variatus* tend to roll into a tight ball, which makes seeing the venter of the metasoma difficult. However, this is the only way to determine the sex of a specimen. Sternum III in males is black, and metallic-colored in females. The tarsal claws have two subsidiary teeth.

Genus *Pseudolopyga* Krombein

Generic diagnosis. Facial setae sparse, erect; scapal basin shallow, with short zone of cross-ridging; malar space less than 1 MOD long; pronotum gently curved without distinct anterior declivity or carinae, with faint anteromedial pit; notauli deep, sulciform; mesopleuron rounded with short scrobal sulcus, without omaulus or scrobal carina; fore and midtarsal claws with two subapical teeth; female hindtarsal claw with one subapical tooth, male with two; scutellar wing fossa simple; forewing medial vein straight, arising after cu-a; tergum III margin evenly rounded, although often turned under laterally; volsella with digitus and cuspis.

Hosts. *Pseudolopyga* have been reared from nests of *Solierella* species (Crabronidae) (Carrillo and Caltagirone 1970). Their parasitic behavior is unusual for chrysidids, as they oviposit on juvenile lygaeid bugs provisioned by the true host wasp. The eggs fail to develop unless the bug is stung and provisioned by a *Solierella* female. *Pseudolopyga* females never approach the host's nest.

Distribution. Western USA, in the Sonoran and Transition Zones.

Discussion. The sexually dimorphic tarsal claw dentition is an unusual trait seen in *Pseudolopyga*. Hindtarsal claws are medially bidentate in females (Fig. 37) and apically tridentate in males (Fig. 32). *Pseudolopyga* can also be distinguished by the longer Rs stub (at least half as long as medial vein) and the tarsal claw dentition. There are three species of *Pseudolopyga* worldwide, only two in North America and both are Californian. The third species occurs in Chile. The North American species were reviewed by Kimsey (1982).

Key to the California Species of *Pseudolopyga*

Clypeal apex notched medially; mesopleuron with large circular punctures and little or no microridging; gena usually with little or no microridging ....*carrilloi* (Bohart & Brumley)

Clypeal apex unnotched medially; mesopleuron without discrete circular punctures, covered with microridging; gena with extensive microridging ....*taylori* (Bodenstein)

Pseudolopyga carrilloi (Bohart & Brumley)

Figs. 27, 32, 37, Map 60

Hedychridium carrilloi Bohart & Brumley 1967:232. Holotype ♂; USA: California, Colusa Co., Arbuckle (UCDC).

Geographic Range (Map 60). Southwestern USA, except Oregon, also in southern Idaho; 88 specimens were seen.

California Records. Collected March through September; **Alameda Co**.: Albany; **Colusa Co**.: Arbuckle; **Contra Costa Co**.: Antioch; **Imperial Co**.: Ogilby Rd. near Yuma Junction; Glamis; **Inyo Co**.: Eureka Valley dunes; **Kern Co**.: McKittrick; 10 mi. w McKittrick; **Riverside Co**.: San Timoteo Canyon; White Water; **San Bernardino Co**.: 6 mi. s Ivanpah; **San Diego Co**.: Scissors Crossing; Borrego Valley; **San Luis Obispo Co**.: Paso Robles; **Tulare Co**.: Ash Mt.; 9 mi. w Creighton Ranch Park; **Yolo Co**.: Davis.

Hosts. *P. carrilloi* has been reared from nest cells of the crabronid wasps *Solierella plenoculoides* W. Fox (as *blaisdelli* (Bridwell)) and *S. peckhami* (Ashmead) (Carrillo and Caltagirone 1970, Parker and Bohart 1968).

Discussion. This species can be distinguished from *taylori* by the apicomedially notched clypeus and absence of extensive microridging on the gena and thorax.

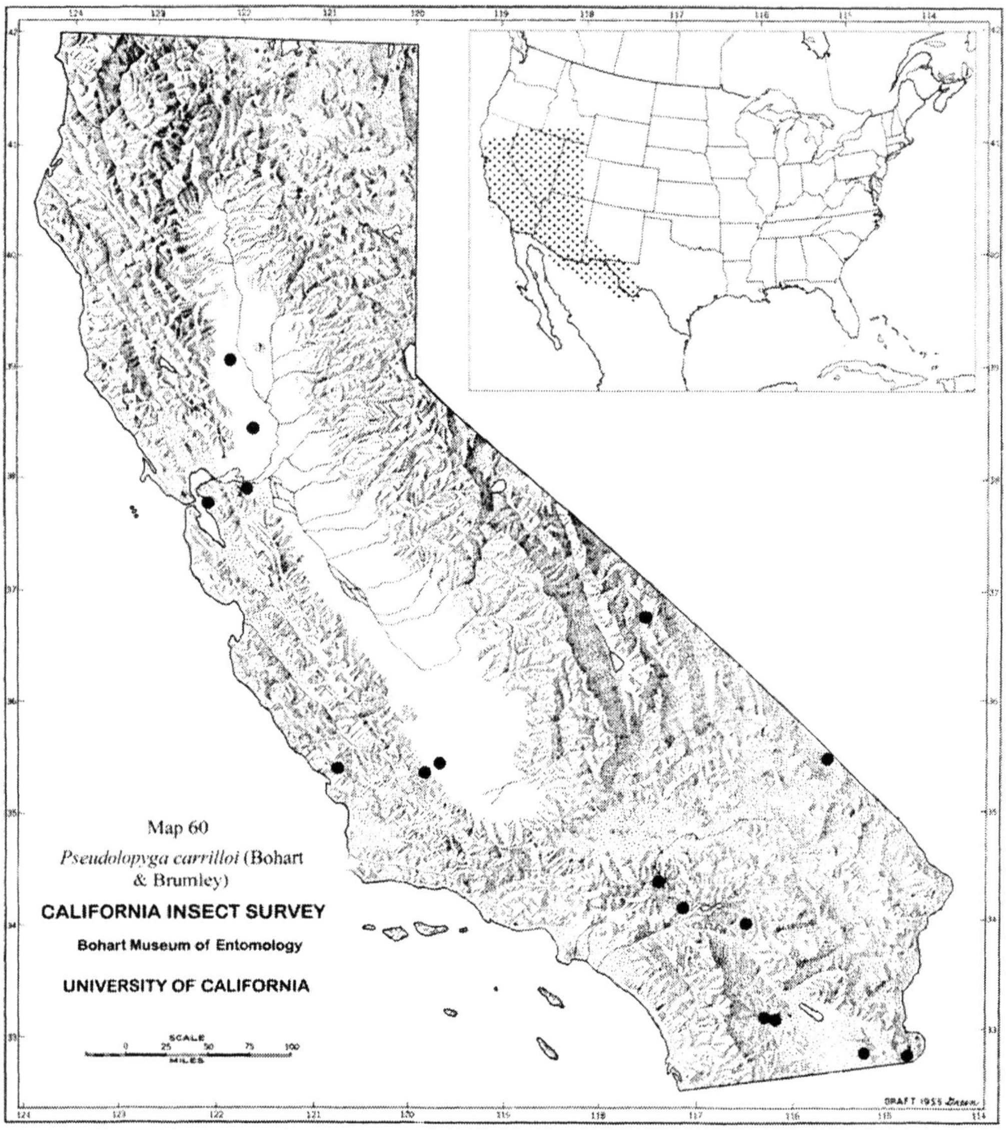

Map 60. California distrubution of *Pseudolopyga carrilloi*. Inset: overall distribution.

***Pseudolopyga taylori* (Bodenstein)**

Map 61

Holopyga taylori Bodenstein 1929:19. Holotype ♂; USA: Idaho, Twin Falls Co., Hollister (USNM).

Geographic Range (Map 61). Western USA; 28 specimens.

California Records. Collected April through October; **Contra Costa Co**.: Antioch; **Inyo Co**.: Deep Springs; **Riverside Co**.: Thousand Palms; Hopkins Well 18 mi. w Blythe; **San Bernardino Co**.: Dagget; 7 mi. sw Kelso; 15 mi. w Baker; **San Diego Co**.: Borrego Valley; **Sierra Co**.: Independence Lake; **Yolo Co**.: Davis.

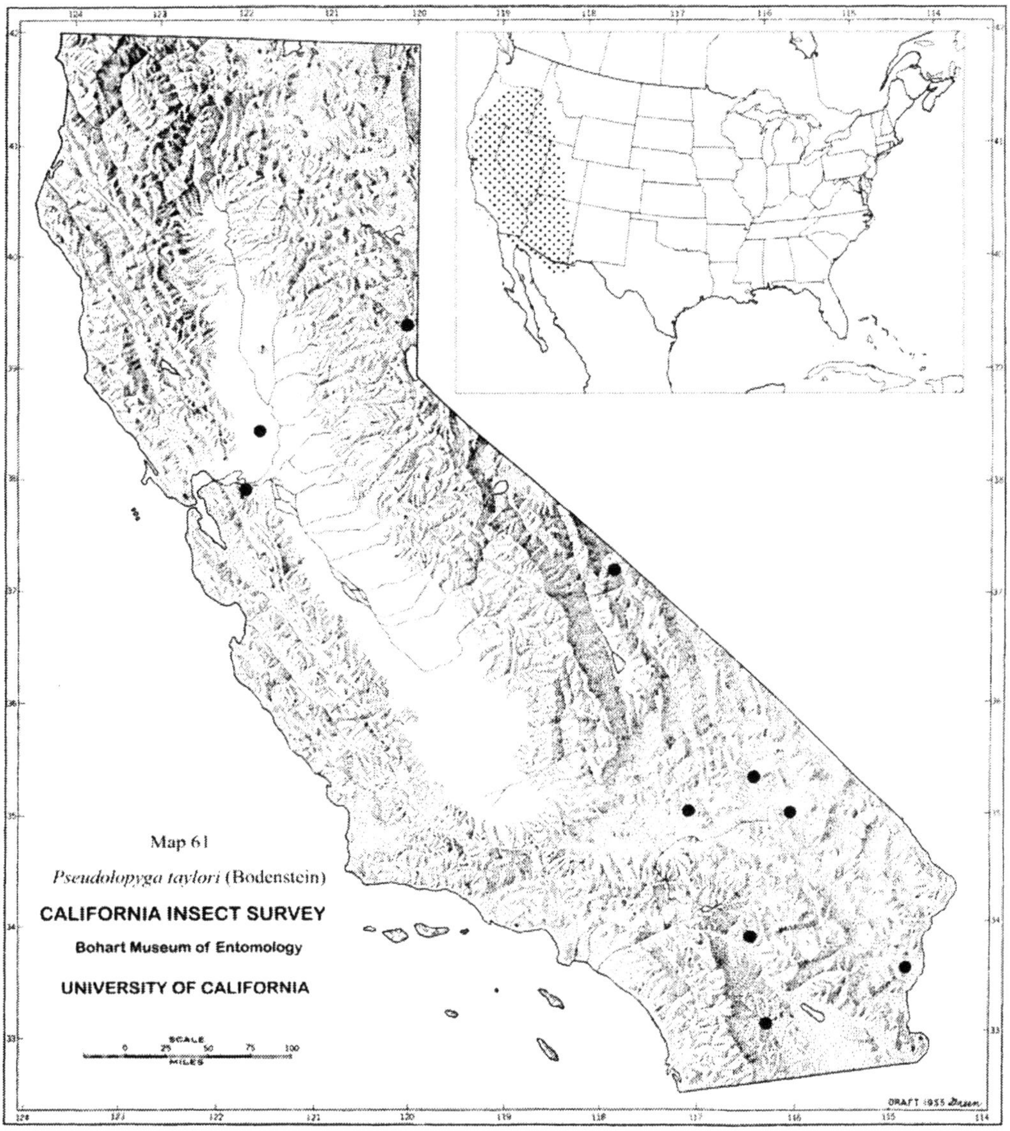

Map 61. California distribution of *Pseudolopyga taylori*. Inset: overall distribution.

Hosts. This species was reared from nests of *Solierella blaisdelli* (Krombein 1967).

Discussion. The clypeus of *taylori* is entire, without a medial notch, and the gena and thorax are extensively microridged.

Genus *Pseudomalus* Ashmead

Generic Diagnosis. Scapal basin deep, smooth, asetose; malar space less than 1 MOD long and bisected horizontally by genal carina; head generally cupped posteriorly, with carinate postocular edge; pronotum deeply concave laterally; scutum with small, scattered punctures, except large dense ones clumped posteriorly between notauli; mesopleuron angulate, with juncture of omaulus, verticaulus and scrobal carina strongly projecting ventrally, scrobal sulcus oblique; scutellum with anterior margin unmodified; metanotum broadly rounded; forewing medial cell asetose, medial vein strongly arched and arising before cu-a, stigma short, broad and apically rounded; forefemur carinate ventrally, often subapically broadened particularly in females; tarsal claws with 3-5 subsidiary teeth; tergum I-III strongly convex, tergum III lateral margin straight or sinuate, usually deeply notched apicomedially, without transparent apical rim.

Hosts. These wasps are nest parasites of members of the crabronid subfamily Pemphredoninae.

Distribution. This is a holarctic genus with the majority of species occurring in the Central Asian Steppes.

Discussion. The most distinctive feature of *Pseudomalus* is the clumping of the large scutal punctures posteromedially between the notauli. Other diagnostic features include the broadly rounded metanotum, long pubescence on the metasoma, and complete notauli that end posteriorly in a pit. There are five North American species and all occur in California. Bohart and Kimsey (1982) reviewed the genus *Pseudomalus*.

Key to species of *Pseudomalus*

1 Metasomal color reddish to greenish or golden and strongly contrasting with thoracic coloration..*auratus* (Linnaeus)

- Metasomal color blue, purple or green, concolorous with thorax.............................2

2 Scutum anterolaterally swollen on either side of notauli, appearing trilobate in anterior view, area between parapsidial line and notaulus polished with small, evenly distributed punctures...........................*trilobatus* (Bohart and Campos)

- Scutum anterolaterally flattened, appearing evenly convex in anterior view, area between parapsidial line and notaulus either finely striate or "scratched", or irregularly and coarsely punctate (as in Fig. 16)..3

3 Mesopleuron primarily green or bluish green; tergum III apical notch usually obtuse; female forefemur about 2.5x as long as broad............*janus* (Haldeman)

- Mesopleuron primarily black with blue, green or purple highlights; tergum III apical notch acute (Fig. 42); female forefemur about 3x as long as broad..........4

4 Scutum, scutellum and metanotum with few, very large, well-separated, deep pits, mesopleuron without omaulus and scrobal carina strongly projecting; tergum III apical notch narrow, largely parallel-sided............................*speculum* (Say)

- Scutum, scutellum and metanotum with dense, nearly contiguous pits; mesopleuron with omaulus and scrobal carina strongly projecting; tergum III apical notch triangular in shape, broadest apically.........................*purpuratus* (Provancher)

Pseudomalus auratus (Linnaeus)

Fig. 36, Map 62

Sphex auratus Linnaeus 1758:572. Holotype ♀; Europe (BMNH).

Hedychrum sinuosus Say 1828:53. Holotype ♀; USA: Indiana (destroyed).

Geographic Range (Map 62). Palearctic Region, eastern USA and scattered in the rest of the USA; 48 specimens were examined.

California Records. Collected in May; **San Joaquin Co.**: Stockton.

Hosts. Hosts of *auratus* are unknown in North America. In the Palearctic Region *auratus* has been reported from nests of *Passaloecus* and *Cemonus* species (Crabronidae: Pemphredoninae) (Grandi 1961, Tsuneki 1970, Mocsáry 1887).

Discussion. *P. auratus* has apparently been introduced into North America from Europe via crabronid nests in hollow stems in plant material. This species is widespread in the Palearctic Region. Although similar to *purpuratus, auratus* can be distinguished by the brightly coppery gold colored metasomal terga, with greenish tints, which contrasts with the blue, green and purple thorax.

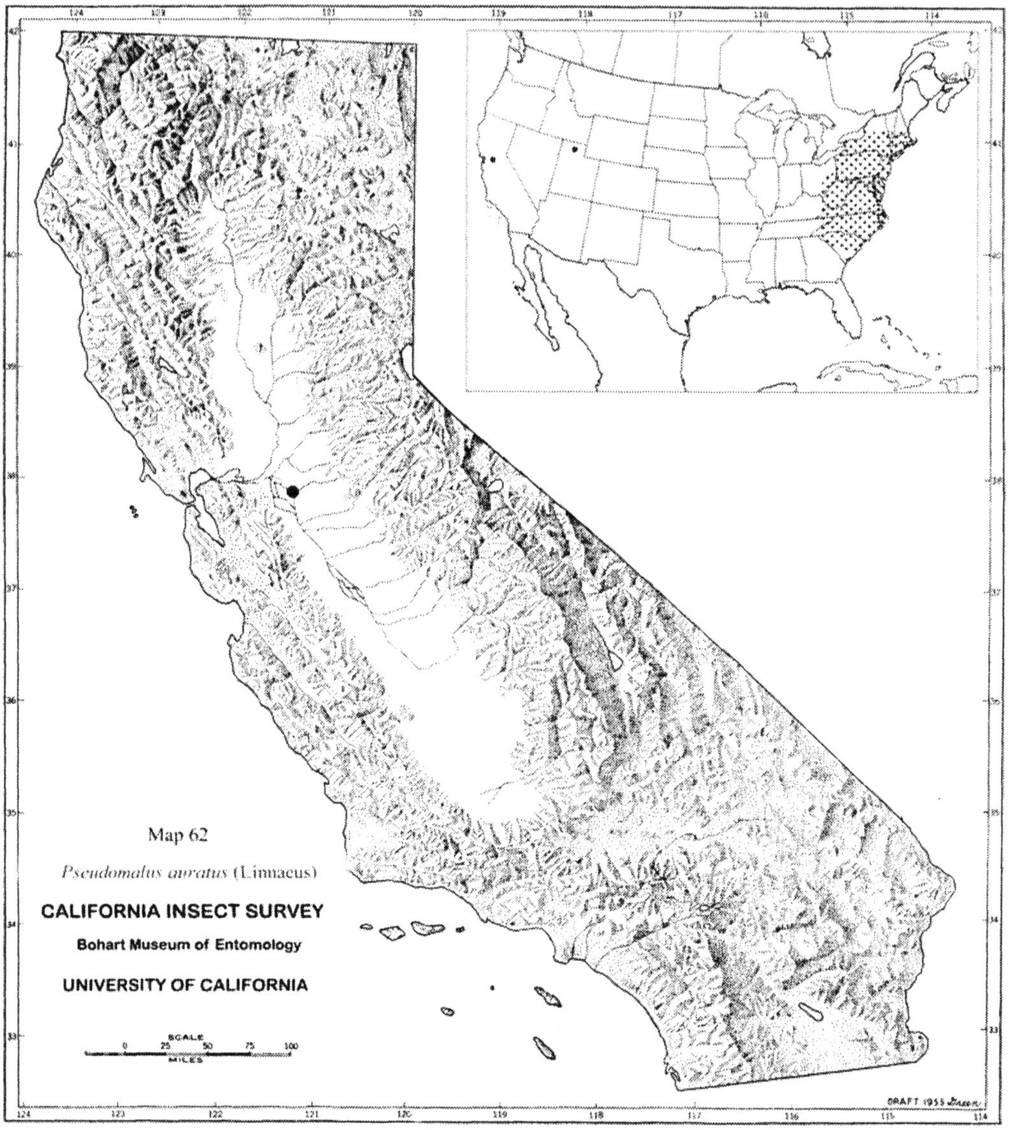

Map 62. California distribution of *Pseudomalus auratus*. Inset: overall distribution.

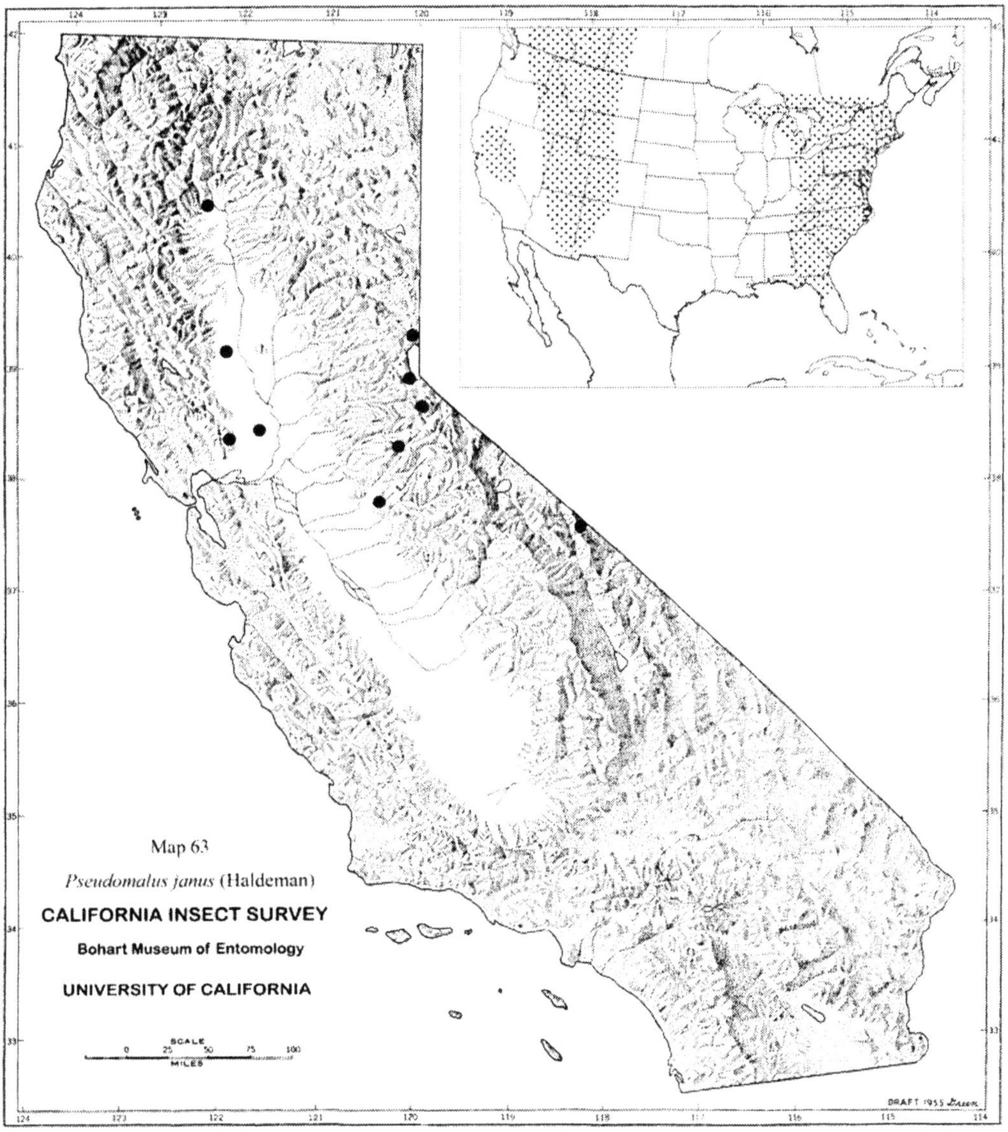

Map 63. California distribution of *Pseudomalus janus*. Inset: overall distribution.

Pseudomalus janus (Haldeman)

Fig. 16, Map 63

Hedychrum janus Haldeman 1844:53. Neotype ♀ (designated by Bohart and Campos 1960); USA: New York, Mt. MacIntire (CU).

Elampus coruscans Norton 1879:234. Lectotype ♀ (designated by Cresson 1928); Canada (ANSP).

Omalus corruscans Aaron 1885:214. Misspelling of *coruscans* (Norton) 1879.

Omalus semicircularis Aaron 1885:215. Lectotype ♂ (designated by Cresson 1928); USA: Colorado (ANSP).

Geographic Range (Map 63). Widespread from coast to coast in North America, including southern Canada, in Transition and Canadian Zones; 184 specimens were studied.

California Records. Collected May through September; **Alpine Co**.: 1 mi. s Monitor Pass; **Calaveras Co**.: Big Trees; **Colusa Co**: 4 mi. w Maxwell; **El Dorado Co**.: Pope Beach Lake Tahoe; **Mono Co**.: Blanco's Corral; **Nevada Co**.: Sagehen Creek; **Shasta Co**.: 25 mi. e Redding; **Solano Co**.: Monticello Dam; **Tuolumne Co**.: Long Barn; **Yolo Co**.: Davis.

Hosts. *Pseudomalus janus* has been reared from nests of *Pemphredon concolor* Say and *P. confertim* W. Fox (given as *errans* Rohwer) by Krombein (1979).

Discussion. The most distinctive features of *janus* are the obtuse apical notch of tergum III, green to blue green mesopleuron, and unusually broad and sharp-edged forefemur.

Pseudomalus purpuratus (Provancher)

Fig. 42, Map 64

Elampus purpuratus Provancher 1881:302. Lectotype ♀ (designated by Bohart and Kimsey 1982); Canada (LU).

Elampus purpurascens Provancher 1881:303. Misspelling of *purpuratus* Provancher.

Omalus macswaini Bohart and Campos 1960:244. Holotype ♂; USA: Wyoming, Teton Co., Jenny Lake (CDFA).

Geographic Range (Map 64). Widespread in the USA, including Alaska, and in Canada, in transition and Canadian Zones; 360 specimens were examined.

California Records. Collected May through August; **Alpine Co**.: Woodfords; **El Dorado Co**.: Grass Lake; Pope Beach Lake Tahoe; **Lassen Co**.: Hallelujah Junction; **Mariposa Co**.; Mariposa; **Mono Co**.: 1 mi. s Tom's Place; **Napa Co**.: Mt. Veeder; **Nevada Co**.: Sagehen Creek; Fuller Lake; Boca; Hobart Mills; **Placer Co**.: Carnelian

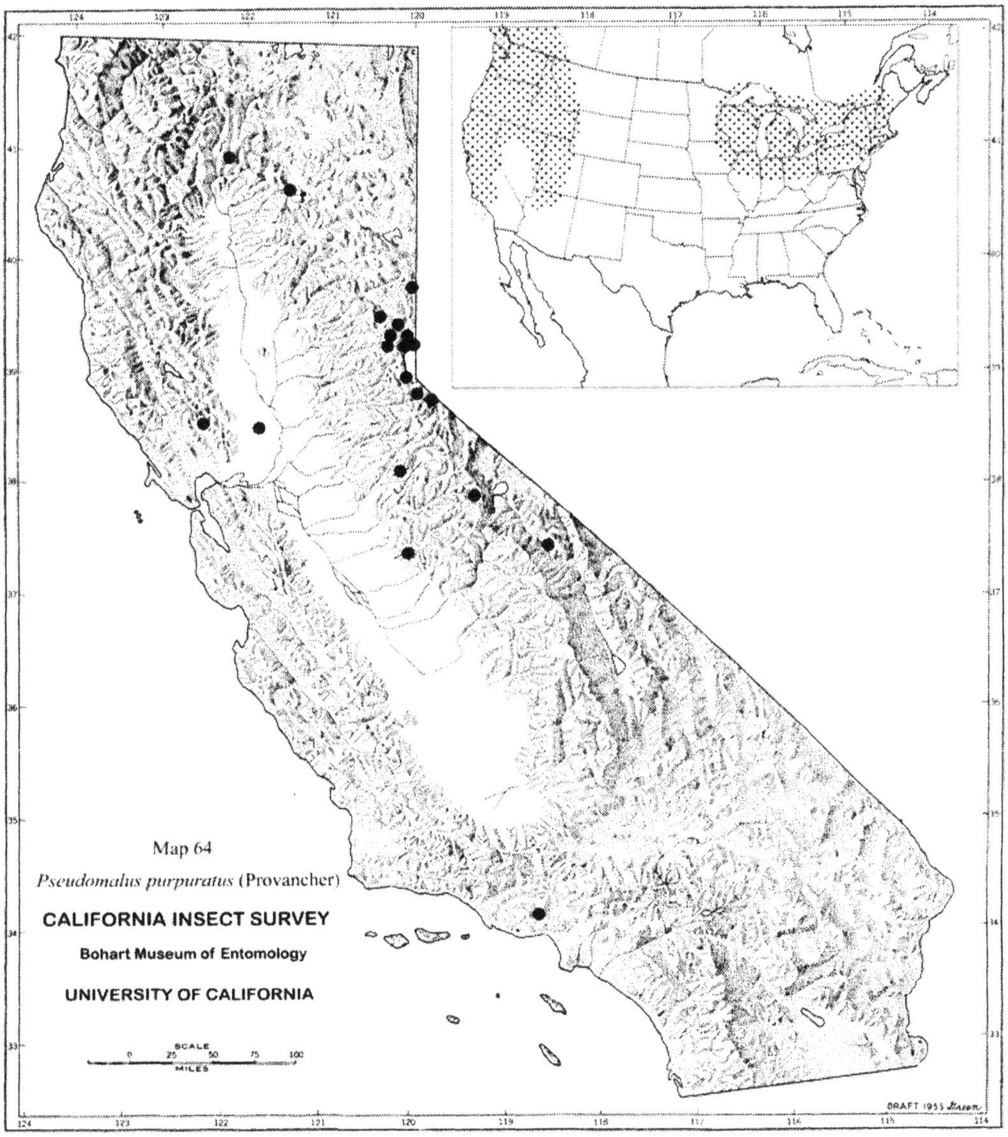

Map 64. California distribution of *Pseudomalus purpuratus*. Inset: overall distribution.

Bay, Lake Tahoe; **Shasta Co.**: Cassel; Hat Creek; **Sierra Co.**: Sierraville; Independence Lake; **Tuolumne Co.**: Tioga Pass; Long Barn; **Ventura Co.**: Sespe Canyon.

Hosts. Hosts include *Pemphredon lethifer* (Shuckard), *Passaloecus cuspidatus* F. Smith, and *Stigmus americanus* Packard (Krombein 1979).

Discussion. The purplish body color, acute apicomedial notch on tergum III and moderately stout forefemur will distinguish *purpuratus* from other *Pseudomalus* species.

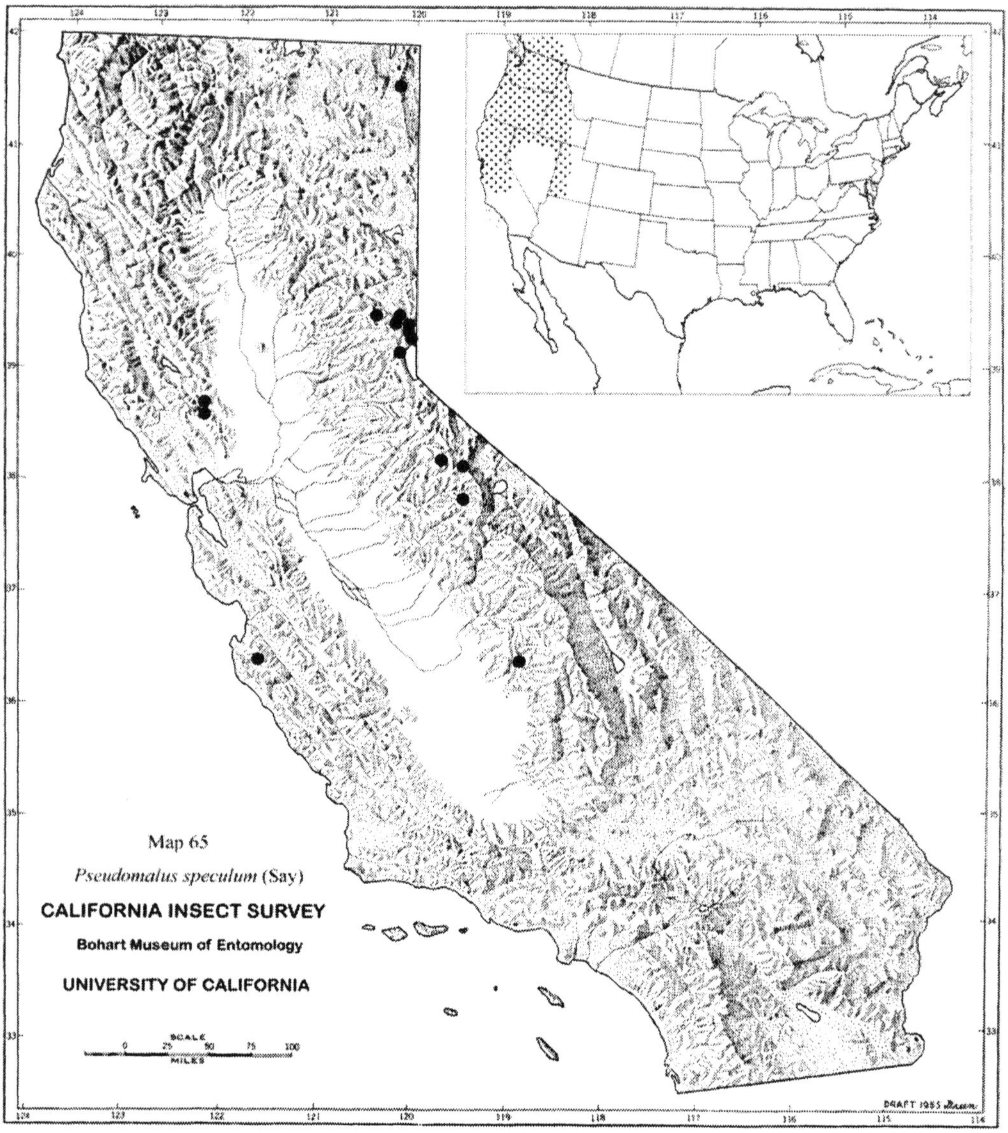

Map 65. California distribution of *Pseudomalus speculum*. Inset: overall distribution.

Pseudomalus speculum (Say)

Figs. 18, 41, Map 65

Hedychrum speculum Say 1836:225. Neotype ♀ (designated by Bohart and Campos 1960); USA: Washington, Rainier National Park, White River (CAS).

Geographic Range (Map 65). Transition and Canadian Zones in western North America, from Colorado to California and Alaska, and in Canada (Manitoba, Northwest Territories); 87 specimens were studied.

California Records. Collected May through August; **Modoc Co**.: Cedar Pass; **Mono Co**.: Sonora Pass; **Monterey Co**.: 1 mi. s Jamesburg Santa Lucia Mts.; **Nevada Co**.: Sagehen Creek; Hobart Mills; **Placer Co**.: Carnelian Bay, Lake Tahoe; **Sierra Co**.: 15 mi. se Sierraville; Independence Lake; **Tulare Co**.: Three Rivers; **Tuolumne Co**.: Chipmunk Flat.

Hosts. Unknown.

Discussion. This species can be recognized by the combination of the brown stained forewing tip, tergum III with apicomedial notch parallel-sided (Fig. 41), and the greatly enlarged punctures on the strongly convex scutum, scutellum and metanotum. Individuals of *speculum* range from 3-4 mm in length.

Pseudomalus trilobatus (Bohart & Campos)

Fig. 44, Map 66

Omalus trilobatus Bohart & Campos 1960:245. Holotype ♂; USA: California: Napa Co., Samuel Springs (beneath Lake Berryessa) (CDFA).

Geographic Range (Map 66). Western USA and British Columbia, Canada, in Transition and Canadian Zones; 331 specimens were studied.

California Records. Collected March through September; **Alameda Co**.: Mt. Hamilton; **Alpine Co**.: Woodfords; **Contra Costa Co**.: Wildcat Canyon; **El Dorado Co**.: Emigrant Gap; **Kern Co**.: 3 mi. w Frazier Park; **Lake Co**.: Kelseyville; Pillsbury Lake; Cobb Mt.; Borax Lake; **Monterey Co**.: Arroyo Seco; **Napa Co**.: Samuel Spring (beneath Lake Berryessa); Butte Canyon 0.5 mi. s Napa; 14 mi. w Monticello Dam; Pope Valley; 4 mi. nw Berryessa (beneath Lake Berryessa); 6 mi. s Pope Valley; **Placer Co**.: 5 mi. w Foresthill; Colfax; 8 mi. e Auburn; **Plumas Co**.: Blairsden; **Riverside Co**.: Idyllwild; **San Benito Co**.: Pinnacles National Monument; 8.5 mi. sw New Idria; **San Bernardino Co**.: Camp Baldy; **Santa Cruz Co**.: Santa Cruz; **Sierra Co**.: Independence Lake; Sardine Lakes; Gold Lake; **Solano Co**.: Mix Canyon summit; Solano Lake; Vacaville; **Tehama Co**.: Red Bluff; **Tulare Co**.: Mineral King; **Tuolumne Co**.: Yosemite National Park; **Yolo Co**.: Cache Creek Canyon; Davis; Putah Canyon; 4 mi. sw Davis.

Hosts. Bohart and Campos (1960) reared this species from *Pemphredon grinnelli* (Rohwer) (as *giffardi* Rohwer) (Crabronidae).

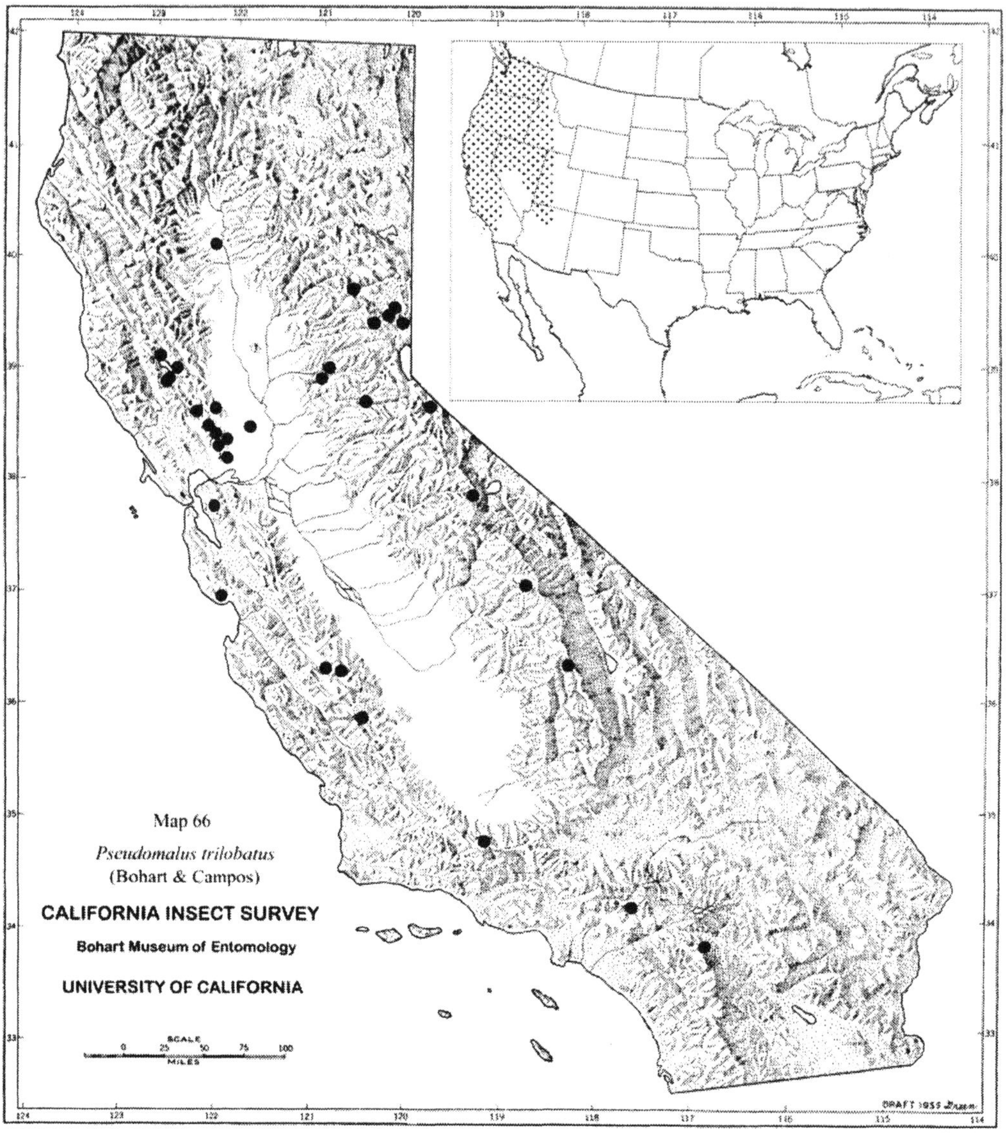

Map 66. California distribution of *Pseudomalus trilobatus*. Inset: overall distribution.

Discussion. The shape of the scutum is diagnostic for this species. When viewed anteriorly the scutum appears to consist of three strong convexities, between the tegula and notaulus and between the notauli. In addition, the tarsal claw has four subsidiary teeth, and tergum III has an acute apicomedial notch, with a basal carina.

Genus *Xerochrum* Bohart

Generic diagnosis. Facial setae sparse, erect; scapal basin indistinct, shallow, with few cross ridges; interocellar sulcus present; malar space about 2 MOD long; mesopleuron rounded, with sharp ventral angle, omaulus straight, weak; notauli sulciform; scutellar wing fossa without projection; forefemur subbasally angulate; mid and hindtibia without pits or depressions on inner surface; tarsal claws edentate; forewing medial vein slightly curved, arising at cu-a, Rs stub less than half as long as medial vein; metasoma primarily nonmetallic red; tergum III evenly rounded; volsella with digitus and cuspis.

Hosts. Unknown.

Distribution. Southwestern USA and northwestern Mexico.

Discussion. This monotypic genus exhibits two features unique among the nearctic Elampini, edentate tarsal claws (Fig. 35) and a non-metallic red metasoma. In addition, the forefemur has a subbasal angle, Rs stub is shorter than the stigma and the forewing medial vein is arched (Fig. 28).

Xerochrum rubeum Bohart

Figs. 14, 28, 35; Map 67

Xerochrum rubeum Bohart 1980:134. Holotype ♂; USA: California, Riverside Co., Thousand Palms (UCDC).

Geographic Range (Map 67). Southwestern USA (AZ, CA, NV, UT) and northwestern Mexico (Baja California Sur); 25 specimens were seen.

California Records. Collected April through June; **Imperial Co.**: 1-2 mi. w Glamis; Kane Springs; **Inyo Co.**: 2 mi. e Lone Pine; 2 mi. w Oasis; **Riverside Co.**: Thousand Palms; **San Bernardino Co.**: Cadiz Dunes; Cronise Valley; **San Diego Co.**: Borrego.

Discussion. Diagnostic characteristics of the genus will serve to identify *rubeum*. This species is rarely collected.

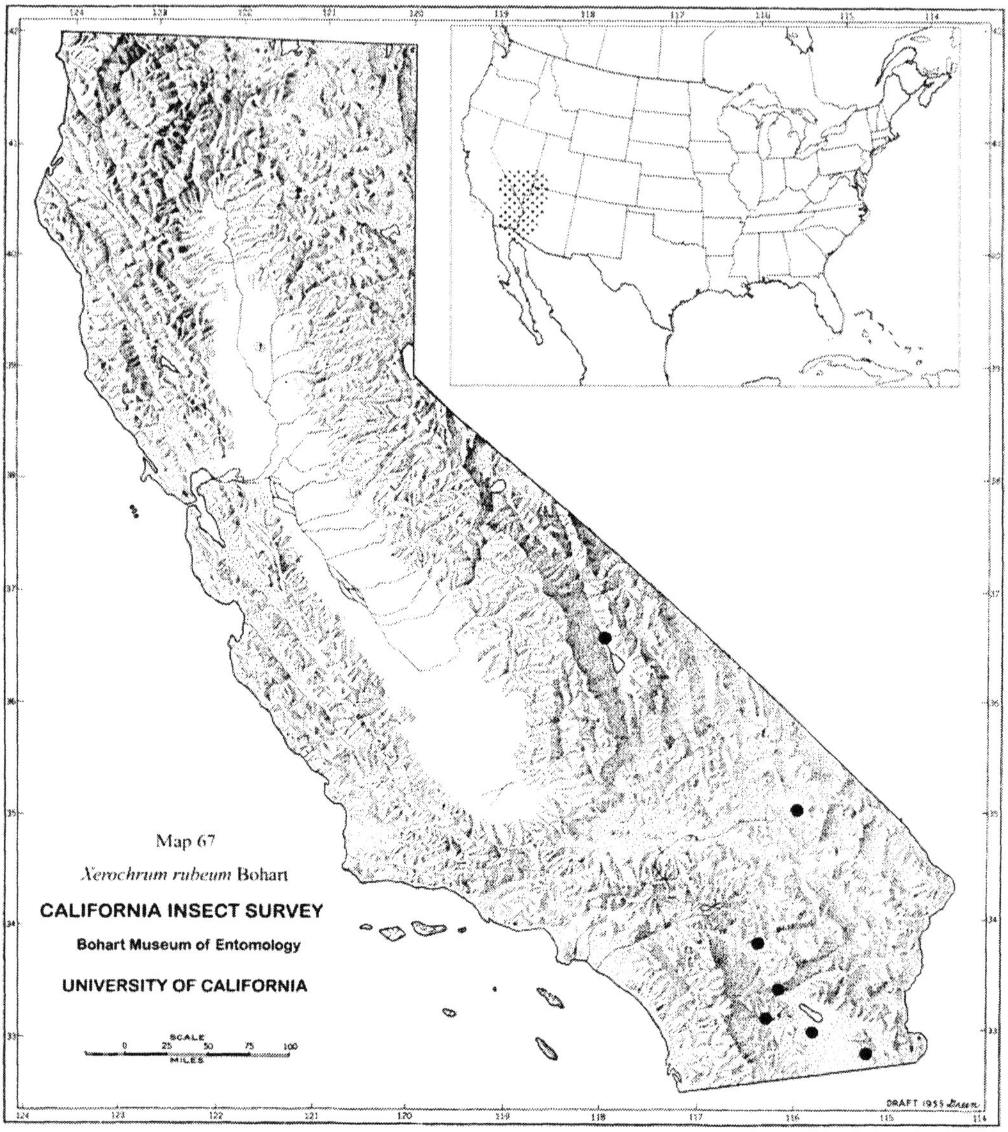

Map 67. California distribution of *Xerochrum rubeum*. Inset: overall distribution.

Tribe Chrysidini

The tribe Chrysidini is the most speciose chrysidid group worldwide, and the majority of species are in the genus *Chrysis*. These wasps are nest parasites of sphecid wasps, solitary Vespidae and Megachilidae. Members are characterized by the subapical transverse pit row of tergum III and the often dentate, lobate or emarginate apical margin. The forewing has the Rs stub sclerotized for more than half the length of the marginal cell and the discoidal and cubital cells are usually complete.

Key to California Genera of Chrysidini

1 Tergum III apical rim with four or six teeth or angles, one species with four weak lobes (as in Figs. 68-76)..*Chrysis* Linnaeus

- Tergum III apical rim with 0-3 teeth or angles (as in Figs. 92-96, 98-103)...........2

2 Tergum III apical rim with a medial notch and lateral angle just beyond pit row (Figs. 54, 55), transverse frontal carina absent or only slightly indicated (Fig. 53) ....*Chrysurissa* Bohart

- Tergum III apical rim, if notched, without lateral angle; transverse frontal carina various...3

3 Tergum III apical rim edentate or rarely with a weak lateral angle, face unusually flattened and completely granulate-punctate without fine median cross-ridging, no transverse frontal carina (Fig. 88)..................................*Chrysura* Dahlbom

- Tergum III apical rim dentate or notched medially, <u>or</u> if not, face with median cross-ridging and/or transverse frontal carina present (as in Fig. 90)...............4

4 Scapal basin with pit in sublateral depression above antennal socket (Fig. 90), pit sometimes weakly indicated in small species with incomplete forewing discoidal cell; tergum III apical rim often tridentate (as in Figs. 95, 101) ....*Caenochrysis* Kimsey & Bohart

- Scapal basin without an unusual depression or pit, forewing discoidal cell complete, tergum III apical rim various but not tridentate.....................................5

5 Tibiae and often tegula and/or tergum III partly whitish (Fig. 103), or metanotum sharply angulate or somewhat projecting backward in profile; vertex without four smooth warts or swellings...*Argochrysis* Kimsey & Bohart

- Tibiae, tegula, and tergum III not partly whitish; metanotum rounded in lateral view; vertex sometimes with four smooth warts or swellings (Fig. 50) ....*Ceratochrysis* Cooper

Genus *Argochrysis* Kimsey & Bohart

Generic Diagnosis. Body usually with white markings on leg joints, white markings also common on male flagellomeres, tegula, tarsi, and post pit area of tergum III; scapal basin shallowly depressed, partly or completely covered with silvery appressed setae, medial area microridged, at least in female; flagellomere I usually longer than II; mandible with single large subapical tooth; malar space 0.5-2.0 MOD long; subantennal space 1.0-2.5 MOD long; genal carina close to eye, without defined subgenal area; transverse frontal carina usually present: biconvex or reverse U-shaped; midocellar area sometimes defined by carina, midocellus not lidded; pronotum shorter than scutellum, medial sulcus weak, lateral depression usually well developed; metanotum rounded, slightly angled in profile, or broadly projecting; mesopleuron simple, with scrobal sulcus, episternal sulcus rarely evident, some species with verticaulus and trace of omaulus; propodeal angle slender, straight or incurved posteriorly; tergum I broad, weakly indented anteromedially; tergum II obtuse posterolaterally; tergum III not depressed before well-developed pit row, apical margin entire or medially emarginate (sometimes medially bilobed); sternum II spots contiguous or narrowly separated and often large; male terminalia: sternum VIII subtriangular to quadrangular, gonocoxa slender and tapering in apical half, cuspis usually broad, aedeagus slender and tapering evenly toward apex.

Distribution. The genus *Argochrysis* only occurs in North America.

Hosts. Members of this genus are nest parasites of sphecid wasps in the genus *Ammophila*.

Discussion. These are distinctively colored wasps. Most species of *Argochrysis* are rainbow colored in various combinations of blue, green, purple, pink and coppery red, and often have the leg joints and antennomeres white. No other chrysidids in North America have this range of color. The markings and coloration of *Argochrysis* species are strikingly similar to those of the Old World genera *Spintharosoma* Zimmerman and *Spintharina* Semenov. However, *Argochrysis* is probably more closely related to *Ceratochrysis* than it is to these two genera, as discussed by Kimsey and Bohart (1991). There are 15 North American species and 12 in California (Bohart and Kimsey 1982)

Key to the California Species of *Argochrysis*

1 Frons without transverse frontal carina, and least interocular distance broader than length of flagellomere I plus pedicel..2

- Frons with well-developed transverse frontal carina or least interocular distance not broader than length of flagellomere I plus pedicel..5

2 Malar space as long as flagellomere I, flagellomere I 1.5x as long as broad; metanotum bulging and irregular but rounded in lateral view; tergum III with apicomedial emargination obtuse to 90°..*alboris* Kimsey

- Malar space shorter than flagellomere I, flagellomere I 1.4x as long as broad; metanotum projecting and angular, or sharp-edged, in lateral view, tergum III with apicomedial emargination variously shaped..4

3 Tergum III apical rim without white markings, terga in both sexes green and blue to purple..*viroris* Kimsey

- Tergum III apical rim with white markings, female terga coppery with some greenish hues, or mostly red, male terga various..*toralis* Kimsey

4 Metanotum distinctly projecting, angulate and/or sharp-edged in lateral view; tergum III apical rim with (*excelsior*) or without whitish markings (*litura* and *inornata*)..5

- Metanotum not projecting, rounded in lateral view; tergum III apical rim with whitish markings, except male *armilla*..7

5 Tergum III apical rim partly to entirely whitish, distinctly notched medially; legs with distinct whitish femorotibial joint marks; male flagellum partly white; punctures on vertex subequal in size to hindocellus..................*excelsior* Bohart

- Tergum III apical rim all green to blue and medially notched or unnotched; legs with femorotibial joints brown to metallic, not whitish; male flagellum entirely brown to black; punctures on vertex considerably larger than hindocellus.........6

6 Tergum III apical margin slightly indented medially in male, entire or nearly so in female; least interocular distance about equal to greatest eye breadth in front view ....*litura* Bohart

- Tergum III apical margin obtusely but clearly notched in both sexes; least interocular distance greater than greatest eye breadth in front view ....*inornata* Bohart

7 Tergum III apical rim produced on either side of medial notch to form distinctive subtriangular lobes; female multicolored in greens, blues and reds; male with front of flagellomeres I-III whitish..8

- Tergum III apical rim without submedial lobes or lobes broadly obtuse; female concolorous or multicolored various; male flagellomeres I-III front not whitish...9

8 Female scapal basin with medial strip all green or coppery, and polished to microridged; male terga multicolored, with coppery or reddish predominating, if mostly green then tergum I much more bluish purple than tergum II; body length averaging 4 mm...*trochilus* (du Buysson)

- Female scapal basin with medial strip red above and punctate hirsute below; male terga blue or green mixed with dark purple; body length averaging 5 mm ....*mesillae* (Cockerell)

9 Flagellum in front nearly all white in male, or at least on flagellomeres I-III in female; female tergum III slightly saddled before pit row, nearly entire apically ....*albicornis* Bohart

- Flagellum dark, with at most flagellomeres I-III white in male; female tergum III evenly convex before pit row, apicomedially notched...............................10

10 Tergum II medially polished between punctures; tergum III apical rim including pits unusually short, half as long as flagellomere I; female frons broader than eye in front view ....*lassenae* Bohart

- Tergum II medially finely sculptured between punctures; tergum III apical rim including pits more than half as long as flagellomere I; female frons equal to or narrower than eye in front view..11

11 Tergum III apical rim distinctly notched, usually without white in male and sometimes in female; frons below frontal carina polished or shiny medially, often with scattered macropunctures; body dorsum blue to green in male, in female often extensively red...*armilla* Bohart

- Tergum III apical rim slightly indented medially, whitish basolaterally; frons below frontal carina closely and finely punctate all across; body dorsum in both sexes blue to green..*secutor* Bohart

Argochrysis albicornis **Bohart**

Map 68

Argochrysis albicornis Bohart, in Bohart and Kimsey 1982:189. Holotype ♂; USA: California, San Diego Co., Borrego Valley (UCDC).

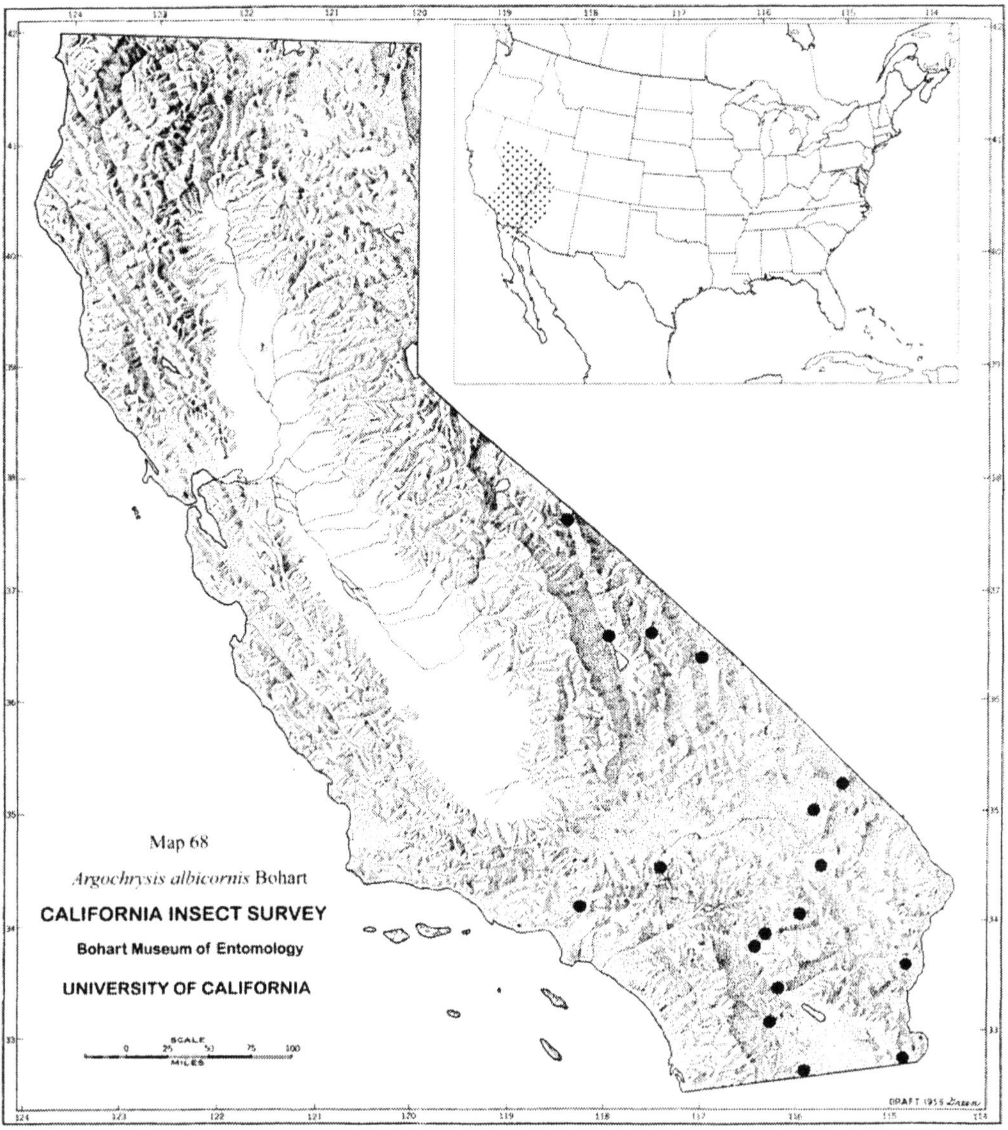

Map 68. California distribution of *Argochrysis albicornis*. Inset: overall distribution.

Geographic Range (Map 68). Upper and Lower Sonoran Zones in desert areas of California, Nevada and Arizona; 58 specimens were seen.

California Records. Collected April through July; **Imperial Co**.: 16 mi. w Winterhaven; 5 mi. w Mt. Signal; 3 mi. w Ocotillo; **Inyo Co**.: Oasis near Brown; Panamint Springs; 13 mi. n Trona; Lone Pine; Death Valley; **Los Angeles Co**.: Angeles Crest Hwy; **Mono Co**.: Benton Crossing; **Riverside Co**.: Palm Springs;

Thousand Palms; 18 mi. w Blythe; **San Bernardino Co.**: Apple Valley; 29 Palms; Amboy Crater; near Kelso; New York Mts.; **San Diego Co.**: Borrego Valley.

Hosts. Unknown.

Discussion. This is an unusually attractive species with snowy white markings. Even the female antennae are extensively whitish. In most *Argochrysis* species white-marked flagellomeres are a male trait. The whitish markings are particularly striking seen against the bluish green, coppery and violet body color. The wings are untinted and both sexes have the face covered with dense silvery setae.

Argochrysis alboris Kimsey

Map 69

Argochrysis alboris Kimsey, in Bohart and Kimsey 1982:185. Holotype ♂; USA: California, Alpine Co., Lake Winnemucca (UCDC).

Geographic Range (Map 69). Western USA (CA, MT, NV, OR, WY), in the Canadian to Sonoran Zones; 31 specimens were examined.

California Records. Collected March through October; **Alpine Co.**: Carson Pass; Ebbetts Pass; Lake Winnemucca; **El Dorado Co.**: Silver Lake; **Imperial Co.**: 3 mi. w Ocotillo; 6 mi. w Mt. Signal; **Inyo Co.**: Mono Pass; 10 mi. s Panamint Springs; 2 mi. e Lone Pine; 2 mi. w Death Valley Junction; Eureka Valley dunes; Brown; **Mono Co.**: Sonora Pass; Leavitt Meadows; Benton; **Nevada Co.**: Sagehen Creek; **Riverside Co.**: Oasis; 18 mi. w Blythe; 7 mi. n Palm Springs; 8 mi. n Blythe; Thousand Palms; Thousand Palms Canyon; **San Bernardino Co.**: New York Mts.; Keystone; **San Diego Co.**: Borrego Valley.

Hosts. Unknown.

Discussion. Diagnostic features of this species include flagellomere I subequal in length to II, metanotum somewhat bulging, no transverse frontal carina, and femoral-tibial joints white.

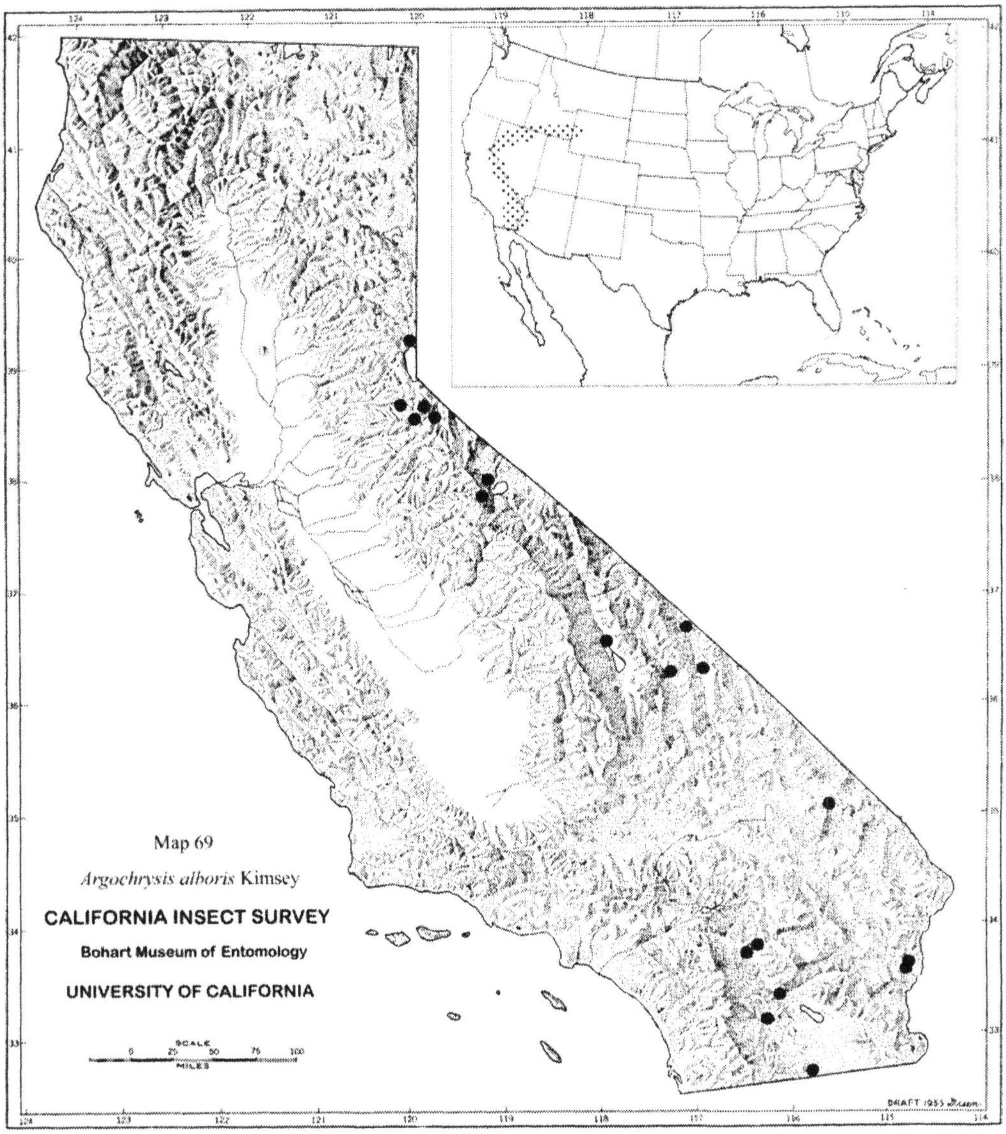

Map 69. California distribution of *Argochrysis alboris*. Inset: overall distribution.

Argochrysis armilla Bohart

Map 70

Argochrysis armilla Bohart, in Bohart and Kimsey 1982:189. Holotype ♂; USA: California, Nevada Co., Sagehen Creek (UCDC).

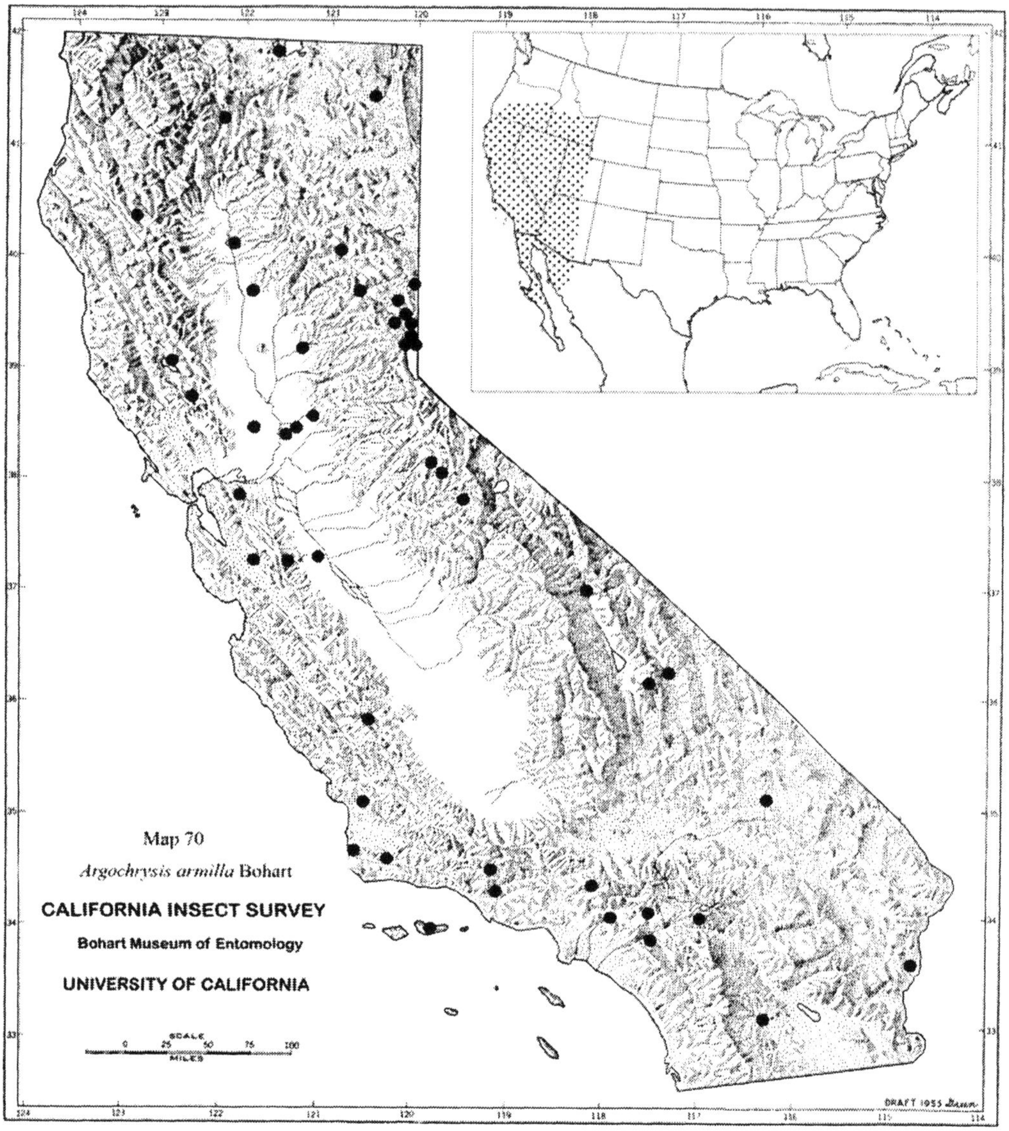

Map 70. California distribution of *Argochrysis armilla*. Inset: overall distribution.

Geographic Range (Map 70). Western USA (CA, ID, OR, NV), and northern Mexico (Baja California), in the Upper Sonoran to Canadian Zones; 552 specimens were studied.

California Records. Collected April through October; **Butte Co.**: 13 mi. n Chico; **Contra Costa Co.**: Mt. Diablo; **El Dorado Co.**: 5 mi. n on Ice House Rd.; 5 mi. s Meyers; Echo Lake; Meyers; **Inyo Co.**: 15 mi. s Big Pine; Darwin Falls; Panamint Springs; **Lake Co.**: Midlake; **Lassen Co.**: Hallelujah Junction; **Los Angeles Co.**:

Claremont; Tanbark Flat; **Modoc Co.**: 30 mi. s Tule Lake; 6 mi. e Alturas; **Monterey Co.**: Arroyo Seco Camp; **Napa Co.**: Samuel Springs (beneath Lake Berryessa); **Nevada Co.**: Boca; Prosser Dam; Sagehen Creek; Hobart Mills; **Placer Co.**: Carnelian Bay, Lake Tahoe; **Plumas Co.**: e Blairsden; Buck's Lake; **Riverside Co.**: 18 mi. w Blythe; Gavilan; **Sacramento Co.**: Carmichael; Sacramento; Folsom; **San Bernardino Co.**: 15 mi. w Baker; Wildwood Canyon 5 mi. e Calimesa; Cajon; **San Diego Co.**: Borrego Valley dunes; **San Luis Obispo Co.**: 5 mi. n Lompoc; Black Mt. 6 mi. ne Pozo; **Santa Barbara Co.**: Bluff Camp, San Rafael Mnts; Santa Cruz Island; **Santa Clara Co.**: 1.5 mi. n San Antone Junction on Mines Rd.; **Sierra Co.**: Independence Lake; Sagehen Creek; Sardine Lakes; Sattley; Sierra Valley; Sierraville; Yuba Pass; **Siskiyou Co.**: Mount Shasta City; **Stanislaus Co.**: Del Puerto Canyon; Denair; **Trinity Co.**: Hayfork; **Tuolumne Co.**: Dardanelles; Kennedy Meadows; Strawberry; **Ventura Co.**: Foster Park; Sespe Canyon; **Yolo Co.**: Davis.

Hosts. Bohart and McLaughlin (1979) and Rosenheim (1987a, b; 1988, 1989, 1993) recorded *armilla* females parasitizing nests of *Ammophila pruinosa* Cresson and *A. marshi* Menke.

Discussion. *A. armilla* is the most abundant *Argochrysis* species in California. Individuals have a transverse frontal carina, a white mark at the base of the hindtibia, and a rounded metanotum. Males tend to be greenish blue to purple. Females are typically reddish coppery.

Argochrysis excelsior Bohart

Map 71

Argochrysis excelsior Bohart, in Bohart and Kimsey 1982:190. Holotype ♂; USA: Arizona, Cochise Co., Portal (UCDC).

Geographic Range (Map 71). Upper and Lower Sonoran Zones in New Mexico, Arizona and California and northwestern Mexico, including Baja California Sur; 19 specimens were seen.

California Records. Collected in April and May; **San Bernardino Co.**: Kelso Wash Cronise Valley; Black Canyon.

Hosts. Unknown.

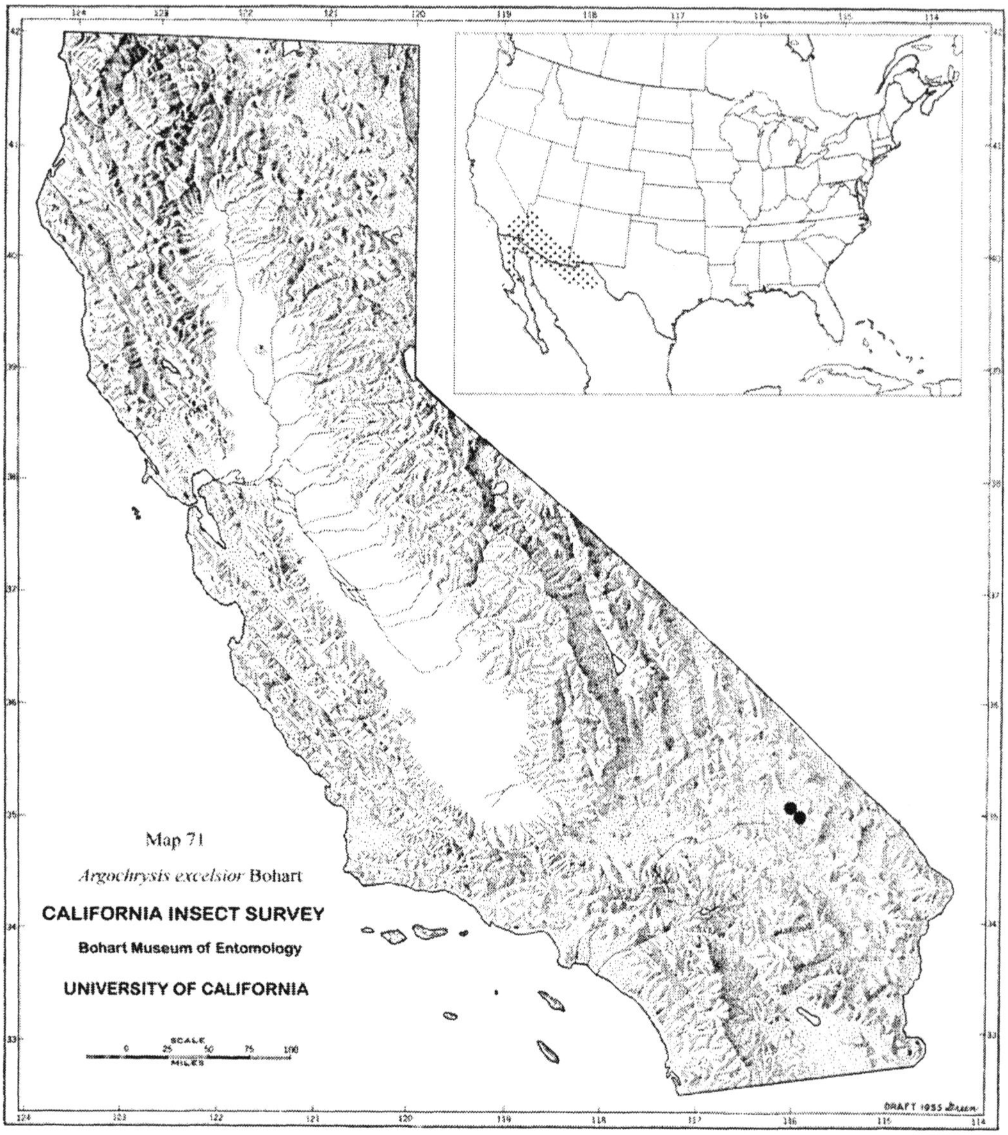

Map 71. California distribution of *Argochrysis excelsior*. Inset: overall distribution.

Discussion. The well-developed transverse frontal carina, reduced ocelli, protruding metanotum, and obtusely notched and white marked rim of tergum III are diagnostic characteristics of *excelsior*.

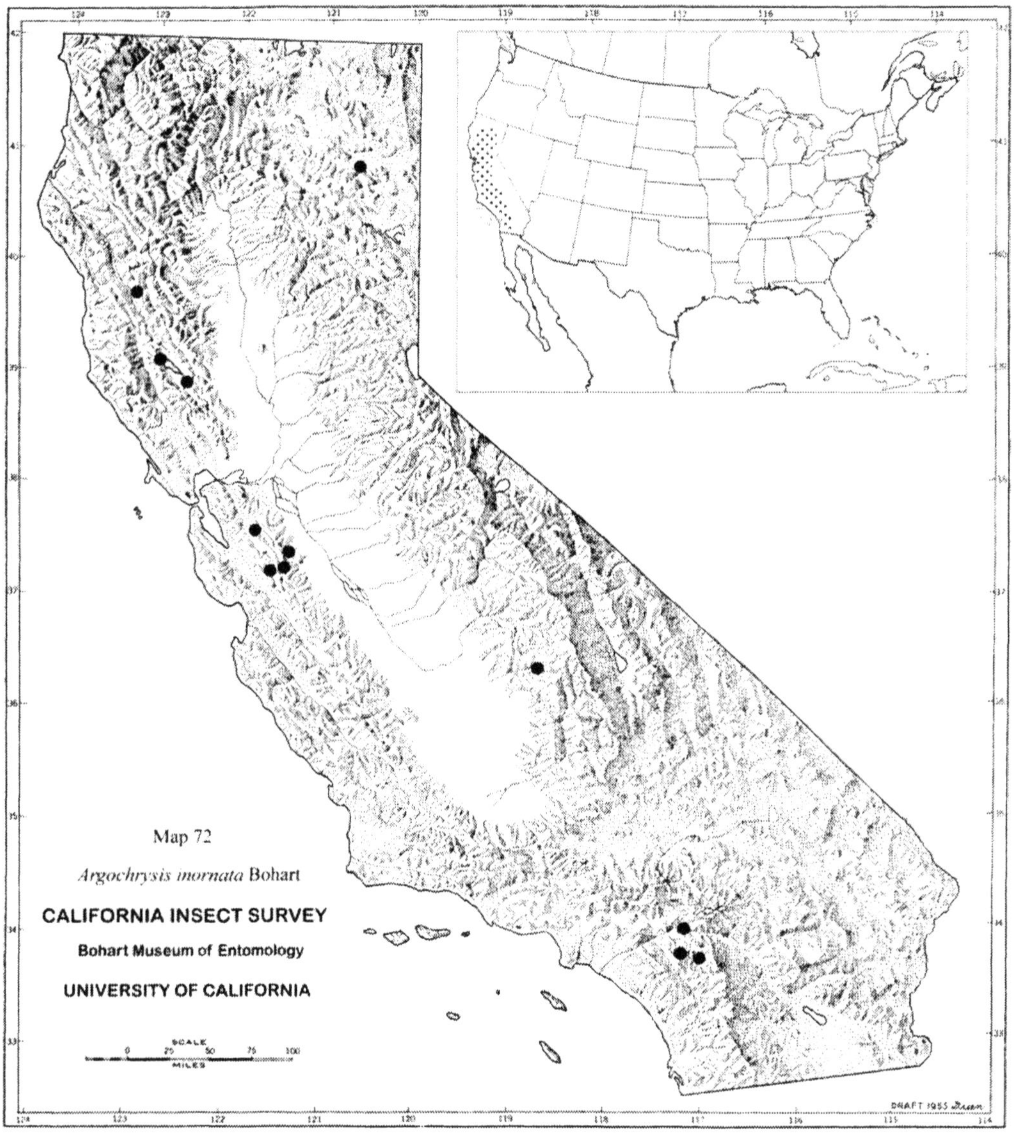

Map 72. California distribution of *Argochrysis inornata*. Inset: overall distribution.

Argochrysis inornata Bohart

Map 72

Argochrysis inornata Bohart, in Bohart and Kimsey 1982:192. Holotype ♂; USA: California, Lake Co., n fork Cache Creek on Hwy 20 (UCDC).

Geographic Range (Map 72). Upper Sonoran Zone in California; 29 specimens were seen.

California Records. Collected in May, June; **Alameda Co**.: Arroyo Mocho; **Lake Co**.: n fork Cache Creek on Hwy. 20; Cache Creek; Blue Lakes; **Lassen Co**.: Madeline; **Mendocino Co**.: Cow Mt.; **Riverside Co**.: Cold Water Canyon, San Jacinto Mts.; **Santa Clara Co**.: Islandobel Creek; Mt. Hamilton; **Stanislaus Co**.: Del Puerto Canyon; **Tulare Co**.: Ash Mt.

Hosts. Unknown.

Discussion. This California endemic species lacks white markings on the antenna, legs and tergum III. Distinctive features include the presence of a transverse frontal carina and distinctly incised apicomedial margin of tergum III.

Argochrysis lassenae Bohart

Argochrysis lassenae Bohart, in Bohart and Kimsey 1982:193. Holotype ♀; USA: California, Lassen Co., Hallelujah Junction (UCDC).

Geographic Range. Upper Sonoran Zone in Lassen Co., California; 6 specimens were seen.

California Records. Collected in June; **Lassen Co**.: Hallelujah Junction.

Hosts. Unknown.

Discussion. This species is another California endemic. Similar to *armilla* and *secutor, lassenae* can be distinguished by the shorter rim on tergum III, the broader female frons, flagellomere I 5x as long as broad, and the rim of tergum III entire. This species is rarely collected.

Argochrysis litura Bohart

Map 73

Argochrysis litura Bohart, in Bohart and Kimsey 1982:193. Holotype ♂; USA: California, Los Angeles Co., Tanbark Flat (UCDC).

Geographic Range (Map 73). Upper Sonoran Zone in California (widespread), southwestern Idaho and western Arizona; 137 specimens were seen.

California Records. Collected March through July; **Contra Costa Co**.: Mt. Diablo; **El Dorado Co**: Pine Hill w Rescue; **Inyo Co**.: Panamint Springs; **Kern Co**.: Glenville; Greenhorn Mts.; Sand Canyon 3 mi. w Brown; **Lassen Co**.: Hallelujah

Junction; **Los Angeles Co.**: Tanbark Flat; **Mendocino Co.**: Cow Mt.; Macayame Mts.; **Monterey Co.**: Arroyo Seco; **Napa Co.**: 25 mi. n Knoxville; **Riverside Co.**:

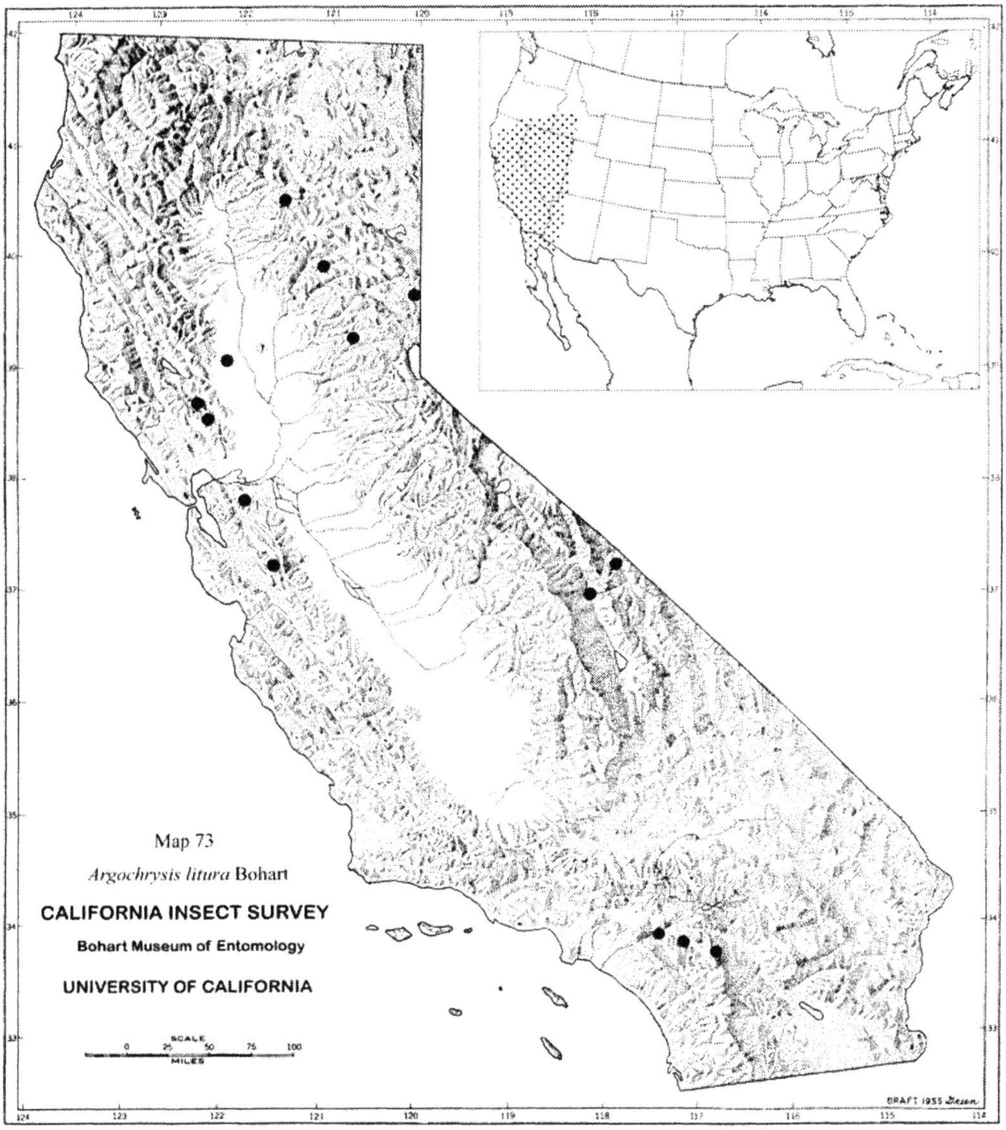

Map 73. California distribution of *Argochryhsis litura*. Inset: overall distribution.

White Water; 5 mi. sw Banning; Thousand Palms Canyon; 7 mi. e Sunnymead; 5 mi. s Hemet; Riverside; Perris; Gavilan; Wildwood Canyon 3 mi. e Yucaipa; **San Bernardino Co.**: Lake Arrowhead; Mill Creek; Joshua Tree State Park; Devore; Mid Hills; Black Canyon near Yucaipa; Summit; **San Diego Co.**: Culp Canyon;

Buchman's Springs; **San Luis Obispo Co**.: Black Mt.; Pozo; La Panza Camp 12 mi. ne Pozo; **Trinity Co**.: Hayfork; **Tulare Co**.: Ash Mt.; **Ventura Co**.: Hungry Valley 5 mi. s Gorman; Santa Susana.

Hosts. Unknown.

Discussion. Diagnostic characteristics of *litura* include the presence of a transverse frontal carina, dark femoral-tibial joints, tergum III without whitish markings on the apical rim and apical rim slightly or not emarginate. Some specimens have a dull, pale, transverse spot on the tegula.

Argochrysis mesillae (Cockerell)

Figs. 5, 103; Map 74

Chrysis mesillae Cockerell 1894a:125. Lectotype ♀ (designated by Cresson 1928); USA: New Mexico, Las Cruces (ANSP).

Chrysis bigeloviae Cockerell 1897:401. Holotype ♂; USA: New Mexico, Las Cruces (USNM).

Chrysis annulipes Mocsáry 1911b:462. Holotype ♂; USA: California (TMB).

Geographic Range (Map 74). Upper Sonoran Zone in western USA and northern Mexico (Baja California, Coahuila), west of 100th meridian; 364 specimens were examined.

California Records. Collected April through September; **Alameda Co**.: Midway Rd.; **Contra Costa Co**.: Antioch; **Inyo Co**.: 5 mi. e Big Pine; Eureka Valley Dunes; Shoshone; **Riverside Co**.: 18 mi. w Blythe; Anza; San Timoteo Canyon; **Sacramento Co**.: Sacramento; **San Bernardino Co**.: Kelso Dunes; **San Diego Co**.: 2 mi. n Warner Springs; **Stanislaus Co**.: Del Puerto Canyon; **Yolo Co**.: Davis.

Hosts. This species is recorded as a nest parasite of *Ammophila californica* Menke (Bohart and McLaughlin 1979), *breviceps* F. Smith and *femurrubra* W. Fox (Krombein 1979).

Discussion. *A. mesillae* is probably one of the best-studied species of *Argochrysis*. The species is characterized by the presence of a transverse frontal carina, and the strongly subdentate rim of tergum III. The rim of tergum III is less clearly subdentate in *trochilus* and *trochilus* is smaller, only 3.5-4.5 mm long versus 5 mm long in *mesillae*.

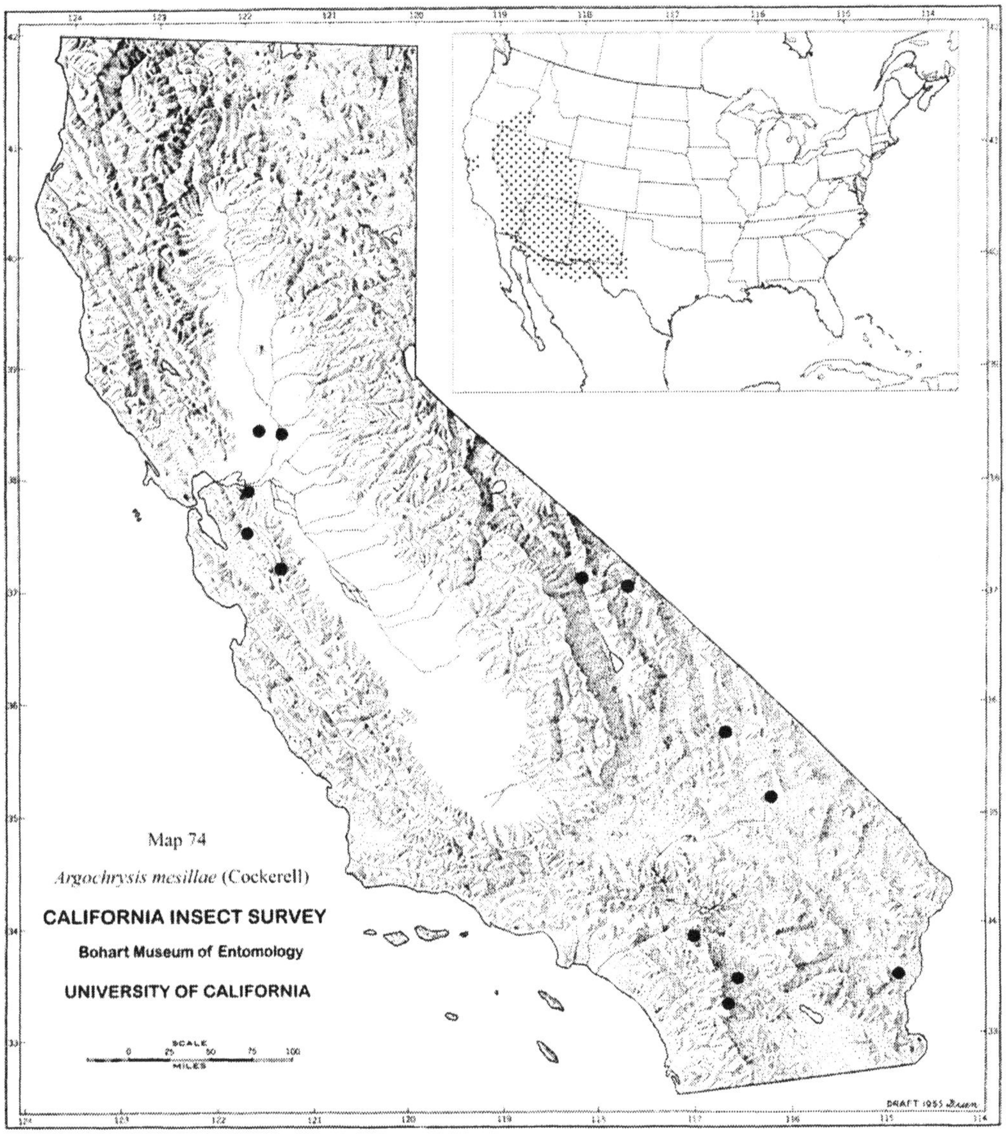

Map 74. California distribution of *Argochrysis mesillae*. Inset: overall distribution.

Argochrysis secutor Bohart

Map 75

Argochrysis secutor Bohart, in Bohart and Kimsey 1982:195. Holotype ♂; USA: California, Inyo Co., Surprise Canyon (UCDC).

Geographic Range (Map 75). Upper and Lower Sonoran Zones in the southwestern deserts of California and Nevada; 15 specimens were seen.

California Records. Collected April though May; **Inyo Co**.: Surprise Canyon; **San Bernardino Co**.: Carson's Well; **San Diego Co**.: Borrego Valley; Chihuahua Valley.

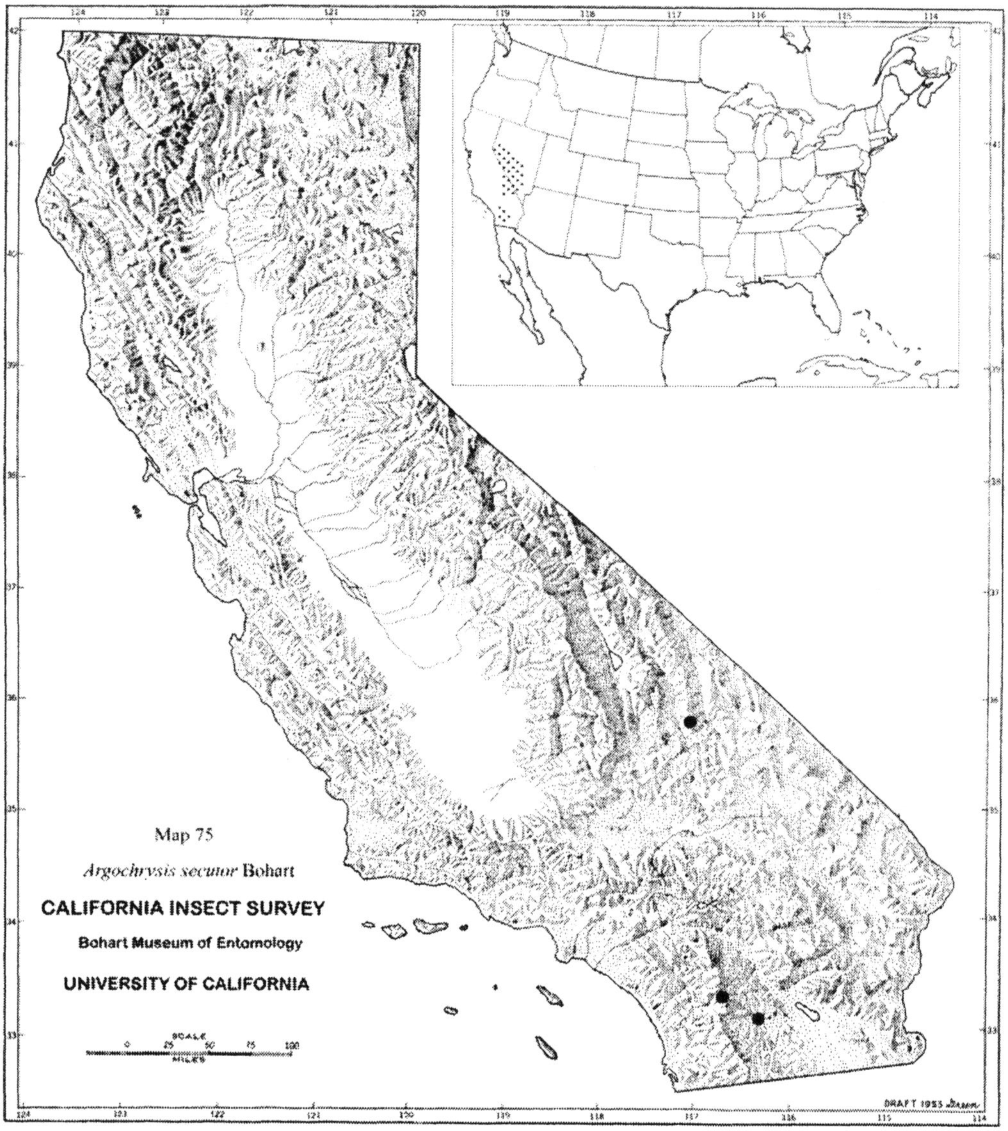

Map 75. California distribution of *Argochrysis secutor*. Inset: overall distribution.

Hosts. Unknown.

Discussion. This species has dark antennae without whitish markings, a rounded metanotum, transverse frontal carina present, and tergum III rim weakly indented medially and whitish laterally. A distinctive feature of *secutor* is the long tergum III rim, which is about two-thirds as long as flagellomere I.

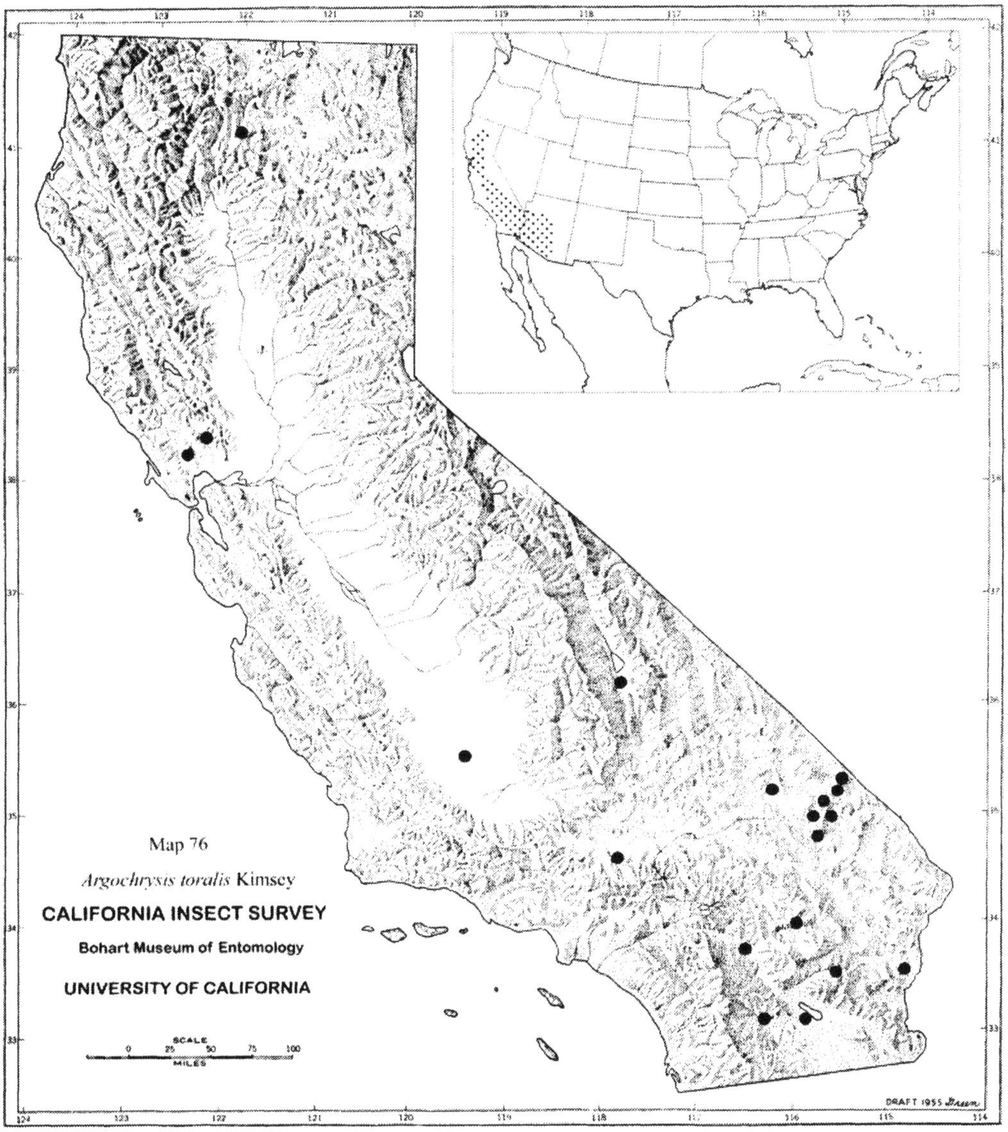

Map 76. California distribution of *Argochrysis toralis*. Inset: overall distribution.

Argochrysis toralis Kimsey

Map 76

Argochrysis toralis Kimsey, in Bohart and Kimsey 1982:187. Holotype ♂; USA: California, San Bernardino Co., Cronise Wash (UCDC).

Geographic Range (Map 76). Upper and Lower Sonoran Zones of California, Nevada, New Mexico and Arizona; 81 specimens were seen.

California Records. Collected April through July; **Inyo Co**.: Olancha; **Kern Co**.: Shafter; **Los Angeles Co**.: 2.5 mi. s Pearblossom; **Napa Co**: Samuel Springs (beneath Lake Berryessa); **Riverside Co**.: Hopkins Well; Thousand Palms; Chuckwalla Springs; 8 mi. w Kane Springs; 18 mi. w Blythe; Hopkins Well; **San Bernardino Co**.: Baker; New York Mts.; Providence Mts.; Granite Mts.; Cronise Wash; Joshua Tree National Monument; Colton Hills; 5.8 mi. nnw Granite Pass; Sacaton Springs New York Mts.; 7 mi. sw Kelso; **San Diego Co**.: Borrego Valley; **Santa Barbara Co**.: Santa Catalina Island; **Siskiyou Co**.: 9 mi. e McCloud; **Sonoma Co**.: Glenn Ellen.

Hosts. Unknown.

Discussion. The lack of a transverse frontal carina, projecting metanotum and nearly flat interocellar area are diagnostic for *toralis*. In addition, the leg joints are whitish, tergum III is apicomedially entire in females or slightly indented in males, the flagellum is dark, and flagellomere I is short (about 1.4-1.5x as long as broad). Males tend to be green, blue and purple, and females have various shades of magenta dorsally.

Argochrysis trochilus (du Buysson)

Map 77

Chrysis trochilus du Buysson 1891:32. Holotype ♀; Mexico (MNHN).

Geographic Range (Map 77). Upper and Lower Sonoran Zones west of the 100th meridian in the USA and northern Mexico (Baja California, Sonora, Chihuahua, Zacatecas); 138 specimens were seen.

California Records. Collected April through August; **Sacramento Co**.: Grand Island; **Imperial Co**.: Glamis; Imperial Palo Verde; **Inyo Co**.: 2 mi. e Lone Pine; Little Lake; Wildrose Canyon; **Riverside Co**.: 18 mi. w Blythe; Thousand Palms; Palm Springs Smoke Tree Ranch; **San Bernardino Co**.: 15 mi. w Baker; Cronise Wash; Needles; **San Diego Co**.: Borrego Valley dunes; **Stanislaus Co**.: Denair; **Yolo Co**.: Davis.

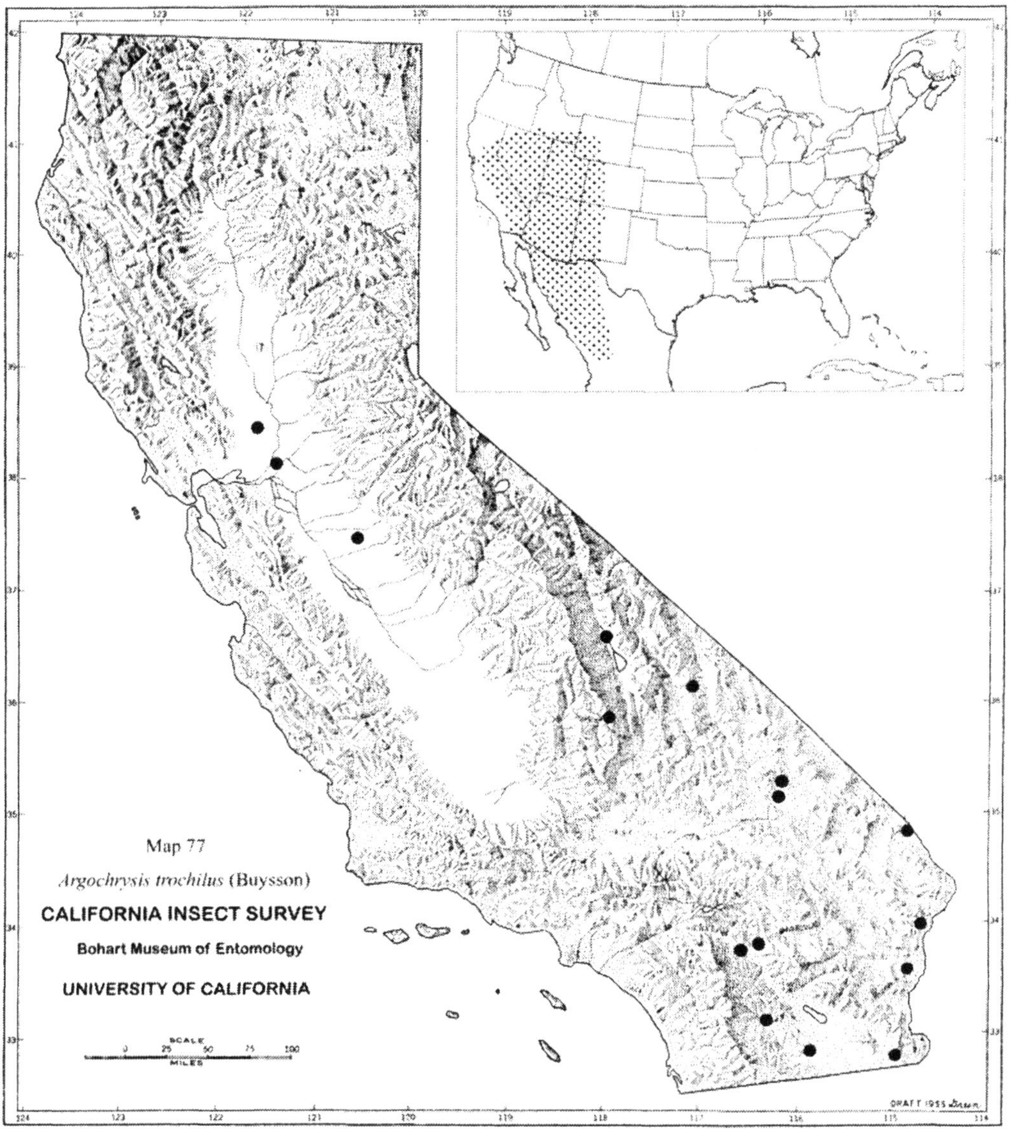

Map 77. California distribution of *Argochrysis trochilus*. Inset: overall distribution.

Hosts. Unknown.

Discussion. *A. trochilus* and *mesillae* are quite similar as described under *mesillae*. However *trochilus* is smaller-bodied, 3.5-4.5 mm long, and the terga are coppery to red.

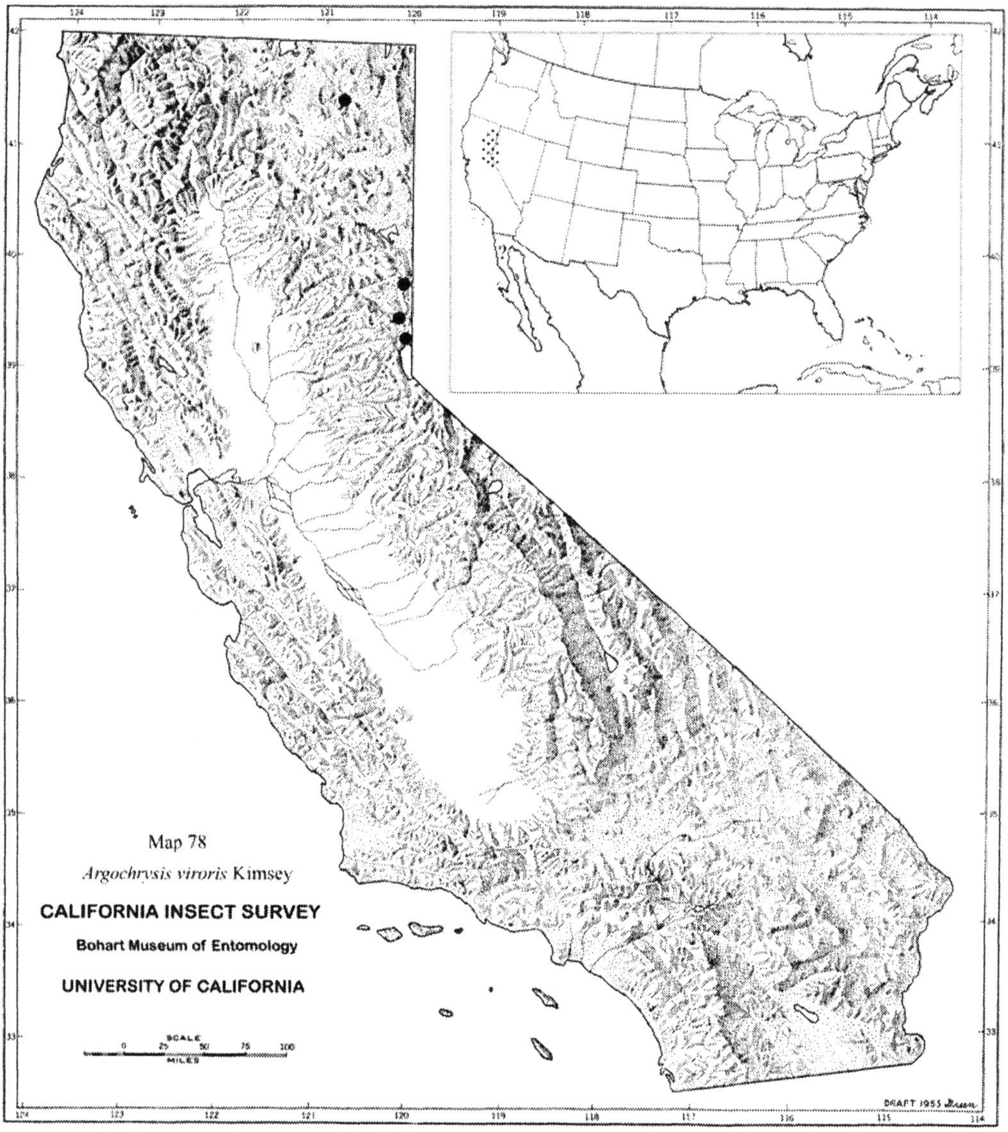

Map 78. California distribution of *Argochrysis viroris*. Inset: overall distribution.

Argochrysis viroris Kimsey

Map 78

Argochrysis viroris Kimsey, in Bohart and Kimsey 1982:188. Holotype ♂; USA: California, Ventura Co., Sespe Canyon (UCDC).

Geographic Range (Map 78). Upper Sonoran to Canadian Zones in California, Nevada and Baja California; 11 specimens were examined.

California Records. Collected in July; **Lassen Co**.: Hallelujah Junction; **Modoc Co**.: Alturas; **Nevada Co**.: Boca; **Sierra Co**.: Independence Lake; **Ventura Co**.: Sespe Canyon.

Hosts. Unknown.

Discussion. The lack of a transverse frontal carina and the projecting metanotum indicate a relationship between *viroris* and *mesillae*. However, *viroris* has no white markings on tergum III, flagellomere I is twice as long as broad, and tergum III is obtusely and weakly notched apicomedially.

Genus *Caenochrysis* Kimsey & Bohart

Generic Diagnosis. Scapal basin with at least narrow medial strip of microridging and ovoid fovea or pit on either side; flagellomere I not more than 3x breadth, shorter than or as long as flagellomere II; malar space 1-2 MOD long; subantennal space usually less than 1 MOD long; subgenal area well defined; transverse frontal carina distinct, convex, biconvex or irregular, often with 2-4 dorsal branches, inner pair often enclosing midocellus; vertex often with distinct posteromedial depression; pronotum as long or shorter than scutellum, medial sulcus weak, lateral depression well-developed and often double; metanotum usually raised or projecting posteriorly; mesopleuron with well-developed scrobal and episternal sulci, omaulus and verticaulus; propodeal angle straight or incurved posteriorly; tergum III pit row distinct, often with a prepit swelling in female, apical margin various: rounded, medially indented, apical rim with 0-3 teeth or angles; sternum II spots usually submedial; male terminalia: sternum VIII elongate triangular, cuspis and digitus slender, aedeagus tapering gradually toward apex.

Hosts. These chrysidids parasitize mud-nest building members of the crabronid tribe Trypoxylonini.

Discussion. Originally confused with the Old World genus *Trichrysis* Lichtenstein, *Caenochrysis* have a unique feature: the facial foveae, and also have a well-developed transverse frontal carina. The vast majority of *Caenochrysis* species are Neotropical. Only a six species of *Caenochrysis* occur in North America, and of these two reach as far west as California. *Caenochrysis* was reviewed by Bohart and Kimsey (1982) as a subgenus of *Trichrysis* Lichtenstein.

Key to the California Species of *Caenochrysis*

Tergum III apically broadly bilobate or rounded (Fig. 96, 102); forewing discoidal cell with one or more outer veins incomplete or lacking...........*doriae* (Gribodo)

Tergum III apically trilobate or tridentate (Figs. 95, 101); forewing discoidal cell complete with outer veins distinctly sclerotized).....................*deversor* Bohart

Caenochrysis deversor (Bohart)

Figs. 95, 101; Map 79

Trichrysis deversor Bohart 1966b:131. Holotype ♂; USA: California, Davis (UCDC).

Geographic Range (Map 79). Western USA, Canada (British Columbia), and northwestern Mexico (Baja California); 103 specimens were seen.

California Records. Collected May through September; **Alameda Co**.: Claremont; **Calaveras Co**.: 3 mi. se Railroad Flat; Railroad Flat; **Contra Costa Co**.: Mt. Diablo; **Del Norte Co**.: s end Illinois Valley; **El Dorado Co**.: 2 mi. n Fairplay; 5 mi. s Kyburz; **Fresno Co**.: Oxalis; **Lake Co**.: Cobb Mt.; **Los Angeles Co**.: Tanbark Flat; **Madera Co**.: Oakhurst; **Mendocino Co**.: UC Hopland Field Station; **Mono Co**.: 11 mi. n Bridgeport; **Napa Co**.: Samuel Springs (beneath Lake Berryessa); **Placer Co**.: Folsom Lake; **Riverside Co**.: Blythe; Riverside; **San Joaquin Co**.: Tracy; **Shasta Co**.: 2 mi. w Shingletown; **Siskiyou Co**.: Mount Shasta City; **Stanislaus Co**.: Del Puerto Canyon; **Tulare Co**.: Wood Lake; **Yolo Co**.: Davis; **Yuba Co**.: Bullards Bar.

Hosts. Parker and Bohart (1966) reared *deversor* from nests of *Trypargilum tridentatum* (Packard) (Crabronidae).

Discussion. In California the two species of *Caenochrysis* can be easily separated, *deversor* is larger bodied, 6-9 mm long (versus 2-4 mm long in *doriae*), and has a sharply tridentate tergum III (*doriae* lacks teeth on the apical rim of tergum III). Additional diagnostic features of *deversor* include the medial ridge of the pit row concave in profile and the transverse frontal carina with four branches extending dorsally.

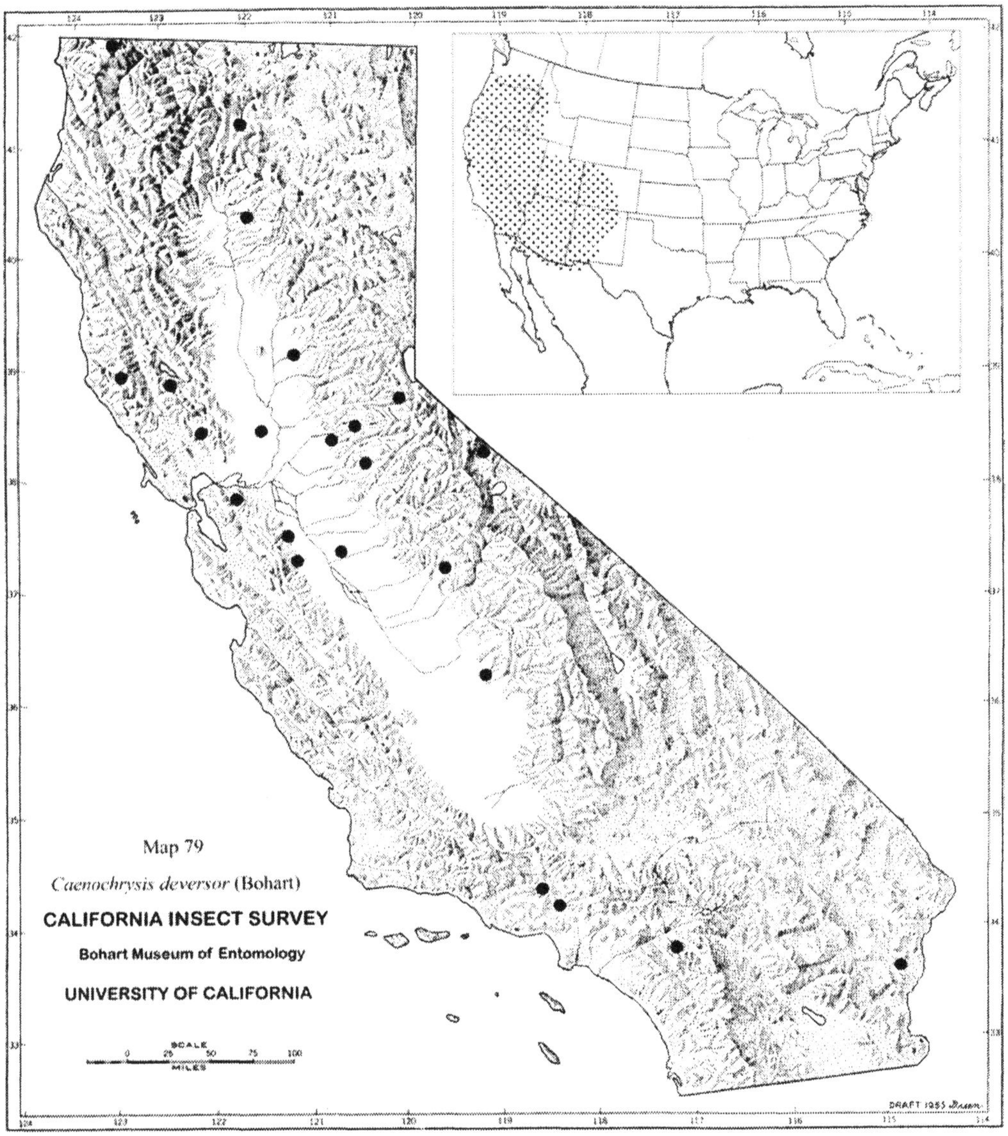

Map 79. California distribution of *Caenochrysis deversor*. Inset: overall distribution.

Caenochrysis doriae (Gribodo)

Figs. 90, 96, 102; Map 80

Chrysis doriae Gribodo 1874:359. Syntype ♀♀; "America Boreali" (MCSN, ZMUC).

Chrysis discreta Aaron 1885:230. Holotype ♀; USA: North Carolina (ANSP).

Chrysis striatella Norton 1879:240. Lectotype ♀ (designated by Cresson 1928); USA: Massachusetts (ANSP).

Chrysis verticalis Patton 1879:67. Holotype ♂; USA: Connecticut, Waterbury (lost ?).

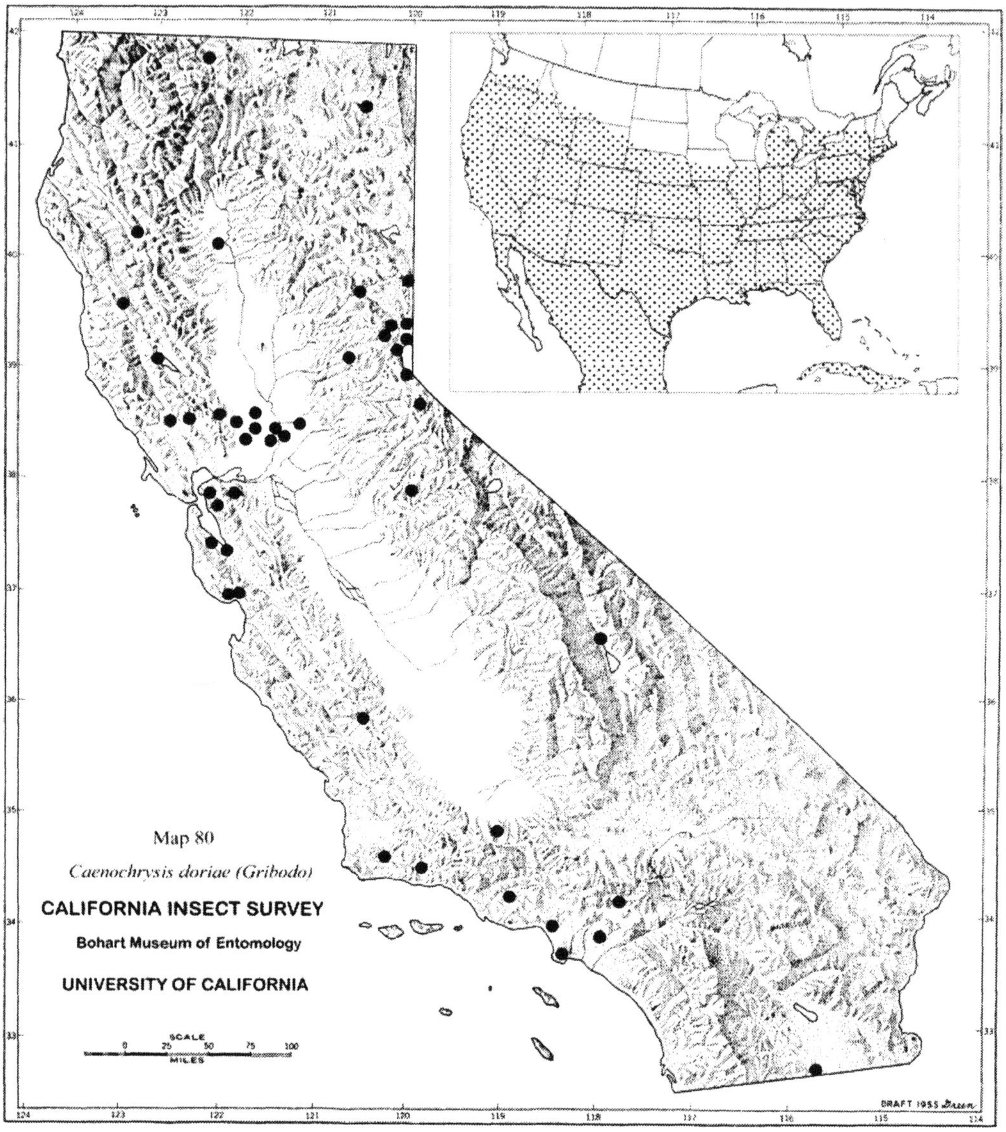

Map 80. California distribution of *Caenochrysis doriae*. Inset: overall distribution.

Geographic Range (Map 80). Widespread in the Nearctic Region, from the Canadian Yukon south into Panama; 1021 specimens were studied.

California Records. Collected April through October; **Alameda Co**.: Wildcat Canyon; Berkeley; **Alpine Co**.: Hope Valley; **Contra Costa Co**.: Moraga; **El Dorado Co**.: Lake Tahoe; **Imperial Co**.: Calexico; **Inyo Co**.: Lone Pine; **Kern Co**.: Mill Potrero; **Lake Co**.: Lakeport; **Lassen Co**.: Hallelujah Junction; **Los Angeles Co**.: Glendale; mountains near Claremont; Pacific Palisades; **Mendocino Co**.: Ukiah; **Modoc Co**.: 24 mi. sw Alturas; **Monterey Co**.: Arroyo Seco Camp; **Napa Co**.: Calistoga; **Nevada Co**.: Fuller Lake; Sagehen Creek; **Orange Co**.: Fullerton; **Placer Co**.: Carnelian Bay, Lake Tahoe; Emigrant Gap; **Plumas Co**.: Blairsden; **Sacramento Co**.: Carmichael; Courtland; Sacramento; **San Mateo Co**.: San Mateo; **Santa Barbara Co**.: Santa Cruz Island; Sedgewick Reserve, 45 km nw Santa Barbara; Santa Ynez; **Santa Cruz Co**.: Felton Springs; Santa Cruz; **Santa Clara Co**.: San Jose; **Sierra Co**.: Independence Lake; Yuba Pass; **Siskiyou Co**.: Swallows 13 mi. n Yreka; **Solano Co**.: Dixon; **Sonoma Co**.: Boyes Hot Springs; **Tehama Co**.: Redding; **Trinity Co**.: Trinity River Camp; **Tulare Co**.: Mineral King; **Tuolumne Co**.: Strawberry; **Ventura Co**.: Foster Park; Lockwood Valley; **Yolo Co**.: 3 mi. nw Knights Landing; Capay; Davis; Putah Canyon; Putah Creek 3 mi. w Davis; Woodland; Yolo bypass near Bryte.

Hosts. This species has been reared from *Trypoxylon bidentatum* W. Fox (Parker and Bohart 1966), *T. frigidum* F. Smith, *Trypargilum collinum rubrocinctum* (Packard) (Krombein 1967) and *T. tridentatum* (Packard) (Bohart and Kimsey 1982).

Discussion. *C. doriae* is a highly variable and widespread species. It can be distinguished by the transverse frontal carina lacking dorsal branches and tergum III apically medially emarginate and not tridentate. Also, unlike *deversor*, the forewing of *doriae* has an incomplete discoidal cell.

Genus *Ceratochrysis* Cooper

Generic Diagnosis. Face usually with transverse frontal carina, scapal basin defined, with at least some medial cross-ridging; flagellomere I usually 3x as long as broad or longer; tergum III apical rim evenly curved, medially emarginate or sublaterally angulate; forewing discoidal cell complete; paramere with apical, subarticulated telomere; cuspis broad with sclerotized outer edge; digitus sword-like and often dentate on outer margin; sternum VIII shield-like with broad base and setose, trilobed or bluntly pointed apical section.

Hosts. These wasps have been recorded as nest parasites of a variety of genera in the Crabronidae and Vespidae (Eumeninae).

Distribution. *Ceratochrysis* species are endemic to North America.

Discussion. This genus appears to be most closely related to *Caenochrysis*. Both genera share the unique feature of a digitate paramere. *Ceratochrysis* species lack the facial foveae characteristic of *Caenochrysis*. The genus contains 27 species, 20 of which occur in California. *Ceratochrysis* species lack the tergal dentition of *Chrysis*, but some have two subapical teeth or lobes on tergum III. Modifications of the apical tergum resemble those of *Chrysura* species. However *Ceratochrysis* species have a well-developed brow, usually with a transverse frontal carina. Members of this genus can have an unusually long flagellomere I and long malar space, and knobbed vertex. Males can be distinguished by metasomal sternum IV visible beyond III as a fringed, narrow metallic plate. Females have at most only the narrow subtriangular apex of sternum IV visible. Species were reviewed by Bohart and Kimsey (1982).

Key to the California Species of *Ceratochrysis*

1 Tergum III apical margin with three angles or lobes (Fig. 56), flagellomere I about twice as long as broad..*antyga* Bohart

- Tergum III apical margin entire, medially emarginated or less commonly bilobate or bidentate, flagellomere I considerably more than twice as long as broad........2

2 Vertex or interocellar area with tubercles (as in Fig. 50).................................3

- Vertex and interocellar area without tubercles..4

3 Tergum III apex broadly medially emarginate (male) or entire (female) (Fig. 57) ....*enhuycki* (Cooper)

- Tergum III apex narrowly but deeply medially emarginate (Fig. 60) ....*crossata* Bohart

4 Malar space 1.5 MOD long or longer in female, 1 MOD long in male (as in Fig. 49)..5

- Malar space less than 1.5 MOD long in female and less than 1 MOD long in male 11

5 Genal carina ending ventrally in flattened lobe adjacent to mandibular base (Fig. 49); tergum II punctation usually dense and contiguous...............*minata* Bohart

- Genal carina ending simply ventrally, without lobe; tergum II punctation various 6

6 Scapal basin with pubescent and punctate area occupying less than one-fourth of interocular area..*bradleyi* Bohart

- Scapal basin with pubescent and punctate area separated in dorsal half by zone of cross-ridging equal to or greater than one-third of interocular area.................7

7 Tergum III apex narrowly curved under, weakly emarginate in male (Fig. 64), entire in female (Fig. 66); female malar space slightly longer than 1 MOD; female frons coarsely punctured laterally, punctation nearly as course as on pronotum ....*cyanosoma* (Mocsáry)

- Tergum III apex sometimes slightly deflected but not curved under, notched medially, or if entire (female *concava*) malar space less than 1 MOD long; female frons punctation various...8

8 Flagellomere I length 3x breadth or less, tergum II punctation slightly separated medially; mesopleuron not ventrally dentate...........................*sierrae* Bohart

- Flagellomere I length more than 3x breadth; tergum II punctation and dentition of mesopleuron various...9

9 Tergum II punctation unusually dense medially, punctures contiguous, without sparsely punctate polished midline; mesopleuron with one or two distinct ventral teeth..*collega* Bohart

- Tergum II punctation less dense, punctures separated by polished areas, sometimes with polished midline; mesopleuron rough ventrally but not dentate..............10

10 Female frons coarsely punctate laterally (nearly as coarse as on pronotum); tergum III apically notched in both sexes, notch acute in male; largest *Ceratochrysis* species, 7 mm long or longer...................*perpulchra* (Cresson)

- Female frons distinctly swollen on either side of central zone of cross-ridging; tergum III apex obtusely emarginate in males, entire in females; smaller species, less than 7 mm long..*concava* Bohart

11 Subantennal distance 1.9 MOD long or longer; frons with transverse frontal carina evanescent; tergum II punctation unusually dense (contiguous) medially.........12

- Subantennal distance less than 1.8 MOD long; frons with transverse frontal carina and tergum II various...13

12 Least interocular distance greater than length of flagellomere I plus pedicel; flagellomere I 3x as long as broad, with long hairs ventrally in male; malar space 2 (male) to 2.5 (female) MOD long..................................*longimala* Bohart

- Least interocular distance subequal to length of flagellomere I plus pedicel, flagellomere I not pubescent and 5-6x as long as broad; malar space 2.5 (male) to 3.0 (female) MOD long..*grisselli* Bohart

13 Tergum III apicomedial notch flanked on either side by sharply tooth-like projection, pit row weakly developed (Fig. 58); punctation unusually dense (contiguous) over entire tergum.......................................*cavicantha* Bohart

- Tergum III entire or notch flanked by rounded lobe or smoothly receding edge, pit row well-developed (as in Figs. 61, 63, 65), punctation various....................14

14 Mesopleuron ventrally distinctly dentate, body length usually more than 6 mm ....*trachypleura* Bohart

- Mesopleuron ventrally edentate, body length usually less than 5 mm..............15

15 Males, sternum IV usually protruding apically as pale margin beyond sternum III; mid and hindtibiae with distal setae tassels (as in Fig. 51)...........................16

- Females, sternum IV apex usually protruding as narrow triangle beyond sternum III, mid and hindtibiae without distal setae tassels, ovipositor sometimes visible 17

16 Tergum III apicomedially slightly and broadly emarginate or indented (Fig. 67) ....*thysana* Bohart

- Tergum III apicomedially deeply and sharply incised (Fig. 61) ....*nearctica* (Mocsáry)

17 Tergum III apical margin notched medially (Fig. 63)...........*nearctica* (Mocsáry)

- Tergum III apical margin evenly rounded and unnotched (as in Figs. 67)..........18

18 Tergum II nearly as long as broad; tergum III with middorsal line nearly straight in lateral view (female only known)....................................*gracilis* Bohart

- Tergum II broader than long; tergum III with middorsal line strongly saddled in lateral view..19

19 Mesopleuron divided in two by strong longitudinal sulcus...........*thysana* Bohart

- Mesopleuron with sulcus narrow or obsolescent (females only known)...........20

20 Frons with punctures largely obscuring central zone of cross-ridging ....*menkei* Bohart

- Frons with discrete central zone of fine cross-ridging................*parkeri* Bohart

Ceratochrysis antyga Bohart

Figs. 56; Map 81

Ceratochrysis antyga Bohart 1966a:115. Holotype ♀; USA: California, Los Angeles Co., Tanbark Flat, San Gabriel Mts. (UCDC).

Geographic Range (Map 81). Southwestern USA and Mexico (Chihuahua), west of the 100th meridian; 192 specimens were studied.

California Records. Collected April through August; **Los Angeles Co.**: Tanbark Flat; **Riverside Co.**: 9 mi. w Banning; **San Diego Co.**: Mt. Laguna; **Sonoma Co.**: Boyes Hot Springs.

Hosts. Bohart (1966a) gave *Trypargilum tridentatum* (Packard) and *Pisonopsis birkmanni* Rohwer (Crabronidae) as hosts of *antyga*.

Discussion. This species is structurally enigmatic and in the past was placed in *Trichrysis* Lichtenstein. Bohart and Kimsey (1982) reassigned all of the American species of *Trichrysis* to *Caenochrysis*. However, reexamination of the type of *antyga* made it clear that this species actually belonged in *Ceratochrysis* (Kimsey and Bohart 1991). The species differs from other *Ceratochrysis* in having an apically trilobate tergum III and relatively short flagellomere I (about twice as long as broad) compared with the longer flagellum seen in other *Ceratochrysis*.

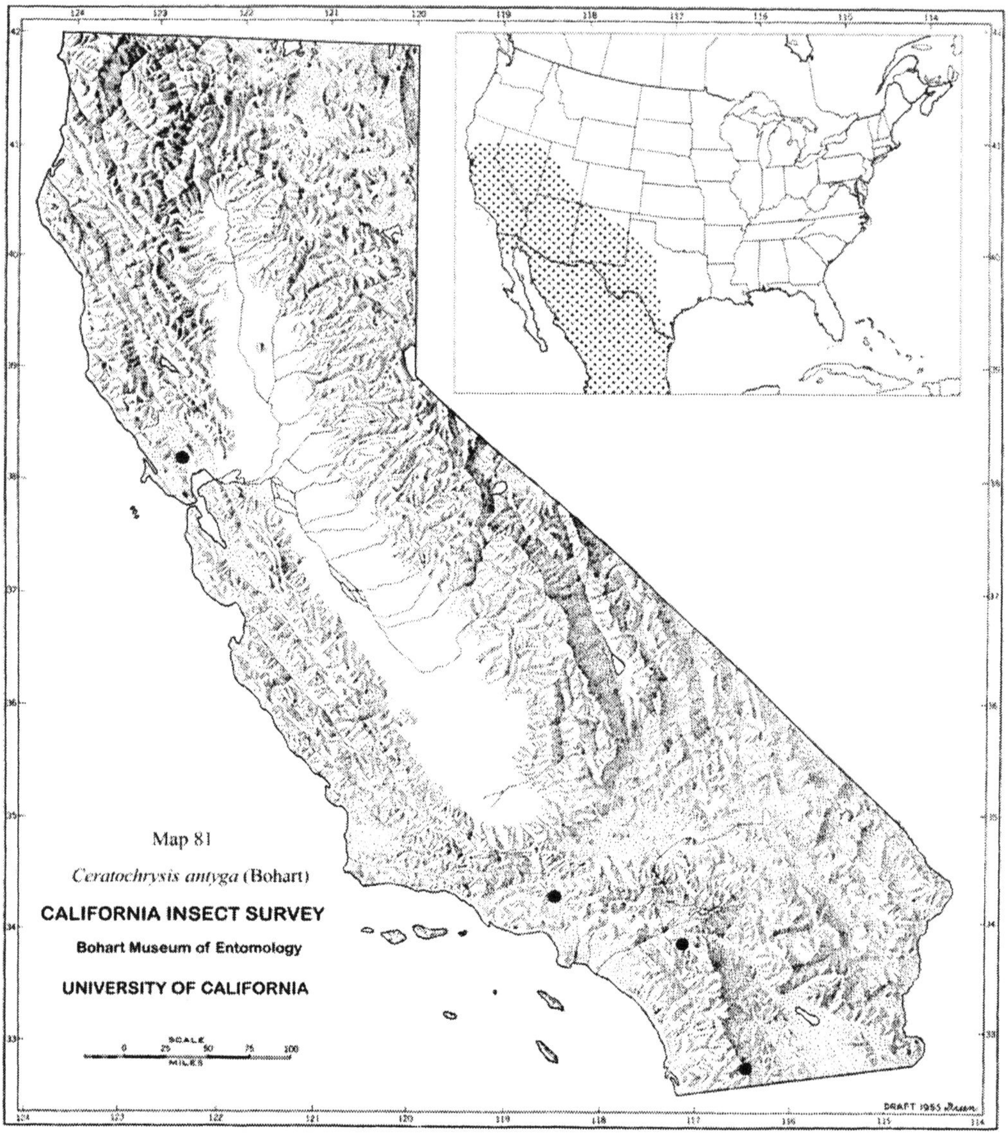

Map 81. California distribution of *Ceratochrysis antyga*. Inset: overall distribution.

Ceratochrysis bradleyi Bohart

Map 82

Ceratochrysis bradleyi Bohart 1982:169. Holotype ♂; USA: Arizona: Yuma (CU).

Geographic Range (Map 82). Lower Sonoran Zone of California and Arizona; 5 specimens were seen.

California Records. Collected May through July; **Riverside Co.**: 6 mi. n Blythe.

Hosts. Unknown.

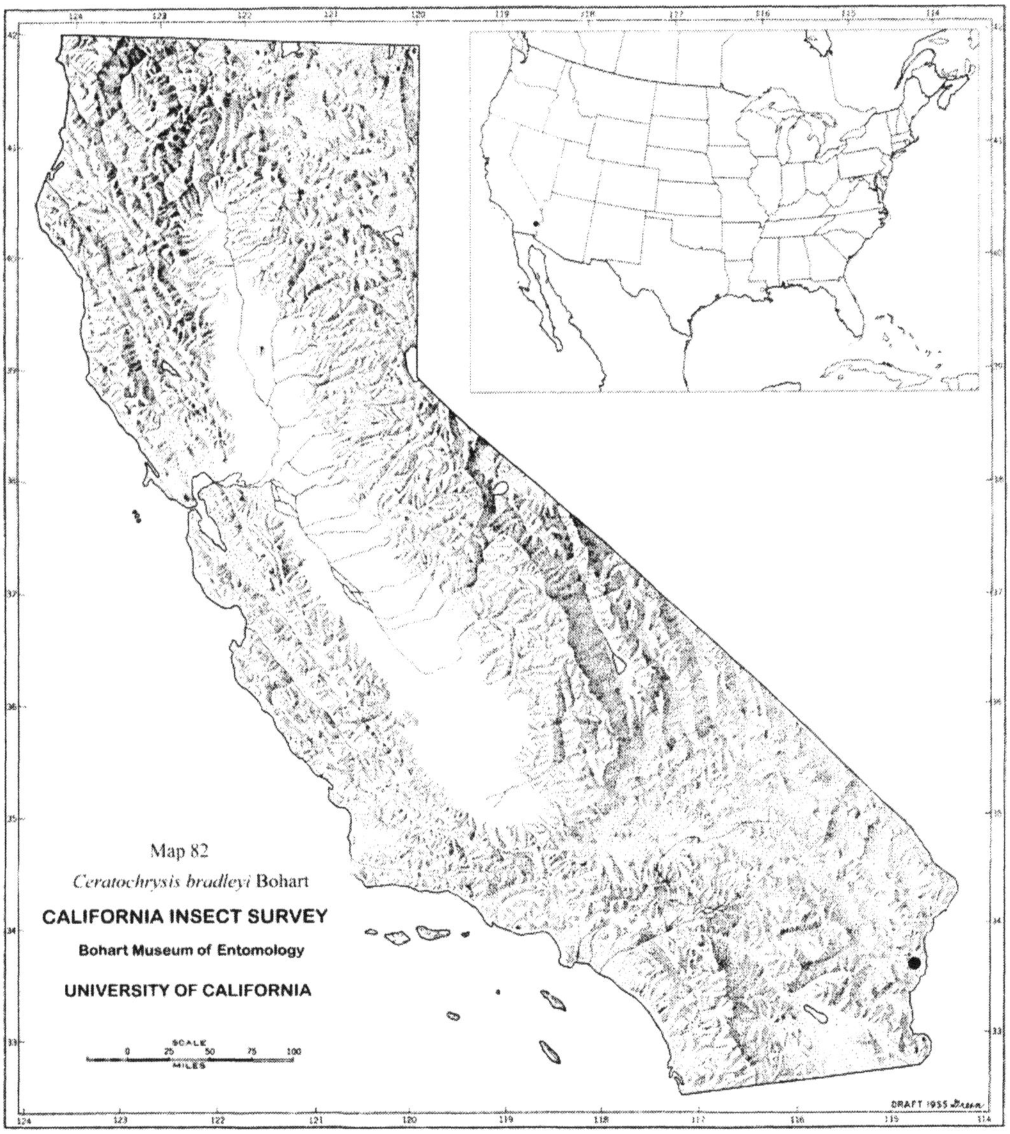

Map 82. California distribution of *Ceratochrysis bradleyi*. Inset: overall distribution.

Discussion. The most distinctive feature of *bradleyi*, seen in both sexes, is the large densely pubescent and punctate area in the scapal basin, leaving only a narrow medial strip of cross-ridging. In addition, the malar and subantennal distances are unusually short and the two sides of the face are unusually flat.

Ceratochrysis cavicantha Bohart

Figs. 58; Map 83

Ceratochrysis cavicantha Bohart 1966a:116. Holotype ♂; USA: California, Inyo Co., Surprise Canyon, Panamint Mts. (UCDC).

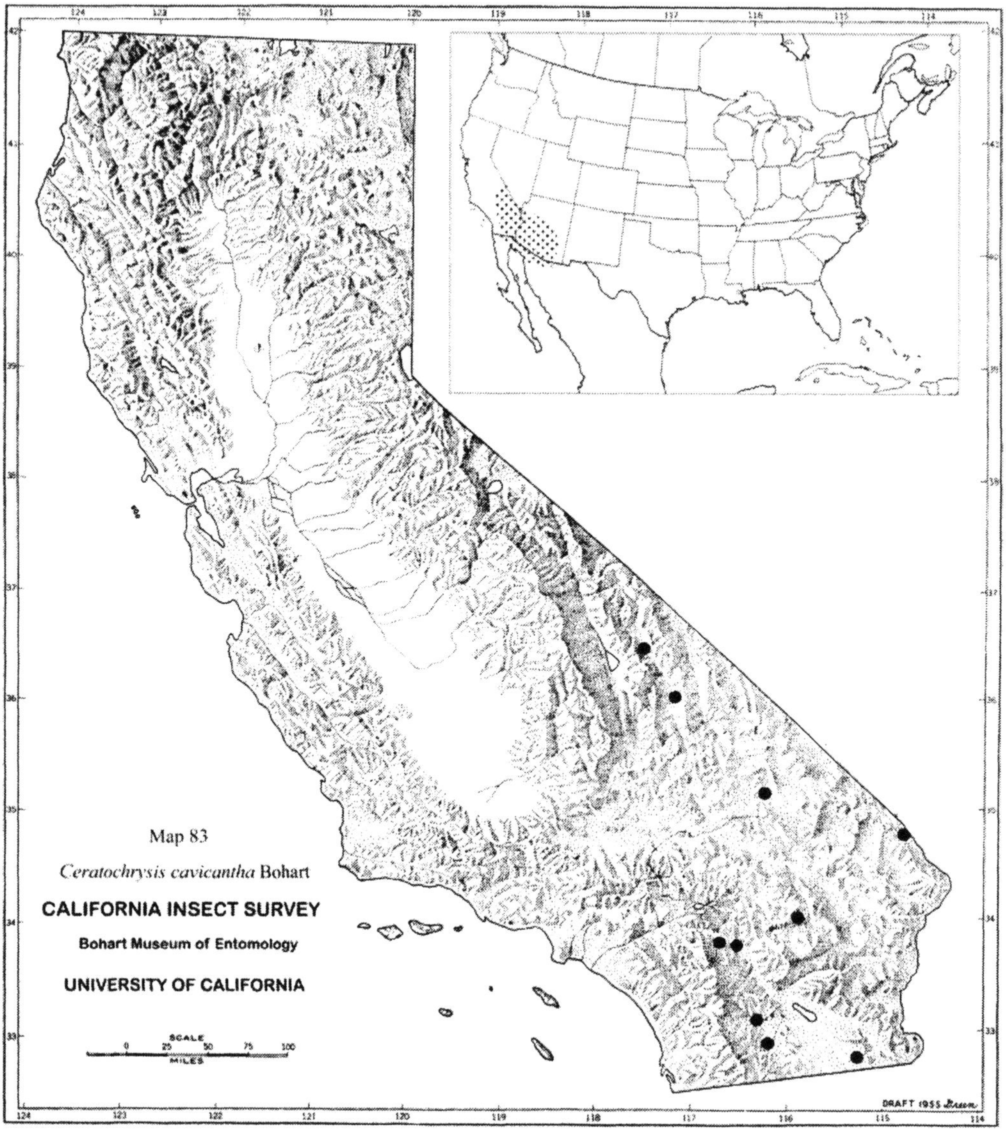

Map 83. California distribution of *Ceratochrysis cavicantha*. Inset: overall distribution.

Geographic Range (Map 83). Lower Sonoran Zone in California and Arizona; 22 specimens were studied.

California Records. Collected in April; **Imperial Co**.: Glamis; **Inyo Co**.: Surprise Canyon; Panamint Mts.; **Riverside Co**.: Palm Springs; Andreas Canyon; **San Bernardino Co**.: 1 mi. w Needles; 5 mi. s Twentynine Palms; Cronise Wash; **San Diego Co**.: Borrego Valley; Palm Canyon

Hosts. Unknown.

Discussion. A number of unusual features characterize *cavicantha*, the apical rim of tergum III has a sharp medial notch, flanked by triangular teeth, the pit row is shallow, tergal punctation contiguous, the face has a weak frontal carina, the scapal basin is broadly V-shaped and lacks cross-ridging, the genal carina is strongly developed and the metanotum is flattened. The color is green with strong purple tints. This is a large-bodied species, averaging 6 mm (males) to 6.5 mm (females) long.

Ceratochrysis collega Bohart

Map 84

Ceratochrysis collega Bohart, in Bohart and Kimsey 1982:170. Holotype ♂; USA: California, Davis (UCDC).

Geographic Range (Map 84). Western USA, Upper Sonoran to Canadian Zones; 52 specimens were studied.

California Records. Collected April through October; **Colusa Co**.: Colusa; **Contra Costa Co**.: Antioch; Bethel Island Harbor; **Nevada Co**.: Boca; **Riverside Co**.: 1 mi. s Temecula; Hopkins Well; **Sacramento Co**.: Sacramento; Sacramento River levee; **San Joaquin Co**.: 8 mi. nw Tracy; Tracy; **Solano Co**.: Ryer Island; **Stanislaus Co**.: 7 mi. nw Tracy; Del Puerto Canyon 17 mi. w Patterson; **Sutter Co**.: Nicolaus; **Ventura Co**.: Sespe Canyon; **Yolo Co**.: Davis.

Hosts. Unknown.

Discussion. This is another species with both the malar space and subantennal distance short, similar to the condition in *concava, fusilis* and *sierrae*, although *collega* is probably most easily confused with *sierrae*. *C. collega* can be distinguished by the contiguous punctation on tergum II and the longer flagellomere I (more than 3.5x as long as broad versus 3x in *sierrae*). In addition, the area immediately below the frontal carina is punctate in *collega* as opposed to irregular and shiny in *sierrae*.

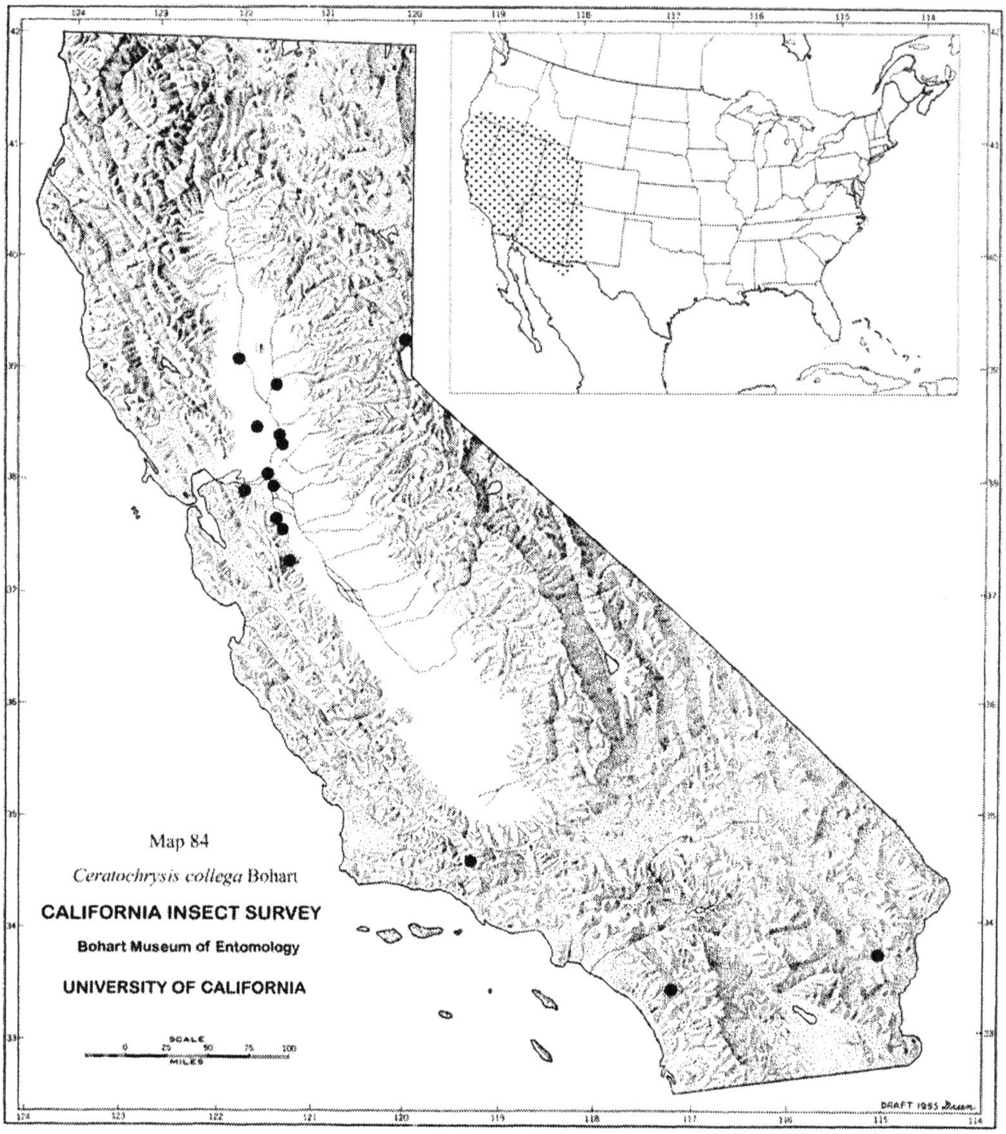

Map 84. California distribution of *Ceratochrysis collega*. Inset: overall distribution.

Ceratochrysis concava Bohart

Map 85

Ceratochrysis concava Bohart 1982:172. Holotype ♂; USA: California, Riverside Co., 10 mi. e Whitewater (UCDC).

Geographic Range (Map 85). Lower to Upper Sonoran Zones in California, Nevada, Arizona and Sonora, Mexico; 62 specimens were examined.

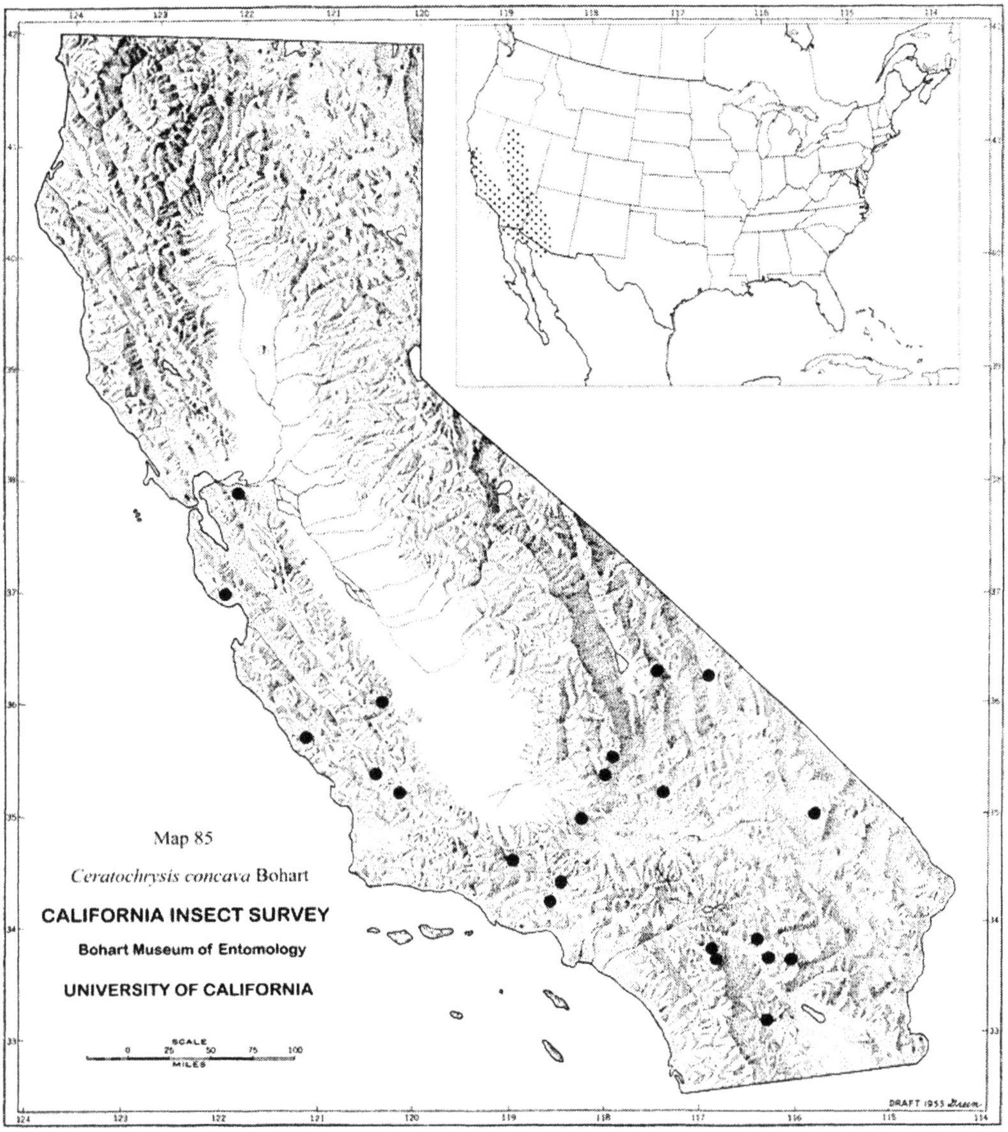

Map 85. California distribution of *Ceratochrysis concava*. Inset: overall distribution.

California Records. Collected March through July; **Contra Costa Co**.: Mt. Diablo; **Fresno Co**.: 12 mi. w Coalinga; **Inyo Co**.: Darwin Falls; 13 mi. s Death Valley Junction; **Kern Co**.: 1 mi. n Mojave Route 466; Indian Wells Canyon; 2 mi. n Inyokern turn-off; Short Canyon; 6 mi. w Inyokern; **Los Angeles Co**.: Santa Monica Mts.; Tanbark Flat; **Monterey Co**.: Bryson; **Riverside Co**.: 10 mi. e Whitewater;

Thousand Palms; Indio; Boyd Desert Res. Center; 10 mi. e Whitewater; 4 mi. s Palm Desert; Indio; San Jacinto River San Jacinto Mts.; **San Bernardino Co.**: Granite Mts.; Kelso Dunes; Kramer Hills; **San Diego Co.**: Borrego Valley; **San Luis Obispo Co.**: 3 mi. s Creston; Black Mt. 6 mi. n Pozo; La Panza Camp 12 mi. n Pozo; **Santa Cruz** Co.: Felton; **Trinity Co.**: Junction City; **Ventura Co.**: 5 mi. s Gorman; Quatal Canyon.

Hosts. Unknown.

Discussion. The swollen sides of the face, with deep medial area (particularly in females) will distinguish *concava* from related species. Additional diagnostic features are flagellomere I 4x as long as broad, the female tergum III entire or nearly so, and the punctation medially on tergum II is less dense (not contiguous) than it is in *collega*.

Ceratochrysis crossata Bohart

Figs. 60, 62; Map 86

Ceratochrysis crossata Bohart 1966a:117. Holotype ♂; USA: California, Los Angeles Co., Tanbark Flat, San Gabriel Mts. (UCDC).

Geographic Range (Map 86). Upper and Lower Sonoran deserts in California, Arizona, New Mexico, Nevada and Mexico (Baja California Sur, Puebla, Mexico D.F.); 121 specimens were seen.

California Records. Collected April through September; **Imperial Co.**: Glamis; **Inyo Co.**: 2 mi. e Lone Pine; Argus Mts.; Darwin Falls; Deep Springs Lake; Little Lake; Lone Pine Creek; Walker Pass; **Los Angeles Co.**: Eagle Rock Hills; Tanbark Flat; San Gabriel Mnts; **Mono Co.**: Chalfant; **Monterey Co.**: Monterey; **Riverside Co.**: 18 mi. w Blythe; Andreas Canyon; Palm Springs; Blythe; Menifee Valley; hills on west end Tahquitz Canyon, near Palm Springs; **San Bernardino Co.**: 12 mi. s Ivanpah; Clipper Valley; Kelso Dunes; **San Diego Co.**: 2 mi. n Warner Springs; Palm Canyon; Borrego Valley; **San Luis Obispo Co.**: Black Lake Canyon.

Hosts. Unknown.

Discussion. This is one of two species of *Ceratochrysis* that have pronounced tubercles on the vertex. Most similar to *quadrituberculata* (not a California species), *crossata* can be distinguished immediately from other species by the setal tassels at the apex of the mid and hindtibiae. The body size is average for *Ceratochrysis* species (5.5 mm in both sexes) and the color is unremarkable.

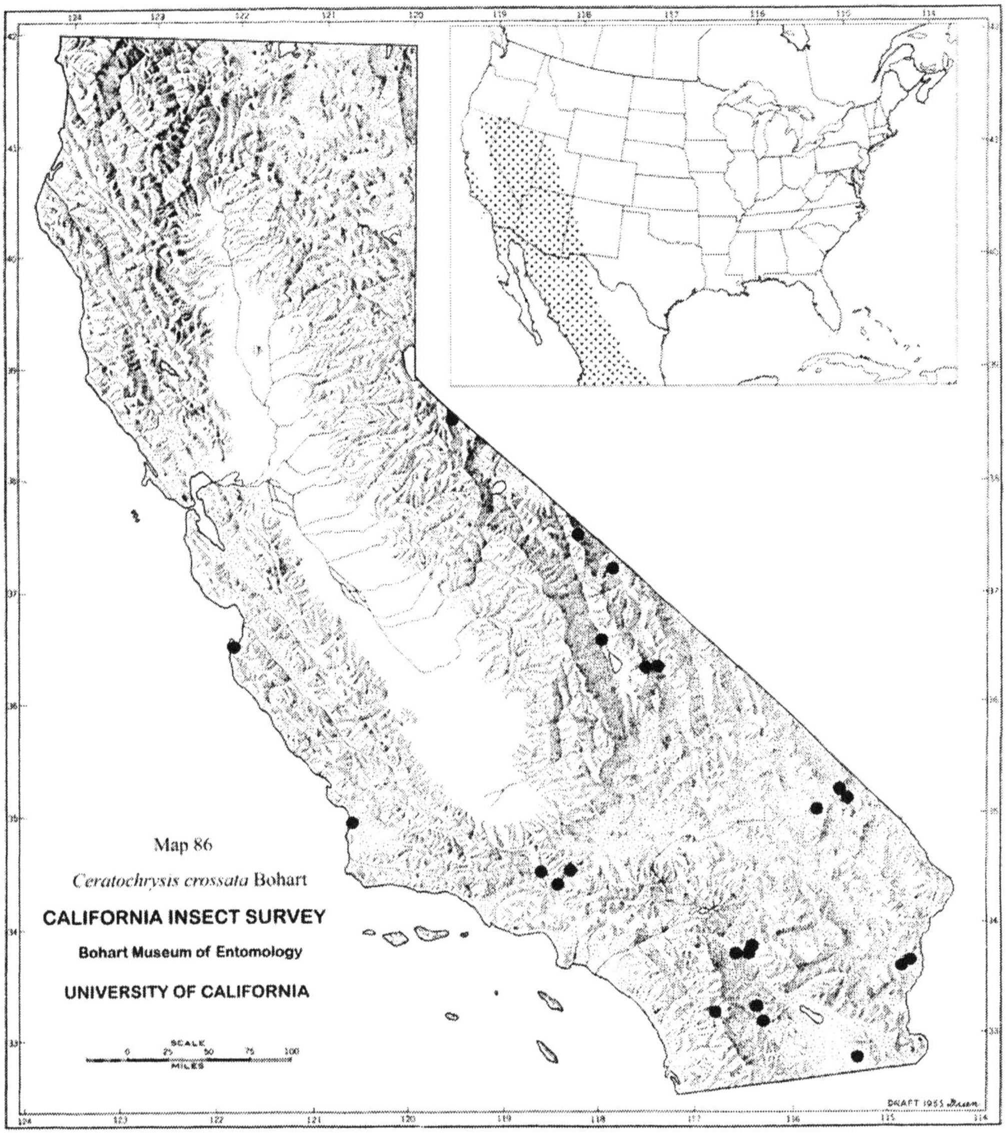

Map 86. California distribution of *Ceratochrysis crossata*. Inset: overall distribution.

***Ceratochrysis cyanosoma* (Mocsáry)**

Figs. 64, 66; Map 87

Chrysis cyanosoma Mocsáry 1914:17. Holotype ♀; USA: California (BMNH).

Geographic Range (Map 87). Upper Sonoran to Canadian Zones in the western USA; 79 specimens were examined.

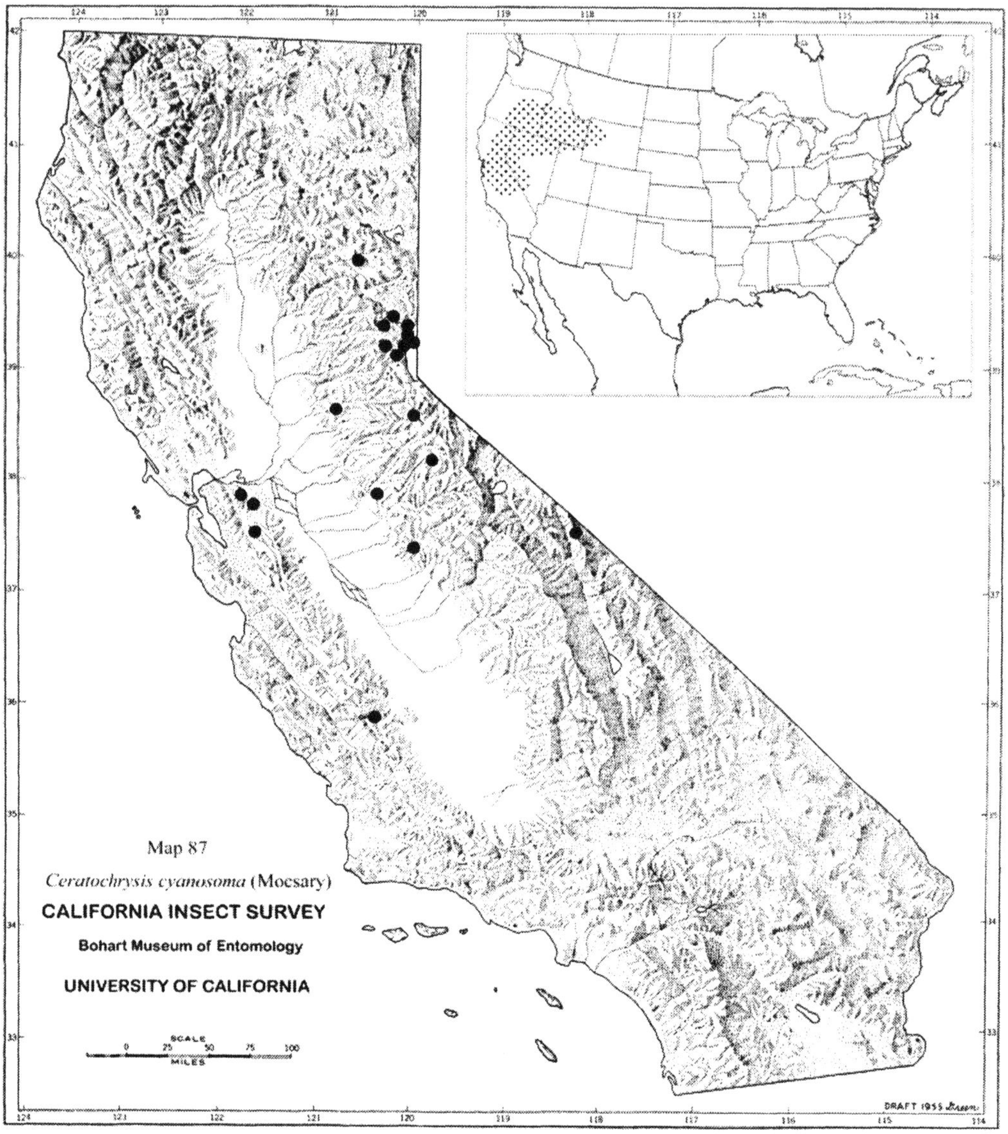

Map 87. California distribution of *Ceratochrysis cyanosoma*. Inset: overall distribution.

California Records. Collected May through September; **Alameda Co**.: Tesla; **Alpine Co**.: Forestdale Meadow; Red Lake; **Contra Costa Co**.: Rock City; Mt. Diablo; **El Dorado Co**.: El Dorado; **Mariposa Co**.: Mariposa; **Mono Co**.: Crooked Creek Laboratory, White Mts.; **Monterey Co**.: Arroyo Seco; **Nevada Co**.: Boca; Sagehen Creek; Hobart Mills; **Placer Co**.: Carnelian Bay, Lake Tahoe; **Plumas Co**.:

Buck's Lake; **Sierra Co.**: Independence Lake; Sattley; Sierra Valley; Yuba Pass; **Tuolumne Co.**: Dardanelles; Sonora.

Hosts. Unknown.

Discussion. This is one of a number of *Ceratochrysis* species with short malar (less than 1 MOD long) and subantennal distances (shorter than 1.5 MOD). *C. cyanosoma* can be distinguished from other similar species by a combination of characteristics: middle third of face with clearly marked finely cross-ridged area, and tergum III with apical margin narrowly curved under, with little or no medial emargination.

Ceratochrysis enhuycki (Cooper)

Figs. 50, 57; Map 88

Chrysis enhuycki Cooper 1952:140. Holotype ♀; USA: New Jersey, Princeton (USNM).

Ceratochrysis tuberella Bohart 1966a:122. Holotype ♂; USA: California: Mono Co., Paradise Camp (UCDC).

Geographic Range (Map 88). Widespread in the USA and Mexico (Morelos); 102 specimens were seen.

California Records. Collected April through August; **Mono Co.**: Paradise Camp; **San Bernardino Co.**: Apple Valley; **San Joaquin Co.**: Tracy; **Sierra Co.**: Sardine Lakes; **Ventura Co.**: Lockwood Creek near Stauffer; **Yolo Co.**: 3 mi. nw Knights Landing, Davis, Woodland.

Hosts. The eumenine vespids *Leptochilus republicanus* (Dalla Torre) and *L. rufinodus* (Cresson) were given as hosts of *enhuycki* by Krombein (1959).

Discussion. This is the second California species with pronounced tubercles on the vertex. However, *enhuycki* is the only one with tergum III apically entire in females and broadly emarginated in males. Additionally, flagellomere I is short, less than 3x as long as broad. Body length is average, 5.0 mm in males and 5.5 mm in females.

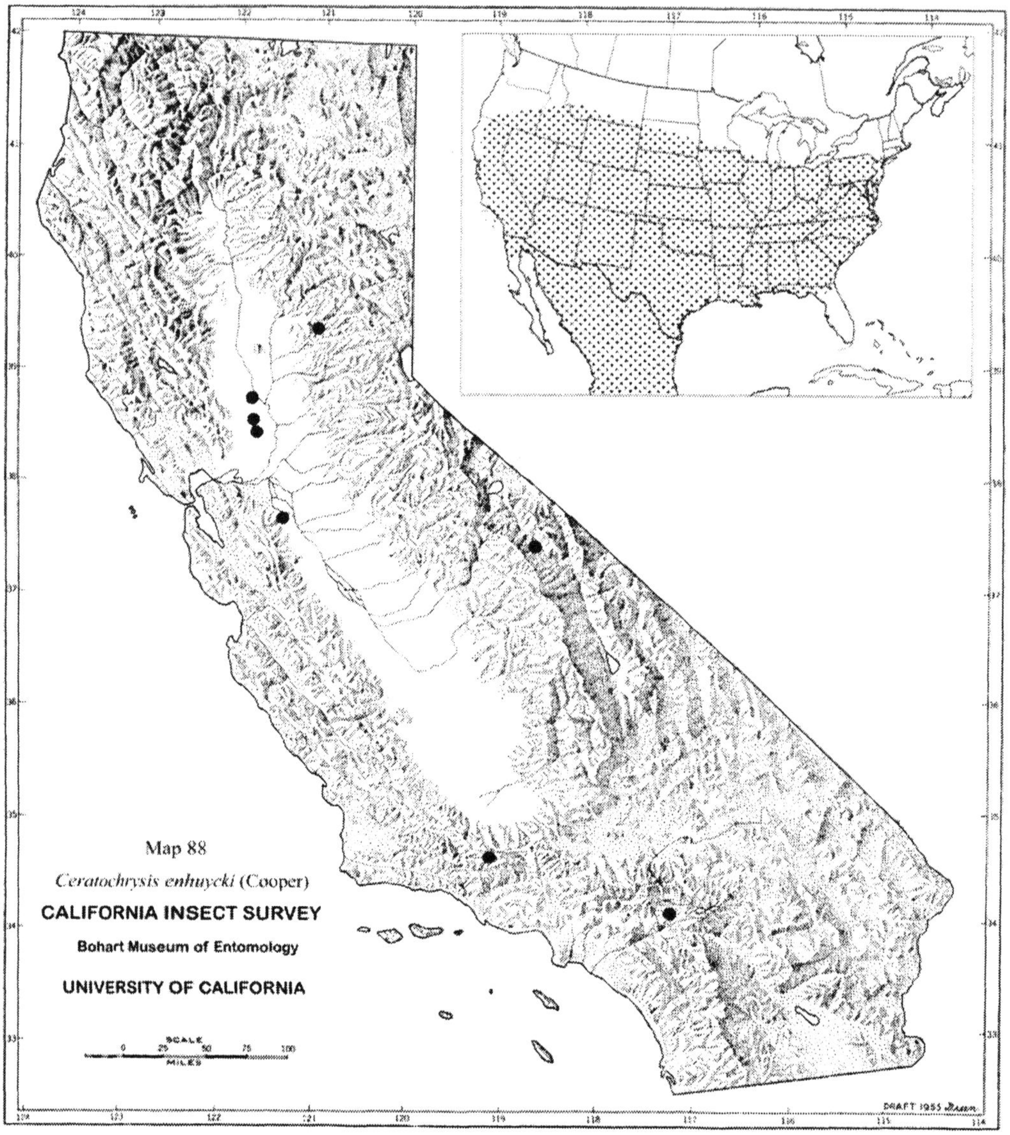

Map 88. California distribution of *Ceratochrysis enhuycki*. Inset: overall distribution.

Ceratochrysis fusilis Bohart

Map 89

Ceratochrysis fusilis Bohart, in Bohart and Kimsey 1982:174. Holotype ♂; USA: Nevada, Churchill Co., 12 mi. ne Stillwater (UCDC).

Geographic Range (Map 89). Upper Sonoran Zone in eastern California, Nevada and western Arizona; 14 specimens were examined.

California Records. Collected in June; **San Bernardino Co.**: Trona.

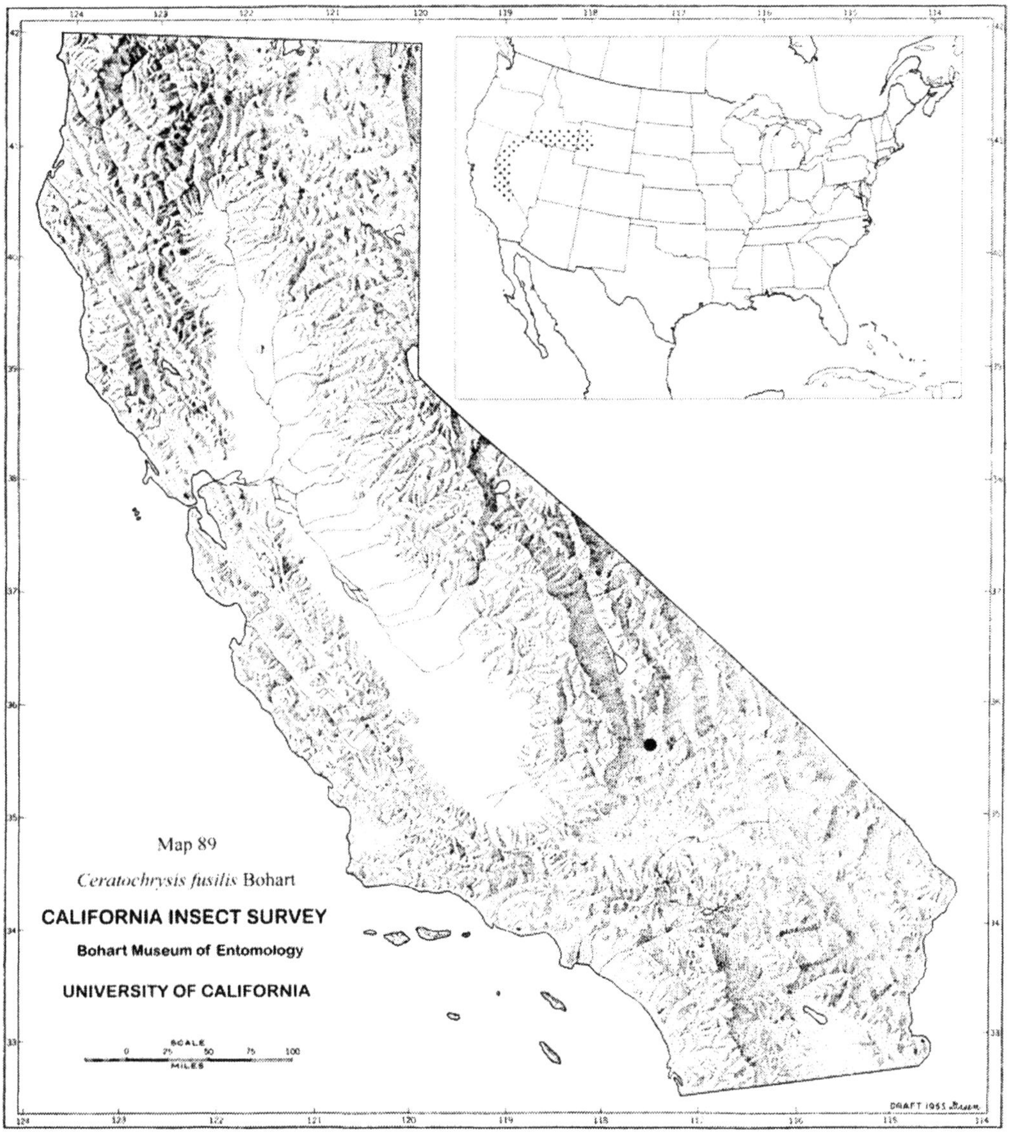

Map 89. California distribution of *Ceratochrysis fusilis*. Inset: overall distribution.

Hosts. Unknown.

Discussion. The most distinctive feature of *fusilis* is the extensive area of fine cross-ridging on the face, which extends up almost to the frontal carina. Additional features are flagellomere I 4x as long as broad, the female tergum III medially emarginate, and punctures on the middle of tergum II separated and not contiguous.

Ceratochrysis gracilis Bohart

Ceratochrysis gracilis Bohart, in Bohart and Kimsey 1982:175. Holotype ♀; USA: California, Kern Co., Rancheria Creek, Piute Mts. (LACM).

Geographic Range. California (Kern Co.); 3 specimens were studied.

California Records. Collected in June; **Kern Co.**: Rancheria Creek; Piute Mts.

Hosts. Unknown.

Discussion. This rarely collected endemic California species is known only from females. Most similar to *thysana* and *menkei*, female *gracilis* can be distinguished by tergum III apical margin entire and the middorsal line (in profile) nearly straight, not saddled, and tergum II nearly as long as broad.

Ceratochrysis grisselli Bohart

Ceratochrysis grisselli Bohart, in Bohart and Kimsey 1982:175. Holotype ♂; USA: California, San Bernardino Co., 11 mi. se Baker (UCDC).

Geographic Range (Map 89). California (San Bernardino Co.); 3 specimens were studied.

California Records. Collected in April; **San Bernardino Co.**: 11 mi. se Baker.

Hosts. Unknown.

Discussion. This California endemic is similar to *longimala*, based on the long subantennal distance (1.9 MOD or longer), and evanescent frontal carina. It can be distinguished from *longimala* by flagellomere I 5-6x as long as broad, without long ventral setae, the least interocular distance about as long as the combined lengths of flagellomere I plus the pedicel, and tergum III apical rim notched medially.

Ceratochrysis longimala Bohart

Ceratochrysis longimala Bohart, in Bohart and Kimsey 1982:176. Holotype ♂; USA: California, Riverside (CAS).

Geographic Range. Upper Sonoran Zone of southern California; 3 specimens were seen.

California Records. Collected in March, May; **Los Angeles Co.**: Gorman; **Riverside Co.**: Riverside.

Hosts. Unknown.

Discussion. The most distinctive feature of *longimala* is the presence of long setae beneath flagellomere I. Other than this feature *longimala* can be separated from the related species, *grisselli,* by flagellomere I 3x as long as broad and the much more extensively developed cross-ridged area of the face. Other distinctions are given under *grisselli*. *C. longimala* is an endemic California species.

Ceratochrysis menkei Bohart

Fig. 59, Map 90

Ceratochrysis menkei Bohart, in Bohart and Kimsey 1982:177. Holotype ♀; USA: California, Inyo Co., 13 mi. s Death Valley Junction (UCDC).

Geographic Range (Map 90). Upper Sonoran Zone in Inyo and Stanislaus Cos., California; 5 specimens were examined.

California Records. Collected in May; **Inyo Co**.: 13 mi. s Death Valley Junction; **Stanislaus Co**.: Del Puerto Canyon.

Hosts. Unknown.

Discussion. This is another rarely collected endemic species known only from the female. Diagnostic features of *menkei* include malar space 2 MOD long, subantennal distance 1.2 MOD long, frontal carina well-developed, with branches extending toward vertex, and tergum III saddled and rounded out apically.

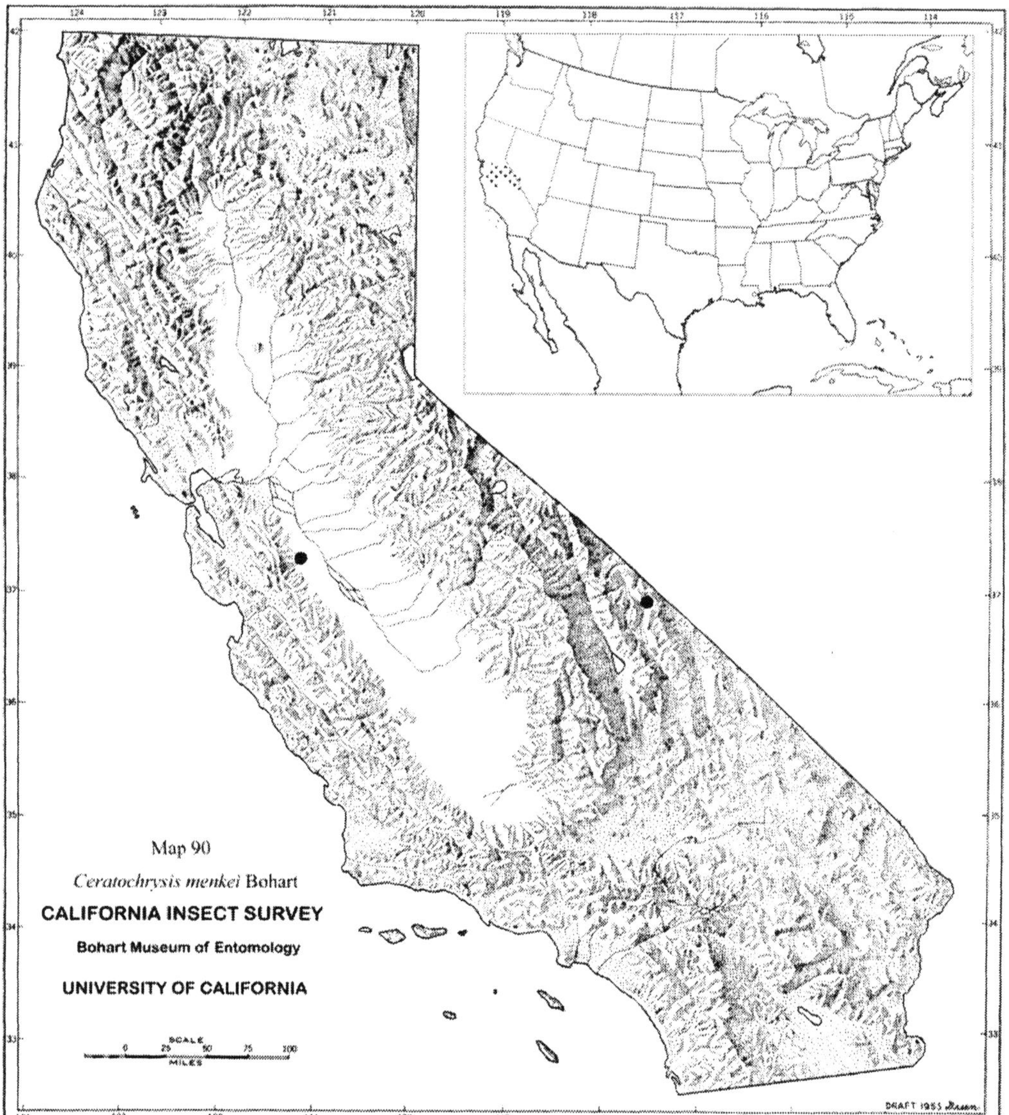

Map 90. California distribution of *Ceratochrysis menkei*. Inset: overall distribution.

Ceratochrysis minata Bohart

Fig. 49, Map 91

Ceratochrysis minata Bohart, in Bohart and Kimsey 1982:177. Holotype ♂; USA: California, Davis (UCDC).

Geographic Range (Map 91). Upper Sonoran to Canadian Zones west of 100th meridian in the USA, from California to Nebraska and Texas, and in Alberta, Canada; 67 specimens were studied.

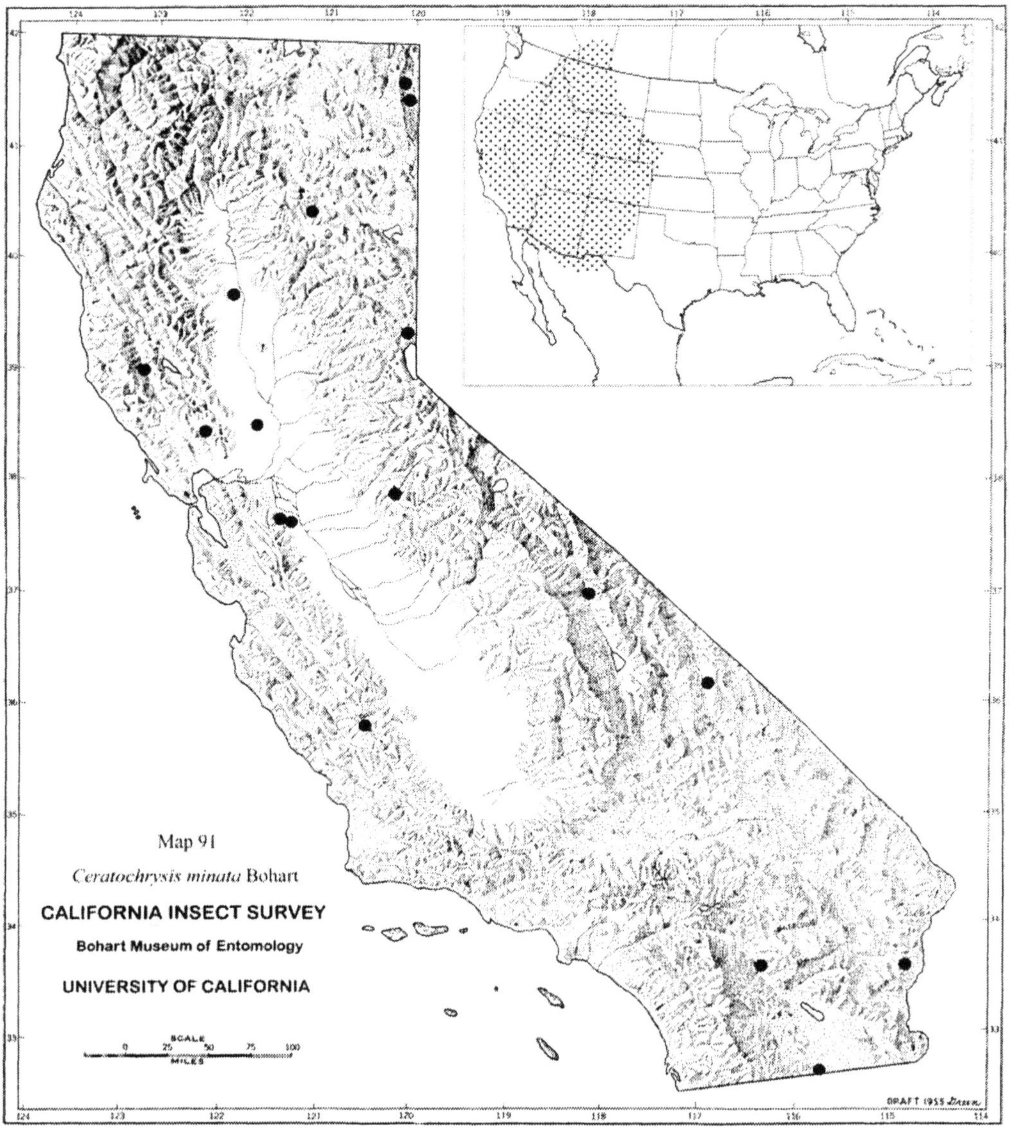

Map 91. California distribution of *Ceratochrysis minata*. Inset: overall distribution.

California Records. Collected April through September; **Glen Co**.: Orland; **Imperial Co**.: Calexico; **Inyo Co**.: 5 mi. e Big Pine; Furnace Creek; **Lassen Co**.: Bridge Creek Camp; **Modoc Co**.: 5 mi. s Lake City; 5.5 mi. e Cedarville; **Monterey Co**.: Arroyo Seco; **Napa Co**.: Samuel Springs (beneath Lake Berryessa); **Nevada**

Co.: Sagehen Creek; **Riverside Co.**: Blythe; Indio; **San Joaquin Co.**: 8 mi. nw Tracy; Tracy; **Stanislaus Co.**: Tracy; Turlock; **Tuolumne Co.**: Tuolumne City; **Yolo Co.**: Davis.

Hosts. Unknown.

Discussion. *C. minata* has a unique diagnostic feature; the gena has a well-developed rounded lobe adjacent to the mandibular insertion, which is pronounced in females and is somewhat less distinct in males. Additional features are the long flagellomere I (4x as long as broad), densely pubescent sides of the face converge above in front view, and the frontal carina with branches extending toward the vertex.

Ceratochrysis nearctica (Mocsáry)

Figs. 61, 63, Map 92

Spintharis nearctica Mocsáry 1911b:462. Holotype ♀; USA: Arizona (TMB).

Ceratochrysis alveata Bohart 1966a:113. Holotype ♂, USA: California, San Diego Co., Borrego Valley (UCDC); 83 specimens were seen.

Geographic Range (Map 92). Lower and Upper Sonoran Zones in the southwestern deserts, from Texas to California, and Sonora and Baja California, Mexico.

California Records. Collected March through October; **Imperial Co.**: Black Mt. 4-5 mi. se Hw 78; Glamis; Palo Verde; Pinto Wash; **Inyo Co.**: 13 mi. s Death Valley Junction; 15 mi. s Big Pine; 3 mi. w Lone Pine; Whitney Portal Rd.; Darwin Falls; Midway Well; Death Valley; Panamint Springs; Big Pine; **Kern Co.**: Short Canyon; **Riverside Co.**: 18 mi. w Blythe; 5.5 mi. nw Indio; hills on west end Menifee Valley; Thousand Palms; **San Bernardino Co.**: Cajon; **San Diego Co.**: Borrego Valley.

Hosts. Unknown.

Discussion. This is a distinctive species with a long malar space (2-3 MOD long), tasseled male mid and hindtibiae, well-developed frontal carina with dorsal branches, and saddled female tergum III.

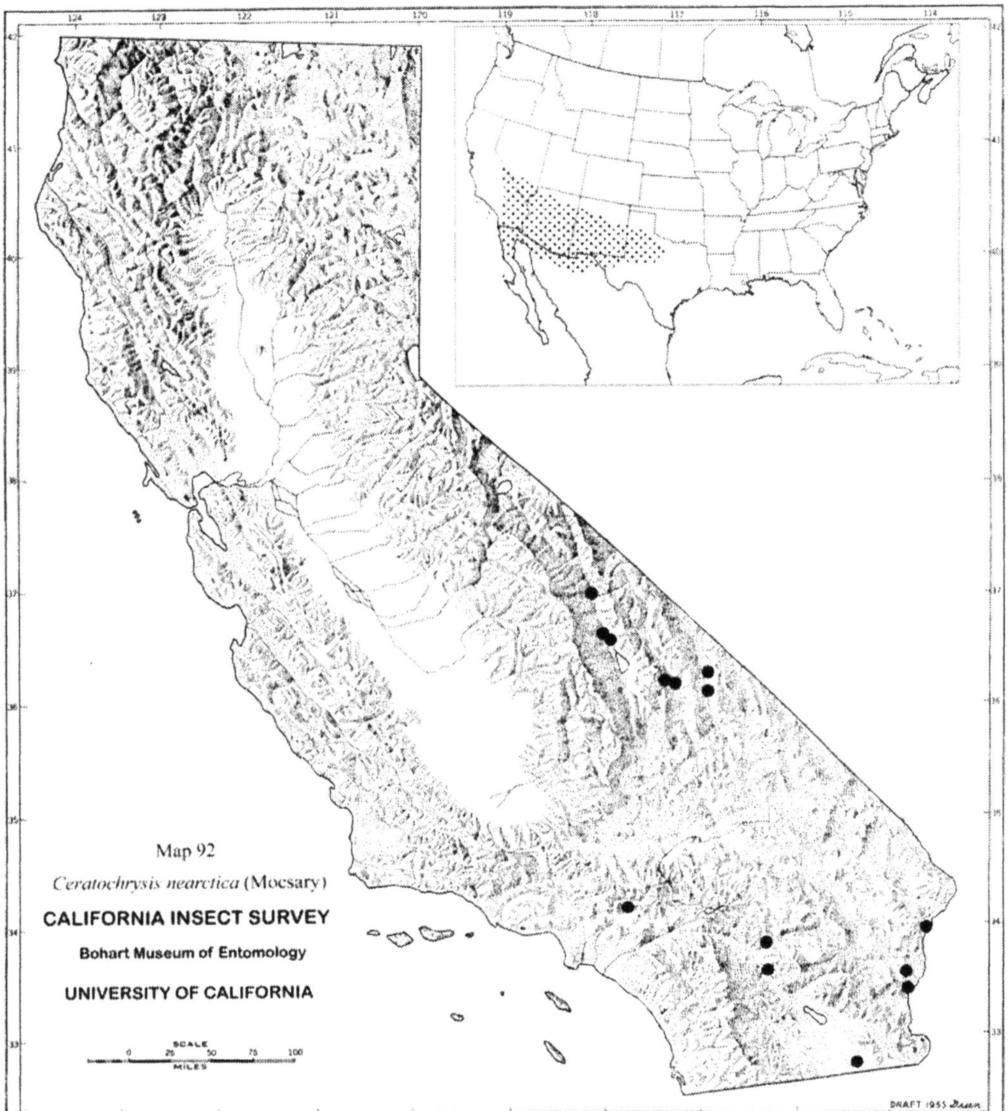

Map 92. California distribution of *Ceratochrysis nearctica*. Inset: overall distribution.

Ceratochrysis perpulchra (Cresson)

Map 93

Chrysis perpulchra Cresson 1865b:308. Holotype ♀; USA: Colorado (ANSP).

Geographic Range (Map 93). USA and southern Canada west of 100th meridian; Upper Sonoran to Hudsonian Zones; mountains of Baja California, Mexico; 650 specimens were examined.

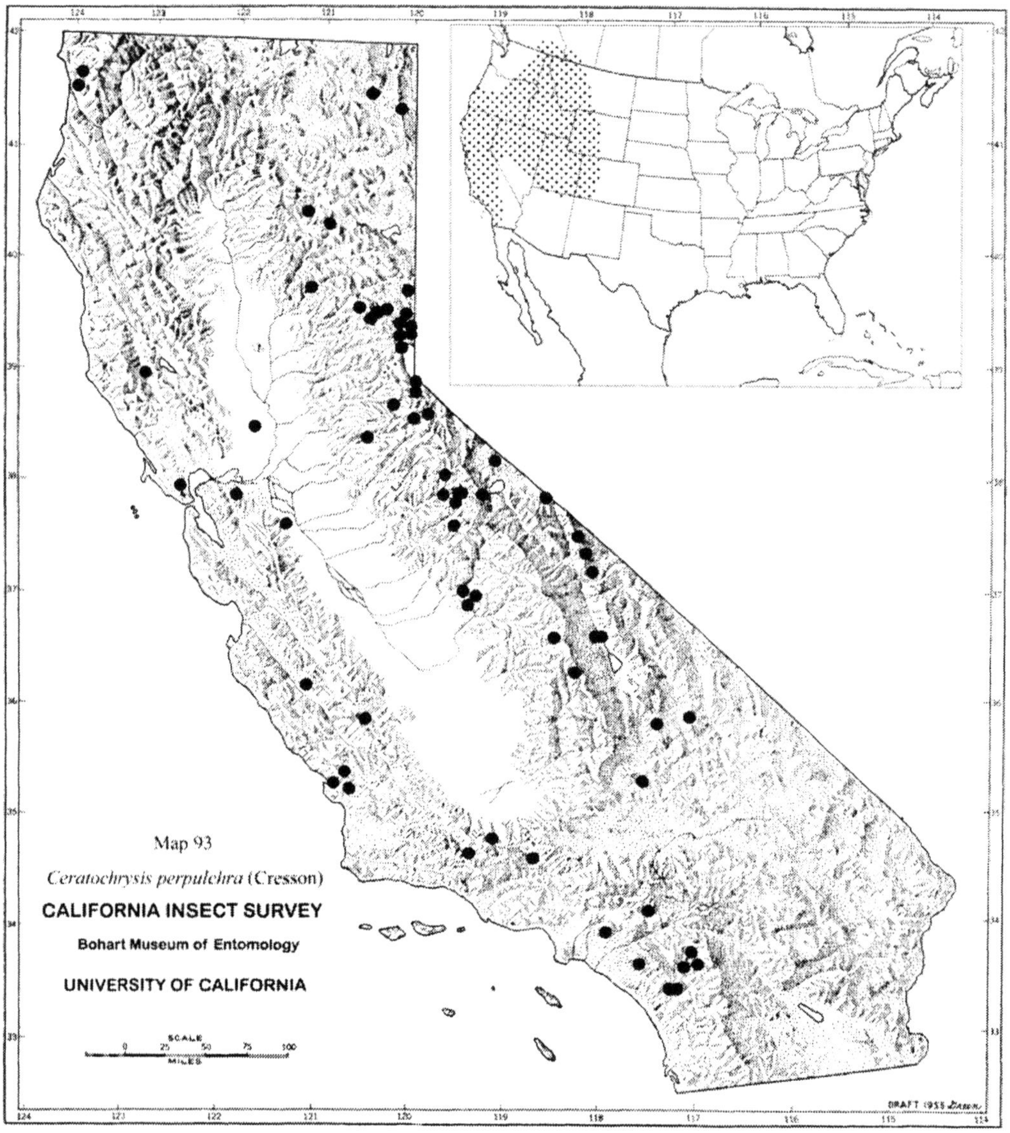

Map 93. California distribution of *Ceratochrysis perpulchra*. Inset: overall distribution.

California Records. Collected April through October; **Alpine Co**.: Carson Pass; Hope Valley; Winnemucca Lake; **Amador Co**.: 2 mi. sw Silver Lake; **Contra Costa Co**.: Mt. Diablo; **El Dorado Co**.: 0.5 mi. w stateline; 5 and 6 mi. w on Ice House Rd.; Grass Lake; **Fresno Co**.: Hammond Ranch; 5 mi. s Auberry; Lone Indian Lake;

Pioneer Basin; **Humboldt Co**.: 2 mi. s Redwood Valley; Kneeland; **Inyo Co**.: Independence; Shulman Grove White Mnts; Surprise Canyon; Whitney Portal; **Kern Co**.: Mt. Pinos; **Lake Co**.: Hopland Grade; **Lassen Co**.: Bridge Creek; Hallelujah Junction; Norvell; **Los Angeles Co**.: 7 mi. w Keen Camp; Claremont; Crystal Lake Rd.; Monrovia Canyon Park; Tanbark Flat; **Marin Co**.: Alpine Lake; **Mariposa Co**.: 1 mi. e Junction Hwy. 120; Big Oak Flat Entrance Station; **Modoc Co**.: 14 mi. sw Alturas; Cedar Pass; **Mono Co**.: 11 mi. n Bridgeport; Mt. Barcroft; 4 mi. e Junction Hwys 120-395; Blancos Corral; Crooked Creek; Mammoth Lakes; Leavitt Meadow; Mt. Barcroft; Pickel Meadow; Sheep Mt.; Sonora Pass; **Monterey Co**.: Arroyo Seco; Soledad; **Nevada Co**.: Boca; Jackson Lake; Sagehen Creek; Hobart Mills; **Orange Co**.: Trabuco Canyon; **Placer Co**.: Carnelian Bay, Lake Tahoe; **Plumas Co**.: Onion Valley; **Riverside Co**.: 10 mi. n Temecula; San Jacinto Mts.; 5 mi. s Hemet; Andreas Canyon; Millard Canyon; The Galivan; White Water; **San Bernardino Co**.: Cajon Junction; **San Joaquin Co**.: Tracy; **San Luis Obispo Co**.: 6 mi. n Santa Margarita; Creston; Morro Bay; **Sierra Co**.: Independence; Kyburz; Sagehen Creek; Sardine Peak; Sattley; Sierra Valley; Sierraville; Yuba Pass; **Tulare Co**.: Deadman Canyon; Kings Canyon National Park; Mineral King Rd.; **Tuolumne Co**.: Boundary Hill Research Area; Yosemite National Park; Emigrant Lake; Leland Meadow; **Ventura Co**.: Sespe Canyon; **Yolo Co**.: Davis.

Hosts. Hicks (1932) reported this species as a parasite of *Ammophila aberti* Haldeman.

Discussion. Of the large-bodied species of *Ceratochrysis, perpulchra* is the one most commonly collected. *C. trachypleura* is similar in size and is largely sympatric with *perpulchra*. They can be separated by the coarser punctation along the sides of the face, purpler body color, and non-dentate mesopleuron of *perpulchra*; *trachypleura* is green, with fine punctation along the sides of the face and the mesopleuron is distinctly dentate. The average body length of *perpulchra* is 6 mm.

Ceratochrysis sierrae Bohart

Map 94

Ceratochrysis sierrae Bohart, in Bohart and Kimsey 1982:181. Holotype ♂; USA: California, Nevada Co., Sagehen Creek (UCDC).

Geographic Range (Map 94). Transition to Canadian Zones west of 100th meridian; the mountains of Baja California, Mexico, and in the Hudsonian Zone of the White Mts. of California; 68 specimens were studied.

California Records. Collected May through September; **Alpine Co**.: Luther Pass; Carson Pass; Winnemucca Lake; **El Dorado Co**.: Grass Lake; Angora Peak; Lake Tahoe; Meyers; **Fresno Co**.: Los Gatos Canyon; **Inyo Co**.: Wyman Canyon; **Lassen**

Co.: Susan River Camp; Hallelujah Junction; Summit Camp Mt. Lassen; **Modoc Co.**: Cedar Pass; Likely; **Mono Co.**: 7 mi. e Bodie; Blanco's Corral; 7 mi. e Tioga Pass; Sardine Creek; 3 mi. n Inyo Co.; Crooked Creek Lab White Mts.; Cottonwood Creek;

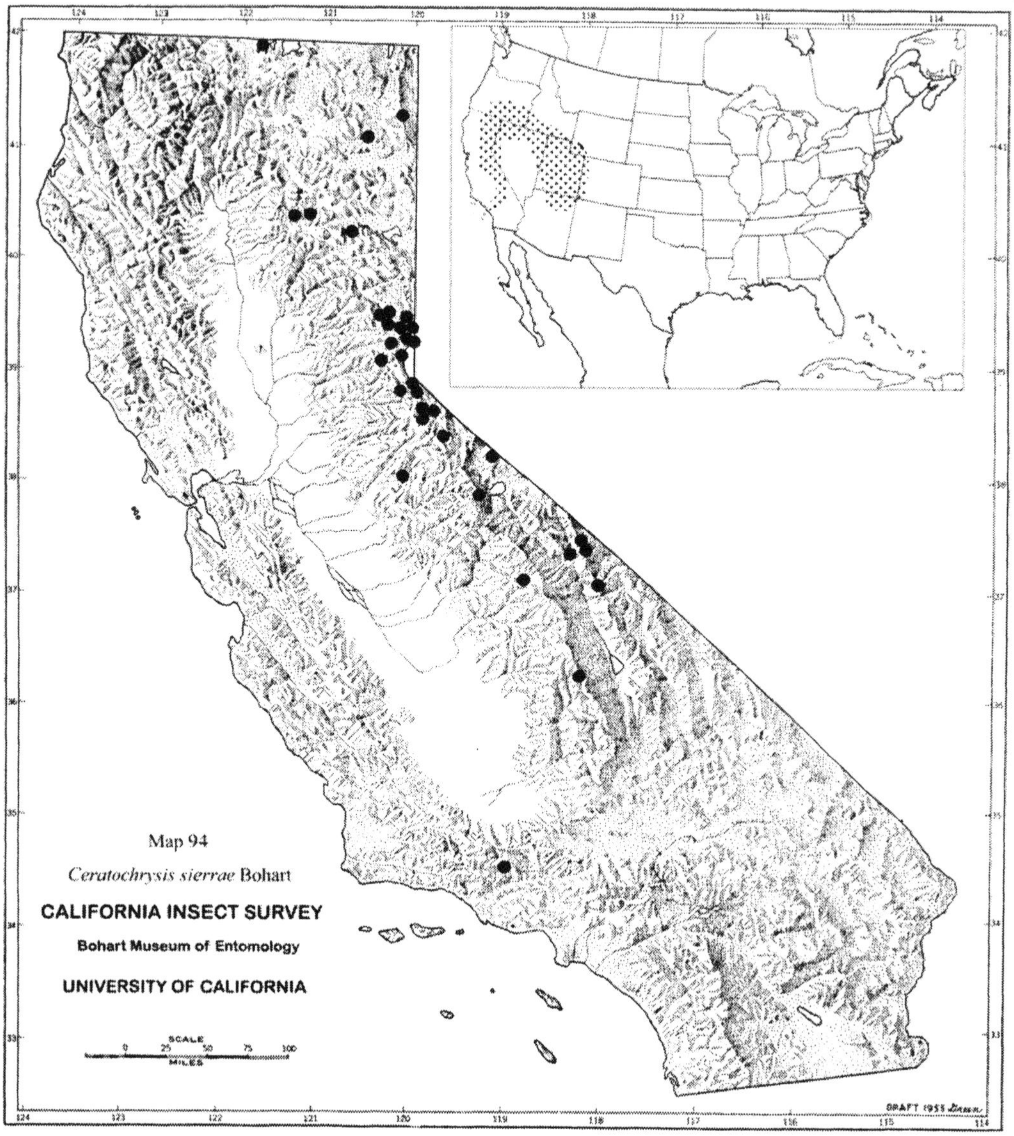

Map 94. California distribution of *Ceratochrysis sierrae*. Inset: overall distribution.

Nevada Co.; Sagehen Creek; Russell Valley; Boca; Hobart Mills; Truckee; **Placer Co.**: Meyers; Carnelian Bay, Lake Tahoe; **Shasta Co.**: Lassen Peak; **Sierra Co.**: Sierraville; Sierra Valley; Sattley; Dog Canyon near Verdi Peak; **Siskiyou Co.**: Lower Klamath Lake; **Tulare Co.**: Mineral King; **Tuolumne Co.**: Strawberry; Chipmunk Flat; **Ventura Co.**: Chuchupate Ranger Station, base Mt. Frazier.

Hosts. Unknown.

Discussion. *Ceratochrysis sierrae* is closely related to *concava, collega* and *fusilis*, as discussed under those species. The diagnostic feature that will distinguish *sierrae* is the shorter flagellomere I (less than 4x as long as broad). This species appears to occur primarily in montane sites at elevations over 1,500 m. (over 3,000 m. in the White Mts.).

Ceratochrysis thysana Bohart

Figs. 51, 67, Map 95

Ceratochrysis thysana Bohart 1966a:120. Holotype ♂; USA: California, Mono Co., Paradise Camp (UCDC).

Geographic Range (Map 95). Upper and Lower Sonoran Zones in California, Nevada, Arizona and Oregon; 67 specimens were studied.

California Records. Collected May through July; **Alpine Co**.: Red Lake; **Inyo Co**.: Townes Pass 18 mi. e Panamint Springs; **Mono Co**.: Paradise Camp; **Riverside Co**.: Deep Canyon 4 mi. s Palm Desert; Riverside.

Hosts. Parker (pers. comm.) reared *thysana* from *Leptocheilus periallis* Parker (Vespidae: Eumeninae).

Discussion. This species shares a number of features with *nearctica*, including the tasseled male mid and hindtibiae, apical rim of tergum III entire or nearly so in both sexes, and female tergum III saddled in profile. *C. thysana* is distinguished by the long malar space (2 or more MOD long), middle third of the scapal basin finely cross-ridged and punctate, mesopleuron divided by a deep longitudinal sulcus, flagellomere I 3.8x as long as broad, and body length 3-5 mm.

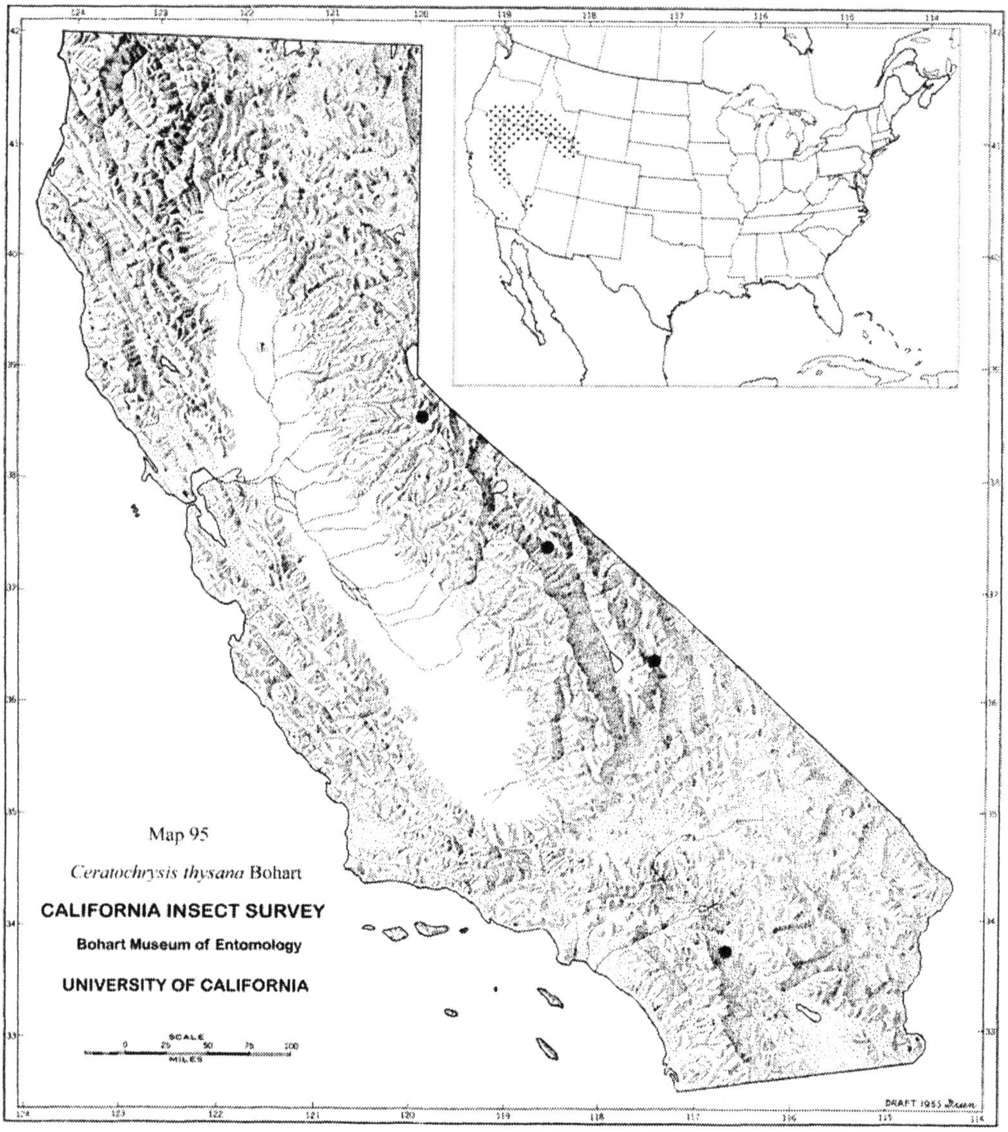

Map 95. California distribution of *Ceratochrysis thysana*. Inset: overall distribution.

Ceratochrysis trachypleura Bohart

Fig. 65, Map 96

Ceratochrysis trachypleura Bohart 1966a:121. Holotype ♂; USA: California, San Diego Co., Warner Springs (UCDC).

Geographic Range (Map 96). Upper Sonoran to Hudsonian Zones from California to Nebraska, Manitoba and British Columbia, Canada, and mountains of Baja California, Mexico, mostly west of the 100th meridian; 432 specimens were examined.

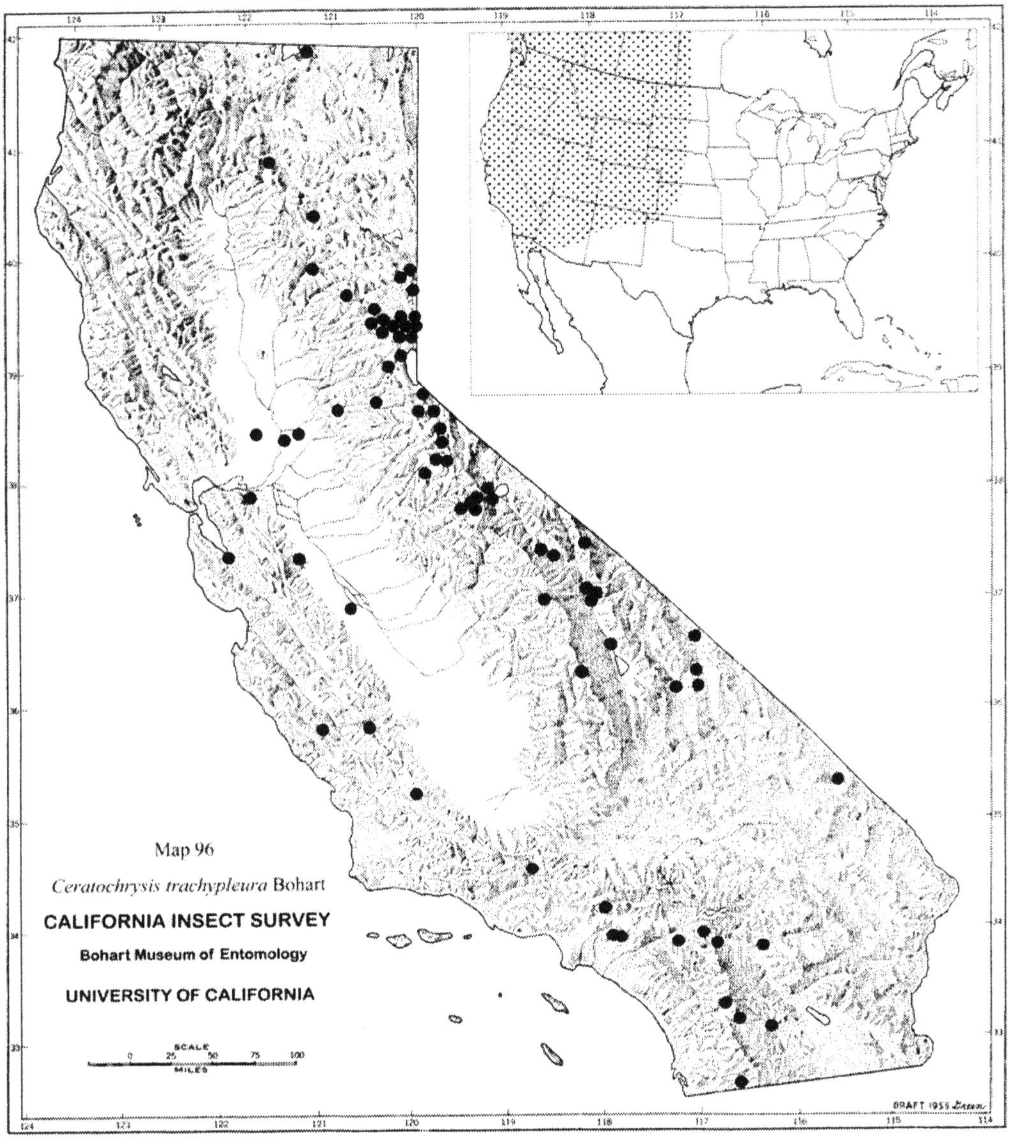

Map 96. California distribution of *Ceratochrysis trachypleura*. Inset: overall distribution.

California Records. Collected March through September; **Alpine Co.**: Shot Rock; Carson Pass; Hope Valley; Lake Winnemucca; **Contra Costa Co.**: Antioch; **El Dorado Co.**: 6-8 mi. n Ice House Rd.; Echo Lake; Grass Lake; Pino Grande; Pollock

Pines; **Fresno Co**.: Pioneer Basin; **Inyo Co**.: Mammoth Lake; 15 mi. s Big Pine; 18 mi. e Panamint Spring; Townes Pass; 6 mi. w Big Pine; e side Panamint Mnts; Eureka Valley dunes; Independence; Lone Pine Creek; **Lassen Co**.: Bridge Creek Camp; Hallelujah Junction; **Los Angeles Co**.: Johnstone Point; La Crescenta; San Gabriel Canyon Rd., 11 mi. n Azusa; Tanbark Flat; Claremont; **Madera Co**.: Agnew Meadows; **Merced Co**.: Dos Palos; **Modoc Co**.: Newell; **Mono Co**.: Lee Vining Meadows; Mono Basin, Sulfur Spring Rd.; 10 mi. s Hwy 167 dunes; Paradise Camp; Sonora Pass; White Mts.; **Monterey Co**.: Arroyo Seco; **Nevada Co**.: 3 mi. n Boca; Boca; Carpenter Ridge; Fullers Lake; Sagehen Creek; Hobart Mills; **Placer Co**.: Carnelian Bay, Lake Tahoe; Brockway Summit; **Plumas Co**.: 8 mi. sw Johnsville; Chilcoot; Hawley Lake; Nelson Creek Rd.; Onion Valley; Vinton; **Riverside Co**.: Banning; Box Springs Mnts; Riverside; Thousand Palms; **Sacramento Co**.: Carmichael; Rio Vista; Sacramento; **San Benito Co**.: Clear Creek R. A.; **San Bernardino Co**.: 12 mi. se Ivanpah; **San Diego Co**.: 2 mi. n Warner Springs; 9 mi. s Warner Springs; Aquanga; Borrego Springs; Borrego Valley; Sunshine Summit near Dodge Valley; **San Luis Obispo Co**.: Black Mt.; 6 mi. n Pozo; La Panza Camp 12 mi. n Pozo; **Santa Clara Co**.: Alum Rock Park, San Jose; **Shasta Co**.: Moose Camp; **Sierra Co**.: Gold Lake; Independence Lake; Sagehen Creek; Sardine Lakes.; Sardine Peak; Sattley; Sierra Buttes; Sierraville; Weber Lake; Yuba Pass; **Stanislaus Co**.: Del Puerto Canyon; **Tulare Co**.: Mineral King; **Tuolumne Co**.: Chipmunk Flat; Leland Meadow; Pinecrest Lake; Strawberry; **Ventura Co**.: Sespe Canyon; **Yolo Co**.: Davis.

Hosts. Hicks (1932) reported the sphecid, *Ammophila aberti* Haldeman as the host of *trachypleura* (given as *perpulchra*).

Discussion. This species most closely resembles *perpulchra* as discussed under that species. The distinctly dentate lower mesopleuron is the most diagnostic feature of *trachypleura*.

Genus *Chrysis* Linnaeus

Generic Diagnosis. Face usually with transverse frontal carina, scapal basin with at least some medial cross-ridging; mesopleuron with scrobal sulcus; tergum III with subapical pit row followed by apical rim with 4 or 6 apical teeth or angles; forewing discoidal cell and cubital cells complete; paramere simple, without flexible apical appendage; body without whitish markings.

Hosts. *Chrysis* species are nest parasites of bees and wasps in the families Crabronidae, Sphecidae, Megachilidae and Anthophoridae, and solitary Vespidae (Eumeninae).

Distribution. The genus *Chrysis* is the most speciose and widely distributed group in the family Chrysididae, occurring on nearly every continent, and accounting for about half of the species in the family.

Discussion. *Chrysis* is a huge, diverse, paraphyletic genus, which appears to have no definitive diagnostic features other than the lack of those features characteristic of other chrysidine genera. There are 50 species of *Chrysis* in California (77 in North America), four of which are endemic to the state. One species, *angolensis*, is introduced from Africa. Important taxonomic features can be seen in the sculpturing of the face and mesopleuron, shape of the lateral propodeal tooth or projection and shape and dentition of the apical margin of the last metasomal tergum. The North American species were revised by Bohart and Kimsey (1982).

Key to the California Species of *Chrysis*

1 Tergum III apically with six teeth or sharp angles (as in Fig. 68)........................2

\- Tergum III apically with four teeth or angles (as in Figs. 69-76)......................10

2 Flagellomere I longer than lateral margin of tergum III from base to apex of basal tooth; propodeal projection posterior margin simple, not dentate or lobulate (Fig. 86)..*tenuicornis* Taylor

\- Flagellomere I shorter than lateral margin of tergum III from base to apex of basal tooth; propodeal projection posterior margin variable..................................3

3 Tergum II posterolateral corner obtusely rounded; flagellomere I longer than II; propodeal projection with posterior margin simple not dentate or lobulate ....*aridula* Bohart

\- Tergum II posterolateral corner sharp, right-angled or acute; flagellomere I shorter than II in some males; propodeal projection dentate or lobulate in some species 4

4 Malar space 1 MOD long or shorter, and shorter than subantennal distance.........5

\- Malar space more than 1 MOD long, and equal to or longer than subantennal distance...6

5 Tergum III lateral tooth broadly obtuse (Fig. 81)..........................*tularensis* Bohart

\- Tergum III lateral tooth a right angle or acute (Fig. 75)....................*serrata* Taylor

6 Malar space as long or longer than 2 MOD, subantennal distance 1.5-2.0 MOD; flagellomere I usually more than 2.2x as long as broad......*praestans* du Buysson

\- Malar space shorter than 2 MOD; subantennal distance less than 1.5 MOD long; flagellomere I usually less than 2.2x as long as broad..................................7

7 Propodeal tooth slightly convex beneath (Fig. 87)..................*arizonica* Bohart

- Propodeal tooth strongly lobate beneath, lobe about 1 MOD in size (as in Fig. 85) 8

8 Tergum III medial and submedial teeth much longer than broad at base ....*oraria* Bohart

- Tergum III medial and submedial teeth subtriangular and about as long as broad at base...9

9 Tergum III with edge of submedial notch thickened and bicarinate, prepit row swelling forming nearly right angle inside view (Fig. 80)..........*wasbaueri* Bohart

- Tergum III with edge of submedial notch not thickened or bicarinate, pre-pit row swelling forming obtuse angle in side view.....................*inaequidens* Dahlbom

10 Mesopleuron dentate, midocellus enclosed by carina extending dorsally from transverse frontal carina......................................*angolensis* Radoszkowski

- Mesopleuron not dentate, midocellar carina variable....................................11

11 Tergum III apical margin bicarinate at least in medial notch; transverse frontal carina distinct, regular and nearly reaching compound eye; genal carina clearly separated from compound eye...12

- Tergum III apical margin not bicarinate, or transverse frontal carina irregular, and genal carina practically touching compound eye at one point........................18

12 Tergum III with strong longitudinal carina before pit row; male flagellomere II shorter than III and about as long as malar space.................*dugesi* du Buysson

- Tergum III without longitudinal carina, male flagellomere II length various......13

13 Midocellus surmounted by thin eyelid-like fold (as in Fig. 84); male lower frons with sparse semierect setae; female tergum III often with prepit swelling; malar space more than 1 MOD long..14

- Midocellus not lidded or lid incomplete; male lower frons with dense silvery appressed pubescence; female without distinct prepit swelling; malar space less than or equal to 1 MOD long..16

14 Tergum III with ridge between medial pits depressed and pit row appearing notched in lateral view..*venustella* Bohart

- Tergum III with ridge between medial pits not depressed, and pit row not appearing notched in lateral view..15

15 Tergum III ridge between medial pits strong and polished; malar space less than 1 MOD long..*astralia* Bohart

- Tergum III ridge between medial pits thin and punctate; malar space equal to or more than 1 MOD long..*venusta* Cresson

16 Tergum II punctures near posterior margin average less than one puncture diameter apart; male flagellomere I slightly longer than II or III; female tergum III prepit depression barely indicated...................................*krombeini* Bohart

- Tergum II punctures near posterior margin average one puncture diameter apart; male flagellomere I slightly longer than II but shorter than III; female tergum III prepit depression various...17

17 Tergum III medial ridge of pit row narrow and often depressed in lateral view; males and some females with narrow smooth swellings between mid and hindocelli; wing membrane untinted; female prepit depression weak ....*calosella* Bohart

- Tergum III medial ridge of pit row strong and slightly raised in lateral view; no interocellar swellings; female with marked prepit depression ....*antennalis* Mocsáry

18 Transverse frontal carina absent or incomplete, often with branches extending toward ocelli, least interocular distance not or slightly greater than eye height 19

- Transverse frontal carina regular and complete, nearly reaching eye, if partial then lacking branches extending toward ocelli and least interocular distance greater than eye height...34

19 Subantennal distance and malar space both 2 MOD long or longer ....*tripartita* Aaron

- Subantennal distance and malar space not both as long as 2 MOD......................20

20 Malar space as long or longer than subantennal distance; tergum III lateral margin somewhat irregular but not distinctly convex or concave...................................21

- Malar space shorter than subantennal distance; tergum III lateral margin often convex..26

21 Metasoma coppery to gold or reddish, contrasting strongly with blue or greenish blue head and mesosoma...*pompella* Bohart

- Metasoma, mesosoma and head concolorous blue to green..................................22

22 Tergum III with lateral tooth obtuse and blunt, all four teeth usually weakly developed, pit row often shallow (Fig. 74); scapal basin extensively polished medially...*scitula* Cresson

- Tergum III with lateral tooth acute and sharp, rarely obtuse, all four teeth well-developed, pit row deep; scapal basin various..23

23 Tergum III punctation coarse and even, not becoming finer toward base of metallic-colored area, apicomedial notch usually forming more than semicircle; frons often with traces of transverse carina..............................*montana* Aaron

- Tergum III punctation coarse toward pit row, becoming much finer toward base of metallic-colored area, apicomedial notch usually less than semicircle; frons various...24

24 Tergum III with apicomedial pair of teeth almost as sharp as lateral pair (Fig. 73) ....*paradisica* Bohart

- Tergum III with apicomedial pair of teeth blunter than lateral pair (as in Fig. 71) 25

25 Tergum II with punctures across distal third dense, mostly less than one puncture diameter apart; body length 6-7 mm.......................................*dorsalis* Aaron

- Tergum II with punctures across distal third less dense, mostly about one puncture diameter apart; body length 7-8 mm.....................................*knowltoni* Bohart

26 Midocellus partly depressed and surmounted by thin eyelid-like fold (as in Fig. 84); transverse frontal carina partial and/or irregular, with branches extending toward ocelli..27

- Midocellus not depressed and not lidded; transverse frontal carina absent, if present and partial, then without branches extending toward ocelli..........................29

27 Upper genal area just behind eye with polished impunctate spot; subantennal distance 2 MOD long..*vibex* Bohart

- Upper genal area just behind eye evenly punctate without impunctate spot; subantennal distance various...28

28 Tergum III laterally undulate or evenly convex in lateral view, apicomedial notch simple and sharp-edged...*propria* Aaron

- Tergum III laterally expanded and angular or subdentate, apicomedial notch partly filled by thin membrane..*tensa* Aaron

29 Subantennal distance and malar space each at least 2 MOD long; transverse frontal carina well-developed, with branches nearly or completely enclosing midocellus ....*tripartita* Aaron

- Subantennal distance and malar space not both as long as 2 MOD; transverse frontal carina weakly developed or absent...30

30 Flagellomere I clearly less than twice as long as broad; tergum III pit row deep, with many pits confluent...*vagabunda* Bohart

- Flagellomere I about twice as long as broad; tergum III pit row various...........31

31 Tergum II shiny toward apex, punctures separated by nearly one puncture diameter; tergum III pit row shallow, pits sometimes evanescent ....*florissanticola* Rohwer

- Tergum II not shiny toward apex, punctures nearly contiguous; tergum III pit row often deep...32

32 Tergum III punctation finer than on scutum; sternum II with two large nearly contiguous black spots..*rivalis* Bohart

- Tergum III punctures as coarse as on scutum; sternum II markings various.......33

33 Male flagellum ventrally with erect setae as long as flagellum diameter; female sternum II with large dark basal spots or dark basal band; terga usually deep blue ....*coloradica* Bohart

- Male flagellum without obvious erect setae; female sternum II without dark spots or with only vague discolorations; terga more green than blue....*xerophila* Bohart

34 Midocellus partly depressed and surmounted by thin eyelid-like fold (as in Fig. 84), pit row usually confluent...*pellucidula* Aaron

- Midocellus sometimes depressed but not lidded; pit row not confluent..............35

35 Transverse frontal carina incurved below to enclose a flattened, oval platform; malar space 2 MOD long or longer, and much longer than subantennal distance; male flagellomere I slightly longer than II..............................*daedala* Bohart

- Transverse frontal carina not incurved below; malar space and male flagellomere I various..36

36 Male flagellomere I length less than or equal to III, both sexes with medial longitudinal ridge or ridge densely micropunctate......................................37

- Male flagellomere I longer than III; tergum II of females and most males with at least trace of polished medial longitudinal ridge....................................43

37 Malar space less than 1.5 MOD long, much shorter than length of flagellomere II 38

- Malar space at least 2.4 MOD long, subequal to length of flagellomere II.........40

38 Tergum III apicomedial notch shallow, forming less than half circle (Fig. 72); body length about 7 mm........,,...*lucifera* Bohart

- Tergum III apicomedial notch forming half circle, body length usually 5-6 mm...39

39 Scutellum and tergum II medially with punctures equal to or more than one puncture diameter apart..*snowi* Viereck

- Scutellum and tergum II medially with punctures less than one puncture diameter apart...*amala* Rohwer

40 Tergum III apicomedial notch deeper than half circle (Fig. 69) whole body often bright purple...*barri* Bohart

- Tergum III apicomedial notch deeper than half circle, whole body usually mixture of blue and green, with some purple..41

41 Least interocular distance longer than foretibia, tergum III apicomedial notch shallow, forming half circle..*pattoni* Aaron

- Least interocular distance as long or shorter than foretibia; tergum III apicomedial notch less than half circle..42

42 Thorax in dorsal view more than twice as long as broad across anterolateral pronotal corners; body length 7-8 mm................................*derivata* Buysson

- Thorax in dorsal view about twice as long as broad at pronotal corners; body length 5-6 mm .... *irwini* Bohart

43 Tergum III with brilliant violet red reflections across tergum, bordering pit row anteriorly, basolaterally on tergum III (as in Figs. 78, 79), or on tergum I above basal insertion...44

- Terga without brilliant violet red coloration..46

44 Tergum III with violet red band bordering pit row anteriorly (Fig. 79) .... *cessata* Buysson

- Tergum III with violet red sometimes visible in pit row and beyond but not as discrete band above pit row...45

45 Tergum I with violet red above basal insertion; tergum III medial ridge of pit row usually absent, resulting in polished reddish purple medial pit; tergum III without basolateral green and violet red spot................................ *violacuna* Bohart

- Tergum I without violet red above basal insertion; tergum III medial ridge of pit row present, dividing row into two series; tergum III basolaterally with green and violet red spot (Fig. 78) .... *moorei* Bohart

46 Ocellocular distance less than twice least hindocellar diameter; tergum III teeth short, but medial pair sharp....................................... *cembricola* Krombein

- Ocellocular distance equal to or longer than twice hindocellar diameter; tergum III teeth various 47

47 Male sternum II with black spots in male reaching midline or only 1 MOD from it (as in Fig. 83); male flagellomere I more than twice as long as broad; if female sternum II spots are more than one midocellus diameter from midline then pit row is sunken medially in side view with overhang above declivity (as in Fig. 77) 48

- Male sternum II with black spots more than 1 MOD from midline (Fig. 82); male flagellomere I about twice as long as broad; female sternum II with spots more than 1 MOD from midline and pit row not appearing sunken medially in side view, without overhang above declivity..49

48 Sternum II with spots reaching midline, rarely 0.5 MOD from it (Fig. 83); male sternum III extensively blue to green; female pit row divided medially by broad partition without sharp declivity in side view.......................... *parkeri* Moore

- Sternum II in male with black spots 0.8-1.0 MOD from midline; male sternum III with green spot of varying size; female pit row with medial ridge weak, in lateral view with sharp declivity and overhang (Fig. 77)........................*apontis* Bohart

49 Frons with transverse frontal carina well-developed...............*nitidula* Fabricius

- Frons with transverse frontal carina vestigial, with three slight remnants or absent ....*schusteri* Bohart

Chrysis amala Rohwer

Map 97

Chrysis amala Rohwer 1909:91. Holotype ♀; USA: Colorado, Florissant (USNM).

Geographic Range (Map 97). U.S.A west of the 100th meridian; 183 specimens were seen.

California Records. Collected March through September; **Alameda Co**.: Tesla; **El Dorado Co**.: Pollock Pines; **Fresno Co**.: Barton Flat; **Lassen Co**.: Hallelujah Junction; **Nevada Co**.: Fuller Lake; **Placer Co**.: Carnelian Bay, Lake Tahoe; **Riverside Co**.: 6 mi. s Palm Springs; **Sacramento Co**.: Sacramento; **San Francisco Co**.: San Francisco; Lone Mt.; **San Luis Obispo Co**.: Black Lake Canyon; **Santa Barbara Co**.: Santa Cruz Island; Goleta; **Santa Clara Co**.: Milpitas; **Sierra Co**.: Yuba Pass; **Tuolumne Co**.: Strawberry; **Yolo Co**.: Woodland.

Hosts. Unknown.

Discussion. This species is closely related to *snowi* and the two are often difficult to separate, particularly in the females. Males can be readily distinguished by a modification of the paramere, which is broadly emarginate in *amala*. However, *amala* can be also distinguished from *snowi* by the closer punctation on the scutellum and tergum II, which is mostly less than one puncture diameter apart, versus more than one puncture diameter apart in *snowi*.

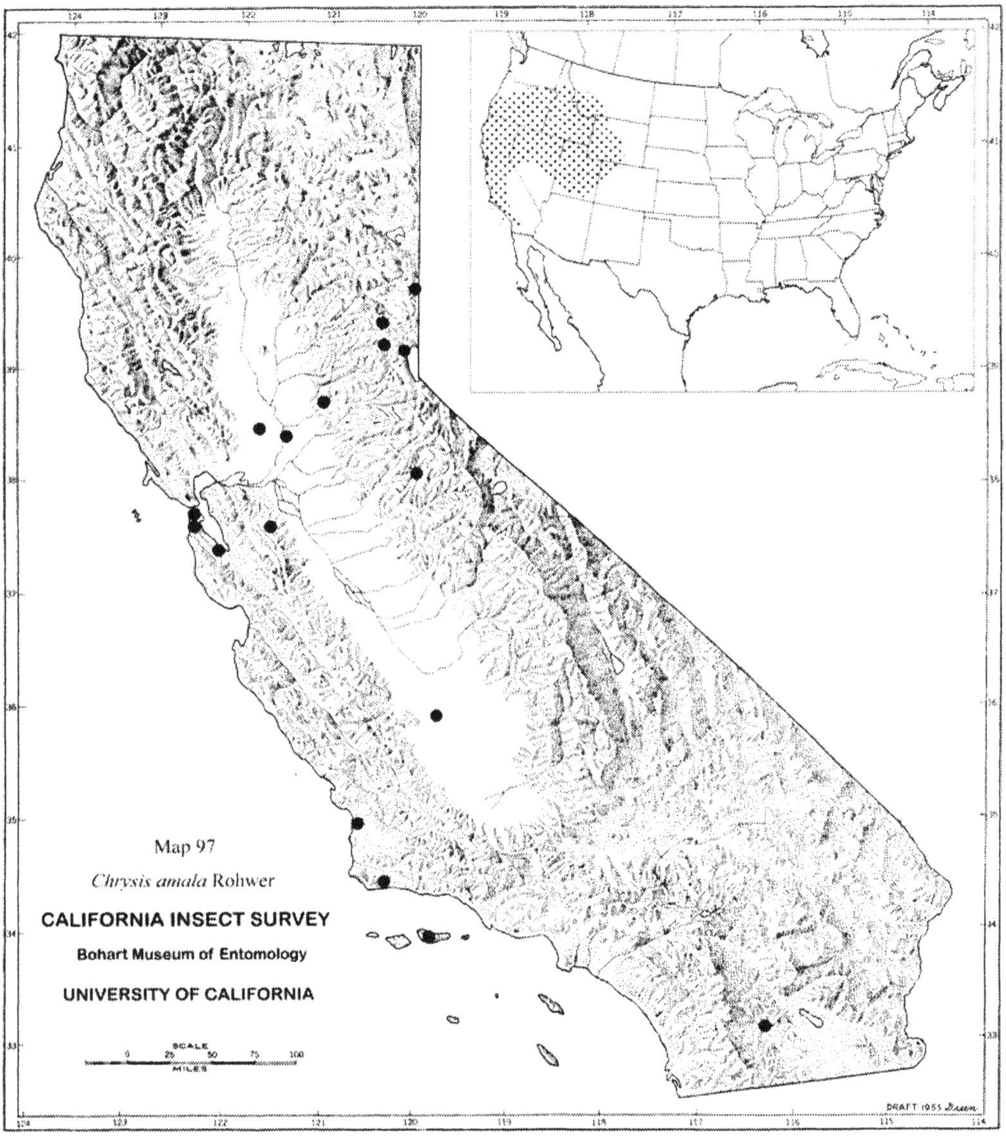

Map 97. California distribution of *Chrysis amala*. Inset: overall distribution.

Chrysis angolensis Radoszkowski

Map 98

Chrysis angolensis Radoszkowski 1881b:219. Holotype sex unknown; Angola (ISZP ?).

Chrysis fuscipennis Brullé 1846:38. Holotype ♀; Philippines (MNHN). Nec Dahlbom 1829. Synonymized by Bohart and Kimsey (1982).

Geographic Range (Map 98). Worldwide in temperate regions, including Ascension Island; 163 specimens were studied.

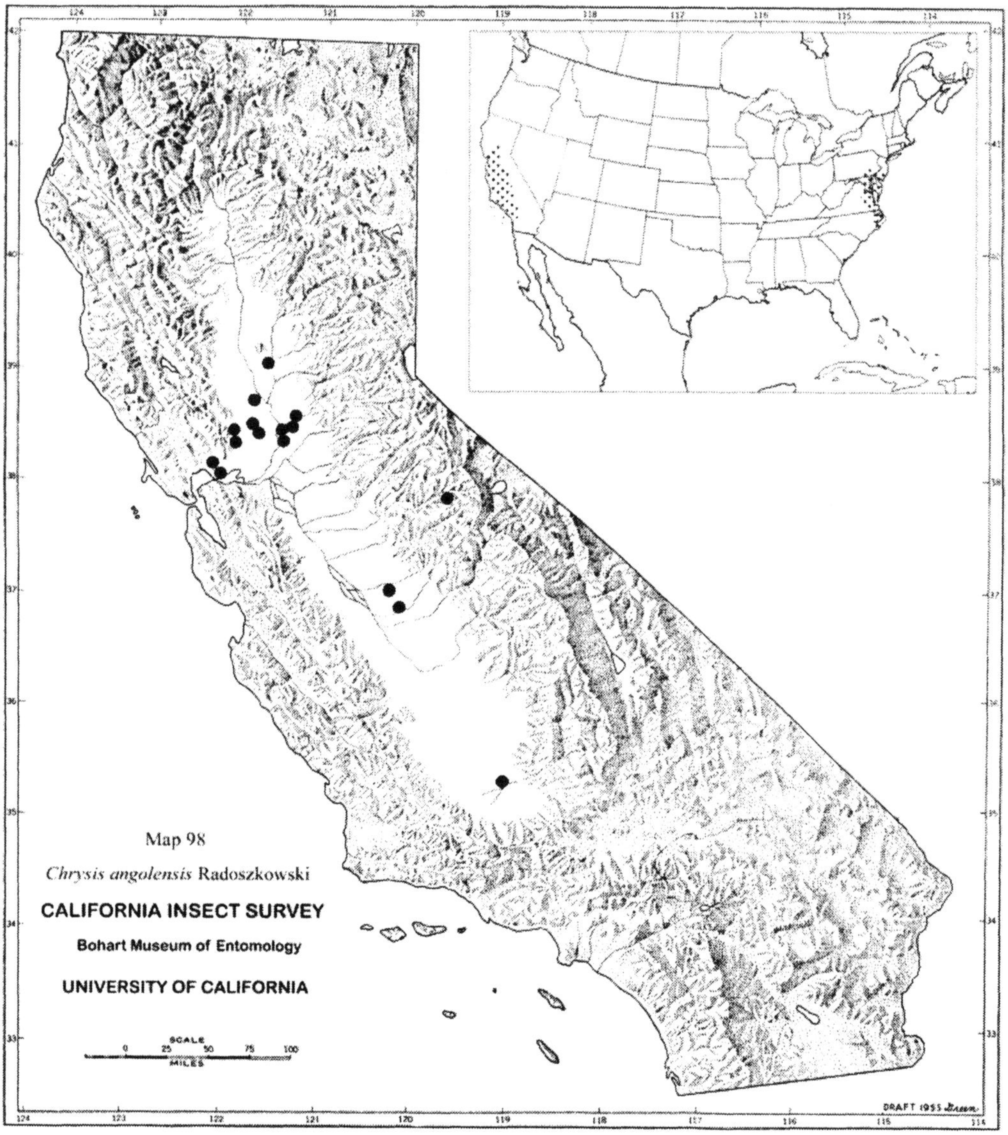

Map 98. California distribution of *Chrysis angolensis*. Inset: overall distribution.

California Records. Collected February through September; **Inyo Co.**: Antelope Springs; **Kern Co.**: Bakersfield; **Madera Co.**: Madera; **Mariposa Co.**: Yosemite Valley; **Merced Co.**: 5 mi. w El Nido; **Placer Co.**: Folsom Lake; **Sacramento Co.**:

American River; Citrus Heights; Sacramento; **San Diego Co**.: Encinitas; **Solano Co**.: Hastings Island; Suisun Marsh 6 km se Suisun City; Vacaville; **Sutter Co**.: Sutter; **Yolo Co**.: 3 mi. w Col. Knight's Lodge; Davis; Putah Canyon; Woodland.

Hosts. In California *angolensis* is a nest parasite of the sphecid genus *Sceliphron* (Stage 1960, Bohart and Kimsey 1982).

Discussion. This introduced species appears to have achieved a nearly cosmopolitan distribution due to its habit of parasitizing sphecids and crabronids that build mud nests on structures, particularly ships. *Chrysis angolensis* can be distinguished from all other North American species by the dark brown-tinted wings, bidentate mesopleuron and obtusely angled teeth on tergum III.

Chrysis antennalis Mocsáry

Map 99

Chrysis antennalis Mocsáry 1912b:564. Holotype ♂; Mexico: Chihuahua, Presidio (TMB).

Chrysis stenodyneri Krombein 1958a:150. Holotype ♀; USA: North Carolina, Dare Co., Kill Devil Hills (USNM).

Geographic Range (Map 102). Widespread from British Columbia to Panama; 381 specimens were seen.

California Records. Collected April through October; **Alameda Co**.: Altamont; **Colusa Co**.: 10 mi. sw Stonyford; **Inyo Co**.: Big Pine Creek; **Lassen Co**.: Hallelujah Junction; **Los Angeles Co**.: Tanbark Flat; **Mariposa Co**.: 1.9 mi. w Mt. Bullion; **Plumas Co**.: east end Lake Almanor; **Riverside Co**.: Blythe; Deep Canyon; Palm Springs; Andreas Canyon; Ripley; **San Benito Co**.: Clear Creek R.A.; **San Bernardino Co**.: Lone Pine Canyon; **San Luis Obispo Co**.: Creston; **Shasta Co**.: Old Station; **Sierra Co**.: Sattley; **Tulare Co**.: Ash Mt.; **Yolo Co**.: 2 mi. n Rumsey.

Hosts. All known hosts are vespids that nest in hollow stems or trap nests, including *Stenodynerus histrionalis* (Robertson), *S. krombeini* Bohart, *S. lineatifrons* Bohart and *Parancistrocerus histrio* Lepeletier (Krombein 1958, 1967), *Microdynerus bakerianus* (Cameron) (Parker 1970), and *Ancistrocerus simulator* Cameron and *Odynerus erythrogaster* Bohart (Bohart and Kimsey 1982).

Discussion. Similar to *stenodyneri, antennalis* can be distinguished by the double-edged apical rim of tergum III, and the edge broader than that of *stenodyneri*. In addition, *antennalis* wings are generally darker colored than in *stenodyneri*. This species is widespread on the East Coast and southwestern U.S.

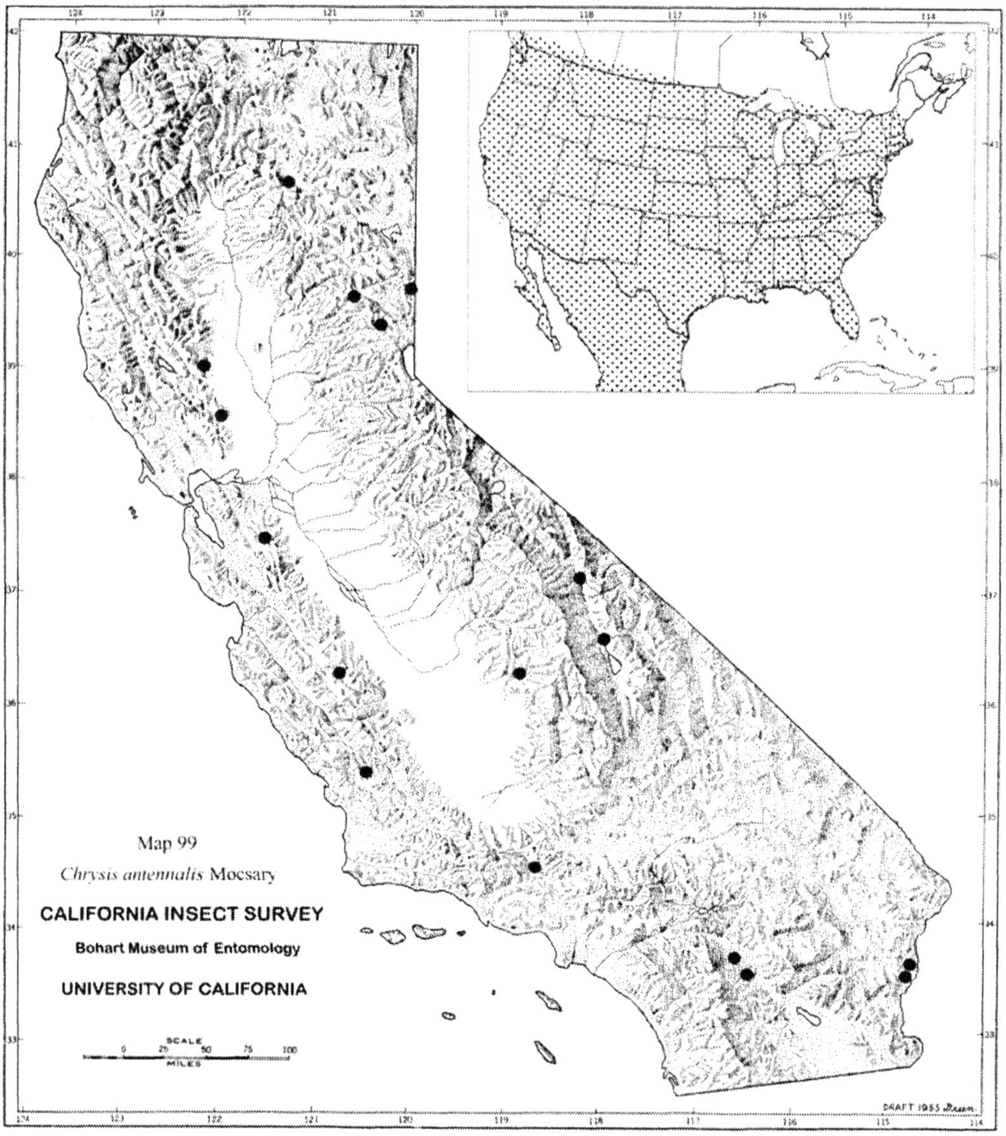

Map 99. California distribution of *Chrysis antennalis*. Inset: overall distribution.

Chrysis apontis **Bohart**

Figs. 1, 77; Map 100

Chrysis apontis Bohart, in Bohart and Kimsey 1982:126. Holotype ♂; USA: California, Mono Co., 5 mi. n Coleville (UCDC).

Geographic Range (Map 100). Western USA (AZ, CA, NV, UT) 99 specimens were examined.

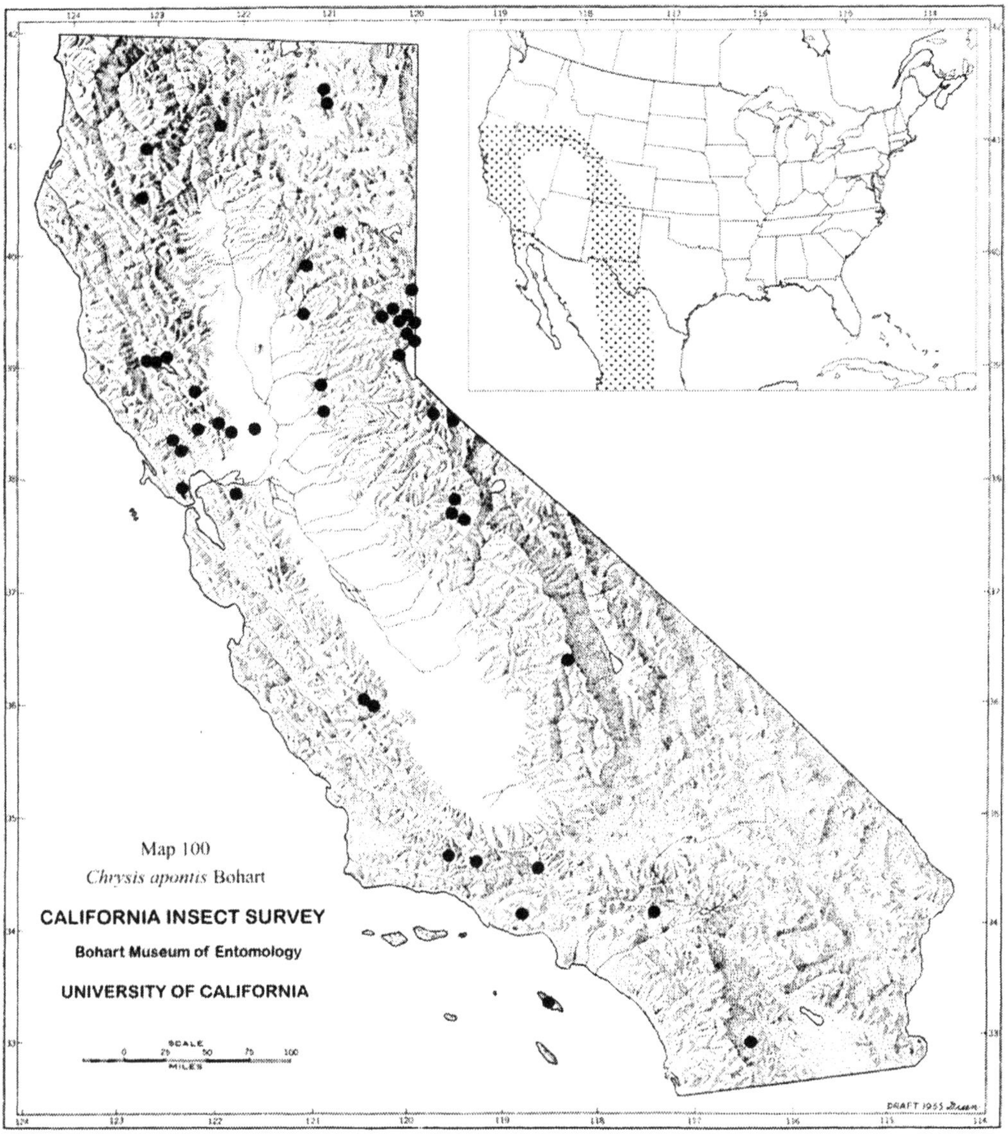

Map 100. California distribution of *Chrysis apontis*. Inset: overall distribution.

California Records. Collected March through October; **Alpine Co**.: Markleeville; **Butte Co**.: Feather Falls; **Contra Costa Co**.: Mt. Diablo; **El Dorado Co**.: Pine Hill, w Rescue; **Fresno Co**.: White Creek; 12 mi. w Coalinga; White Water Stream; 10 mi. w Coalinga; **Lake Co**.: Clear Lake Oaks; Cobb Mt.; Hopland Grade; north fork Cache Creek Hwy. 20; **Lassen Co**.: Black's Mt.; Hallelujah Junction; **Los Angeles Co**.: Crystal Lake; Tanbark Flat; Camp Baldy; **Marin Co**.: Corte Madera; **Mariposa Co**.: Indian Flat; Yosemite; **Modoc Co**.: Hackamore; Inspection Station Canby; **Mono**

Co.: 5 mi. n Coleville; **Napa Co**.: Samuel Springs (beneath Lake Berryessa); **Nevada Co**.: Boca; 17 mi. n Truckee; 3 mi. n Boca; Sagehen Creek; **Placer Co**.: 5 mi. w Foresthill; Carnelian Bay, Lake Tahoe; 5 mi. w Foresthill; **Plumas Co**.: Onion Valley; **San Diego Co**.: Oak Valley; **Santa Barbara Co**.: Los Padres National Forest; Upper Oso Camp, Los Padres National Forest; **Santa Clara Co**.: 10-12 mi. w Coalinga; Silver Creek; **Sierra Co**.: Sierra Valley; Sierraville; Sattley; **Siskiyou Co**.: Junction Springs Rd. and Hwy 97; Mount Shasta City; **Solano Co**.: Monticello Dam; **Sonoma Co**.: Boyes Hot Springs; Glen Ellen; **Trinity Co**.: Bell Creek; 6 mi. sw Denny; Hayfork Ranger Station; **Tulare Co**.: Ash Mt.; **Tuolumne Co**.: Mather turn-off; Yosemite National Park; **Ventura Co**.: Sespe Canyon; **Yolo Co**.: Davis; Putah Canyon.

Hosts. *Chrysis apontis* has been reared from nests of *Euodynerus foraminatus scutellaris* Saussure, *Ancistrocerus* sp. *A. simulator* Cameron, *A. catskill* (Saussure), and *Parancistrocerus* sp. (Vespidae) (Moore 1966, Bohart and Kimsey 1982, Parker and Bohart 1966).

Discussion. Diagnostic features of *apontis* include: sternum II spots close together and 1 MOD from the midline or less, male sternum III with variable sized metallic green spot, female pit row with medial partition weak or absent and forming a deep declivity with overhang in lateral view, and male flagellomere I more than twice as long as broad.

Chrysis aridula Bohart

Map 101

Chrysis aridula Bohart 1962:364. Holotype ♂; USA: Arizona, Bowie (UCDC).

Geographic Range (Map 101). Upper and Lower Sonoran Zones, southern California, southern Arizona and southwestern New Mexico; 10 specimens were seen.

California Records. Collected in May, June; **Riverside Co**.: Andreas Canyon; Morongo Valley; **San Bernardino Co**.: Kingston Mts.; 12 mi. se Ivanpah; New York Mts. Keystone Canyon

Hosts. Unknown.

Discussion. Although *aridula* has six teeth on the apical rim of tergum III, it is more closely related to *propria* than to the other six-toothed *Chrysis* species, based on the lidded ocelli and narrow gena. Tergum III has the apicomedial pits confluent, with no prepit swelling.

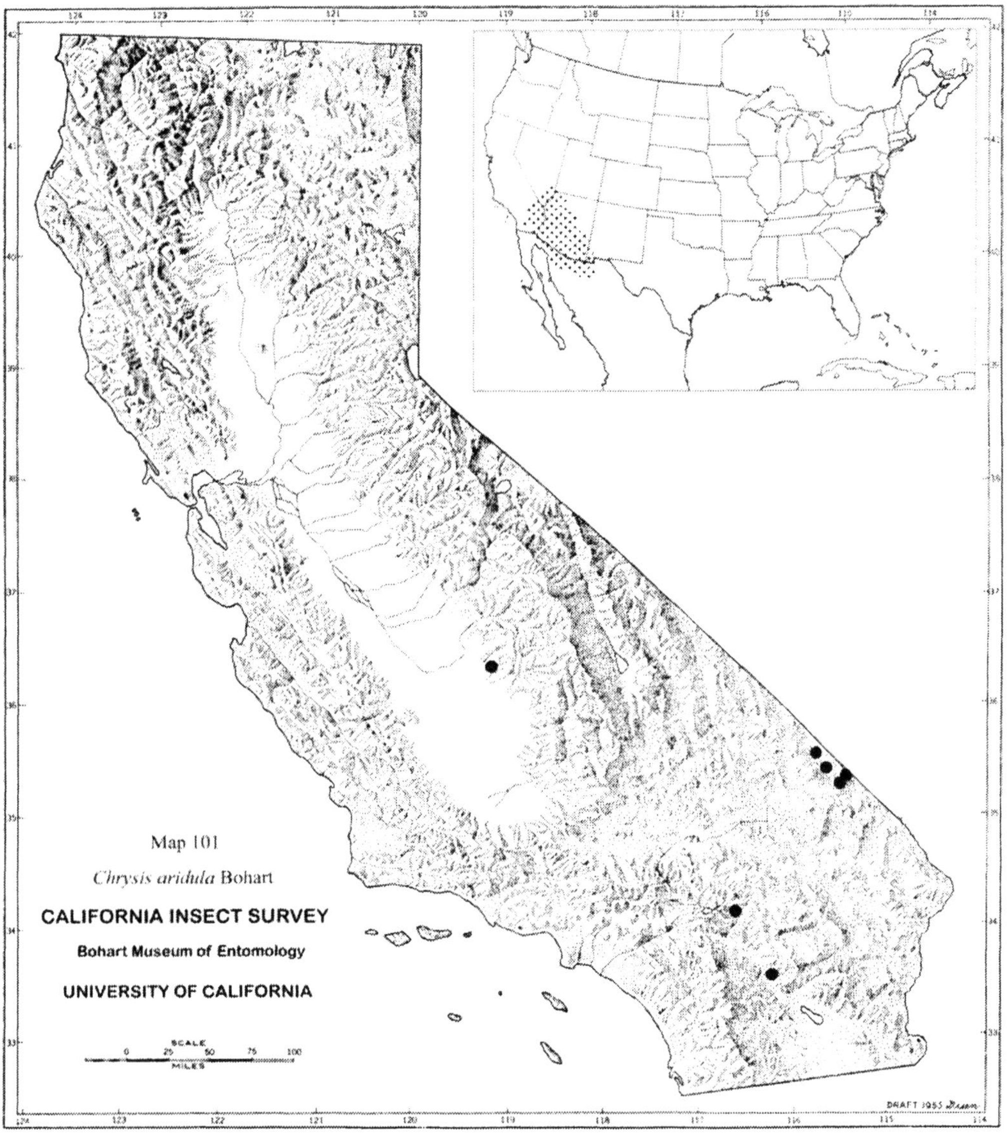

Map 101. California distribution of *Chrysis aridula*. Inset: overall distribution.

Chrysis arizonica Bohart

Fig. 68, 87; Map 102

Chrysis arizonica Bohart 1962:366. Holotype ♂; USA: Arizona, Santa Cruz Co., Ruby (UCDC).

Geographic Range (Map 102). Southeastern California, southwestern Texas, southern New Mexico, Arizona, and northern Mexico (Durango, Sonora and Baja California); 378 specimens were seen.

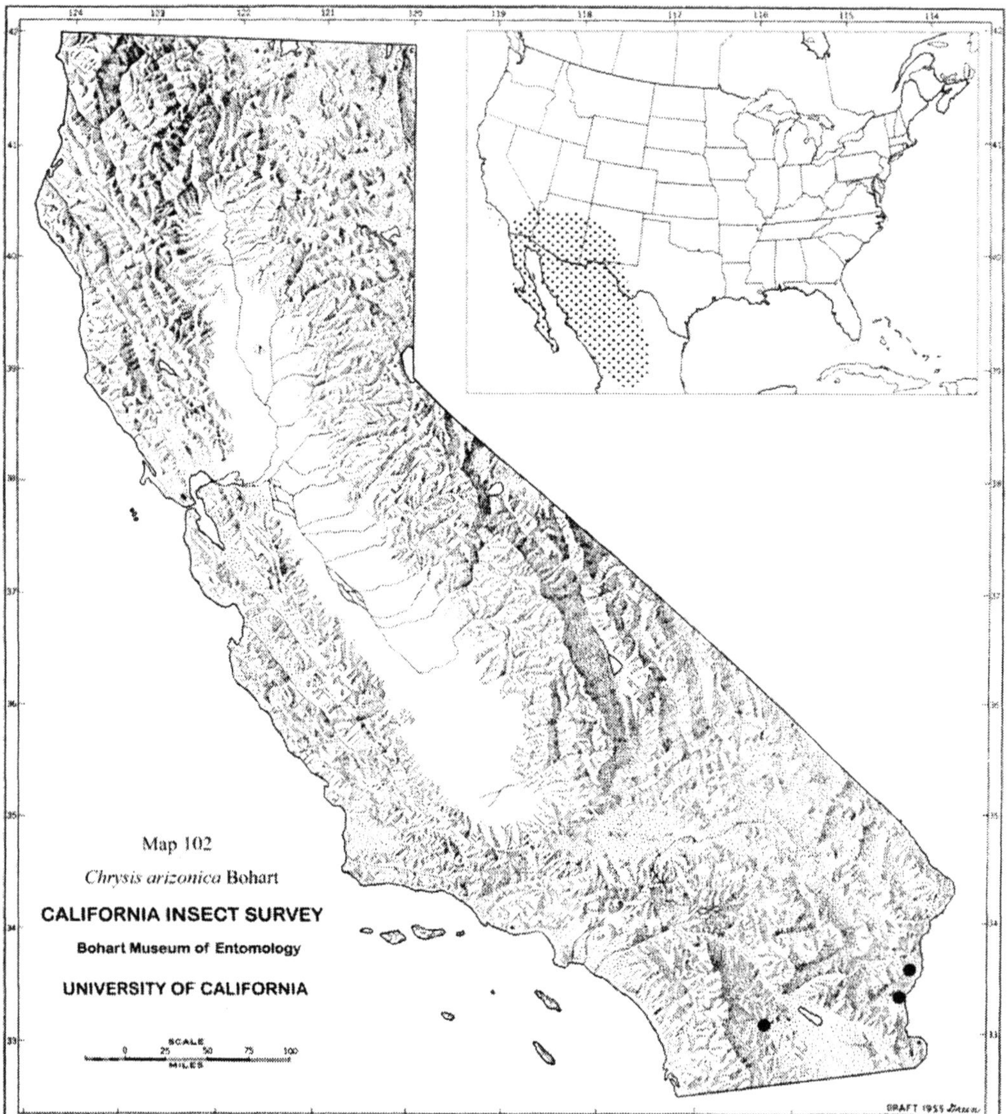

Map 102. California distribution of *Chrysis arizonica*. Inset: overall distribution.

California Records. Collected March and April; **Imperial Co.**: Palo Verde; **Riverside Co.**: 18 mi. w Blythe; **San Diego Co.**: Borrego Valley.

Hosts. Krombein (1967) and F. Parker (pers. comm.) reared this species from *Euodynerus guerrero* (Saussure) and *Parancistrocerus toltecus* (Saussure).

Discussion. This is another six-toothed species. The shape of the teeth are diagnostic for *arizonica*, the medial and submedial teeth are long and prong-like, and the area between is incised up to the pit row. In addition, flagellomere I is usually less than twice as long as broad, sternum II spots are contiguous forming a rectangular black area, the malar space is 1.5 MOD long, and the propodeal tooth has a straight to somewhat convex lower margin.

Chrysis astralia Bohart

Fig. 84, Map 103

Chrysis astralia Bohart 1964:233. Holotype ♂; USA: California, Davis (UCDC).

Geographic Range (Map 103). USA, west of the 100th meridian, Mexico (Baja California, Sonora, Jalisco, Oaxaca) and Costa Rica; in the Lower and Upper Sonoran Zones; 210 specimens were studied.

California Records. Collected June through September; **Contra Costa Co**.: Antioch; **El Dorado Co**.: Pine Hill near Rescue; **Glenn Co**.: Artois; **Kern Co**.: Buttonwillow; 11 mi. n Bakersfield; **Lassen Co**.: Hallelujah Junction; **Madera Co**.: Willow Slough, 26 mi. w Madera; **Marin Co**.: Point Reyes Station; **Monterey Co**.: Arroyo Seco; **Orange Co**.: Trabuco Canyon; **San Bernardino Co**.: 29 Palms; Desert Center; Mojave Narrows Regional Park; **San Joaquin Co**.: Tracy; **Sonoma Co**.: Adobe Rd. near Petaluma; **Tulare Co**.: Wood Lake; **Yolo Co**.: Davis; 2 mi. e Woodland; Elkhorn Ferry; Woodland.

Hosts. Unknown.

Discussion. *Chrysis astralia* is probably most similar to *venusta*. It can be distinguished from *venusta* by the medial carina of the pit row strong and polished, and the malar space less than 1 MOD long.

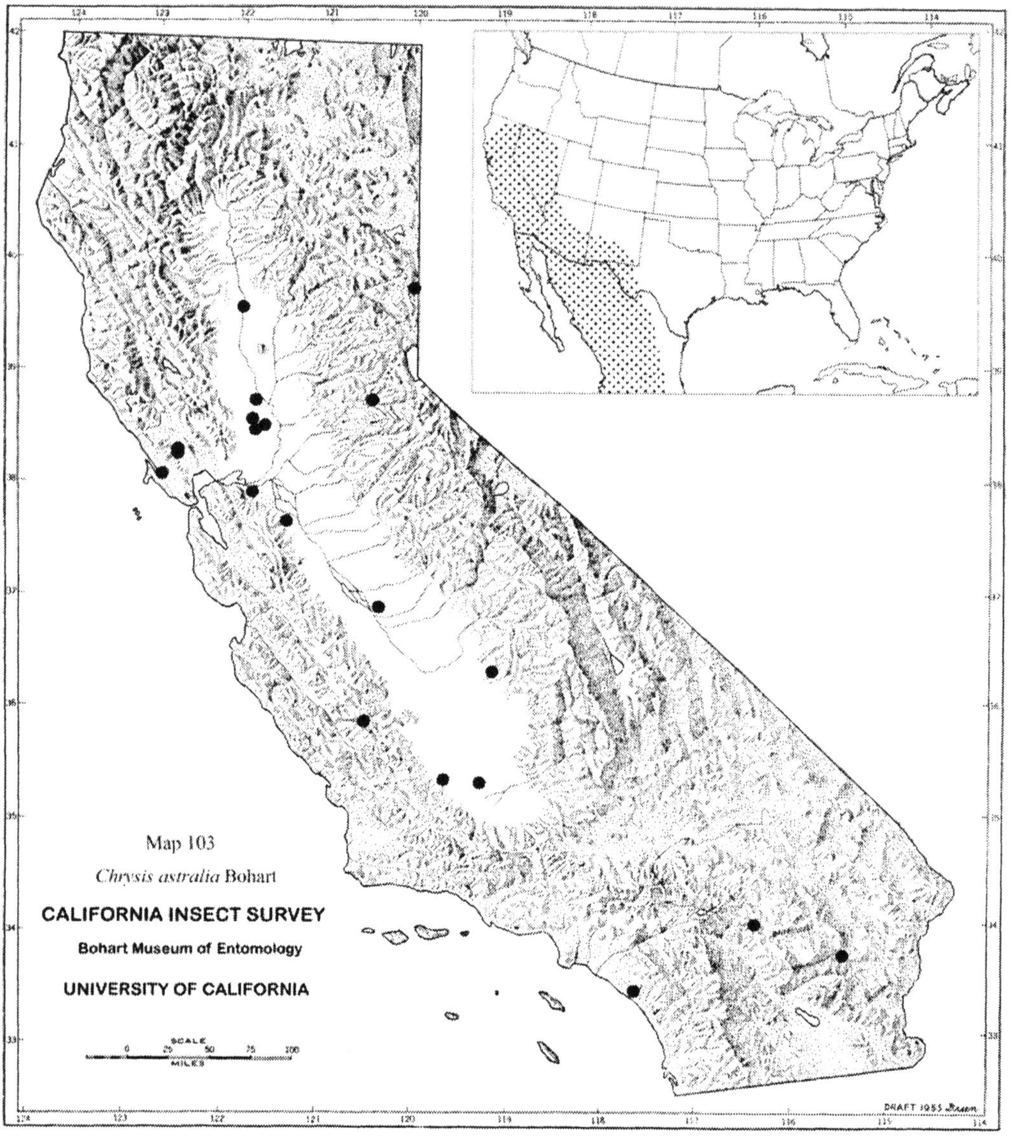

Map 103. California distribution of *Chrysis astralia*. Inset: overall distribution.

Chrysis barri Bohart

Fig. 69, Map 104

Chrysis barri Bohart 1966c:132. Holotype ♂; USA: Nevada, 10 mi. s Wendover (UCDC).

Geographic Range (Map 104). Upper Sonoran Zone from Idaho and Nevada west; 47 specimens were seen.

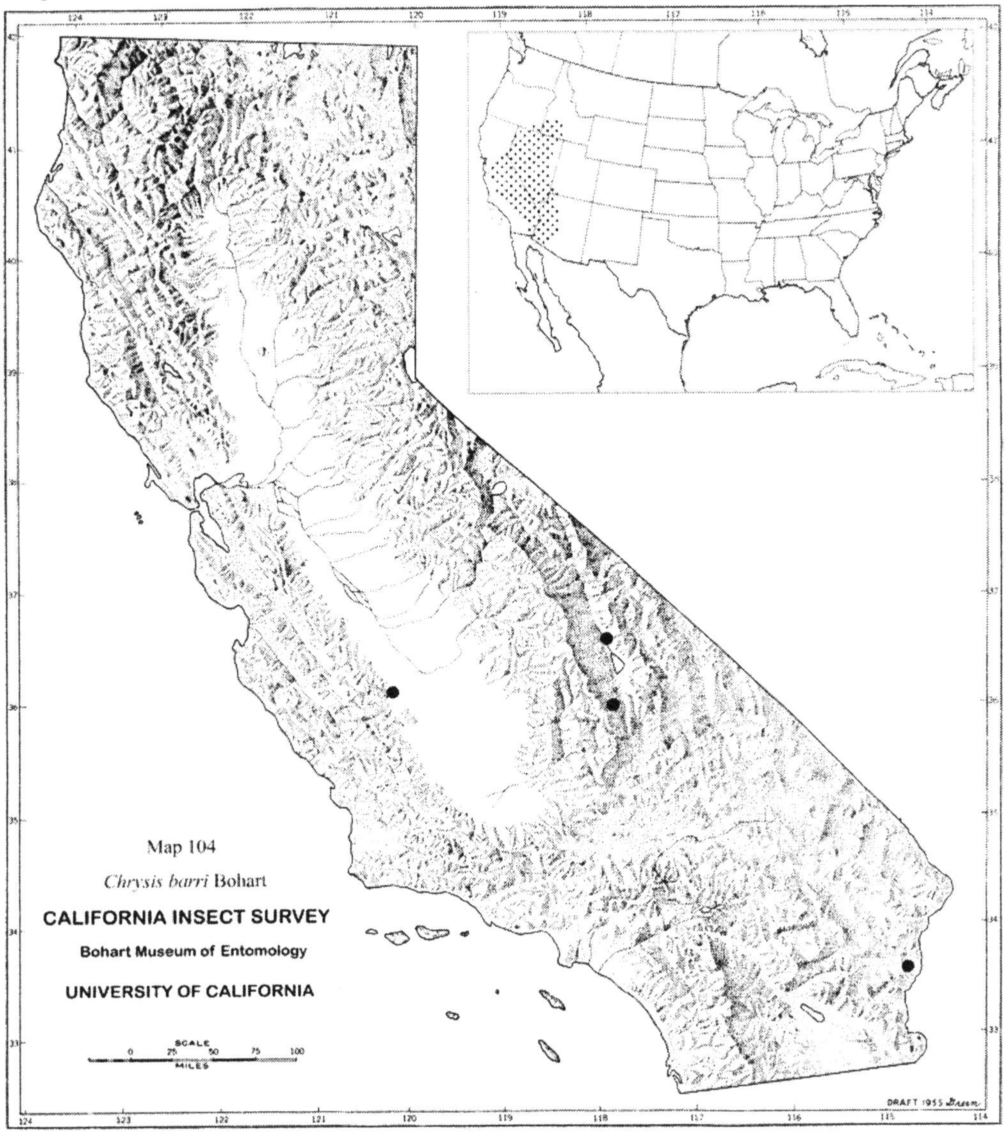

Map 104. California distribution of *Chrysis barri*. Inset: overall distribution.

California Records. Collected May through September; **Fresno Co**.: Jacolitas Canyon; **Inyo Co**.: 3 mi. e Lone Pine; Keeler; Little Lake; **Riverside Co**.: near Blythe; 15 specimens were examined.

Hosts. Parker and Bohart (1966) reared *barri* from nests of *Parancistrocerus toltecus*.

Discussion. Female *barri* in particular are often difficult to separate from those of *amala, lucifera* and *pattoni*. *Chrysis barri* can be distinguished from those species by the sharp apicomedial notch of tergum III, which is deeper than a half circle, the untinted wing membrane, and malar space at least 2.4 MOD long. The body color is often bright purple.

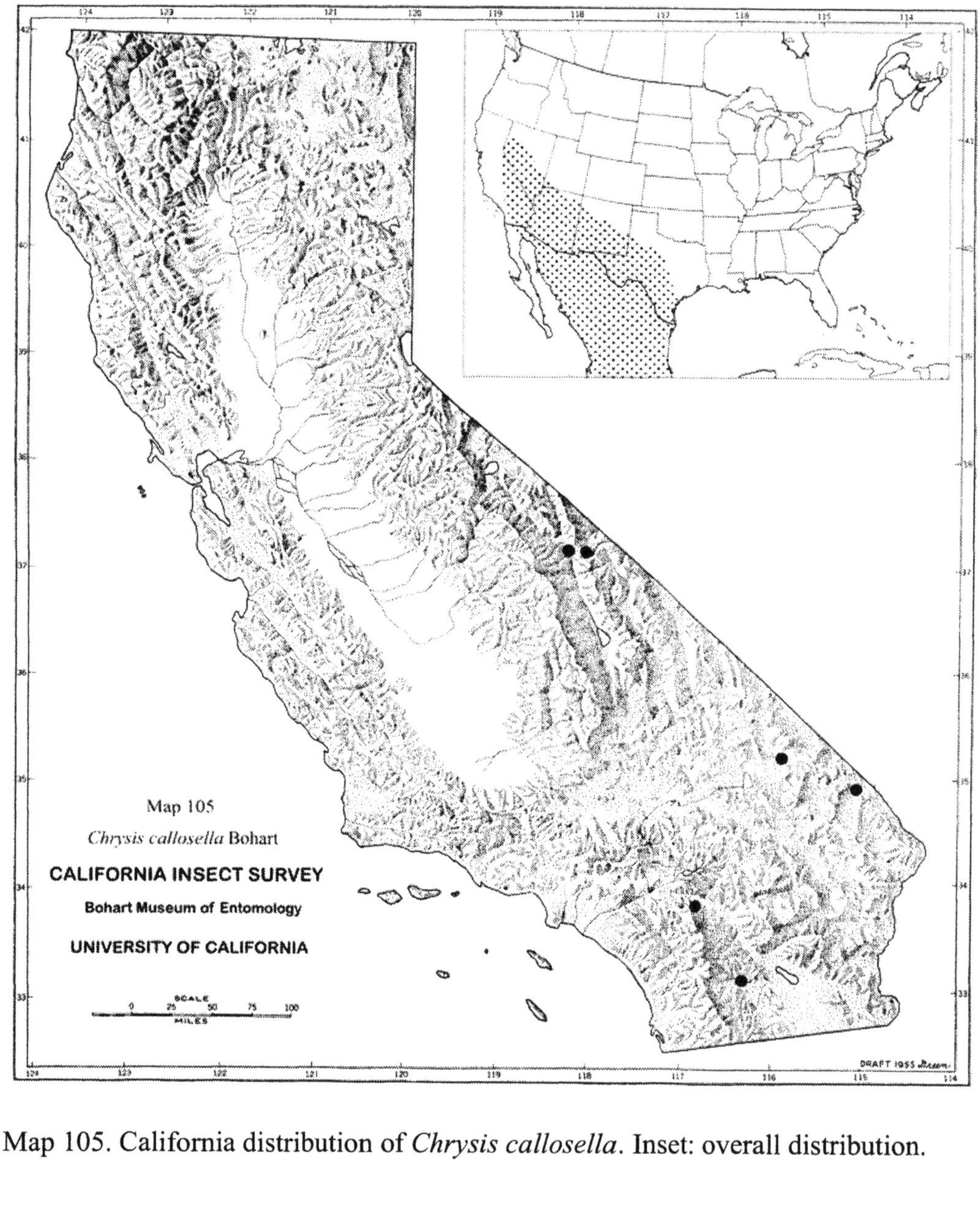

Map 105. California distribution of *Chrysis callosella*. Inset: overall distribution.

Chrysis callosella Bohart

Map 105

Chrysis callosella Bohart 1982:99. Holotype ♂; USA: California, Inyo Co., Antelope Springs (UCDC).

Geographic Range (Map 105). Desert areas west of 100th meridian in the USA and Mexico (Puebla, Chihuahua, Sonora); 47 specimens were examined.

California Records. Collected April through August; **Inyo Co**.: Antelope Springs; Big Pine; **Riverside Co**.: Banning; **San Bernardino Co**.: Granite Mts.; 15 km nne Cima; **San Diego Co**.: Borrego.

Hosts. Unknown.

Discussion. The oblique interocellar carina, when well-developed, easily separates *callosella* from related species. However, it may be weakly developed in some females. This species is generally smaller than *antennalis* and the pit row on tergum III has the medial ridge narrow and often depressed.

Chrysis cembricola Krombein

Map 106

Chrysis cembricola Krombein 1958b:53. Holotype ♀; USA: West Virginia, Lost River State Park (USNM).

Geographic Range (Map 106). Upper Sonoran to Canadian Zones in USA and Canada 247 specimens were studied.

California Records. Collected June and July; **Nevada Co**.: Sagehen Creek; **Sierra Co**.: Independence Lake.

Hosts. Krombein (1958, 1967) reared this species from nests of *Symmorphus canadensis* (Saussure) (Vespidae: Eumeninae).

Discussion. This is a small chrysidid, averaging 6 mm in length. It can be distinguished from other species by the short ocellocular distance, the large black spots on sternum II, the short teeth of tergum III, the apicomedial notch of tergum III unusually shallow in females, sternum III mostly metallic blue or green, and tergum II somewhat polished posteriorly.

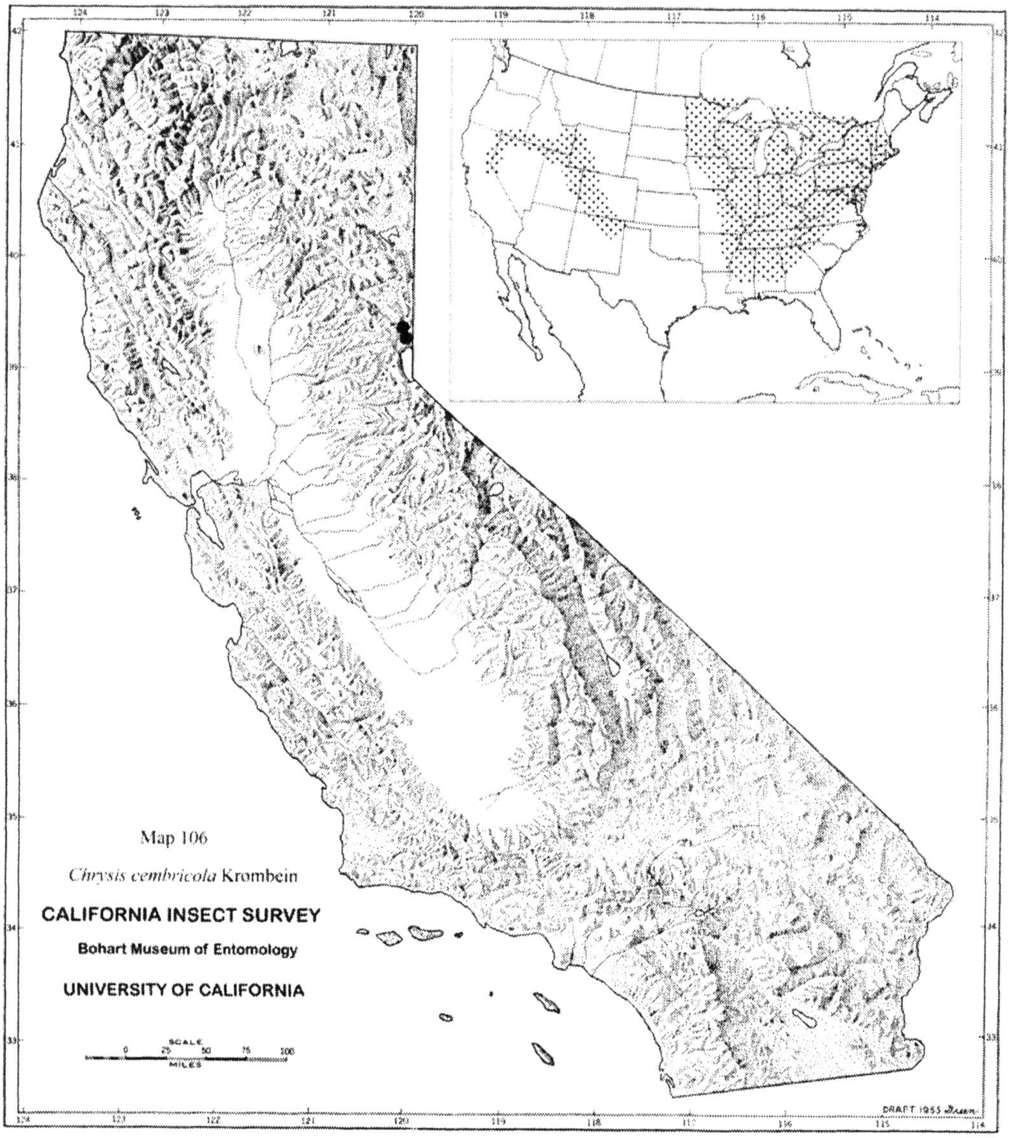

Map 106. California distribution of *Chrysis cembricola*. Inset: overall distribution.

Chrysis cessata du Buysson

Fig. 79, Map 107

Chrysis cessata du Buysson 1891:36. Holotype ♀; "Nat. Bridge, Virginia" (MNHN).

Chrysis chalcopyga Mocsáry 1914:48. Syntype ♀♀; "Colorado" (TMB).

Chrysis nitidula of authors, nec Fabricius 1775.

Geographic Range (Map 107). Transcontinental in both Canada and the U.S., primarily in the Upper Austral and Canadian Life Zones; 672 specimens were seen.

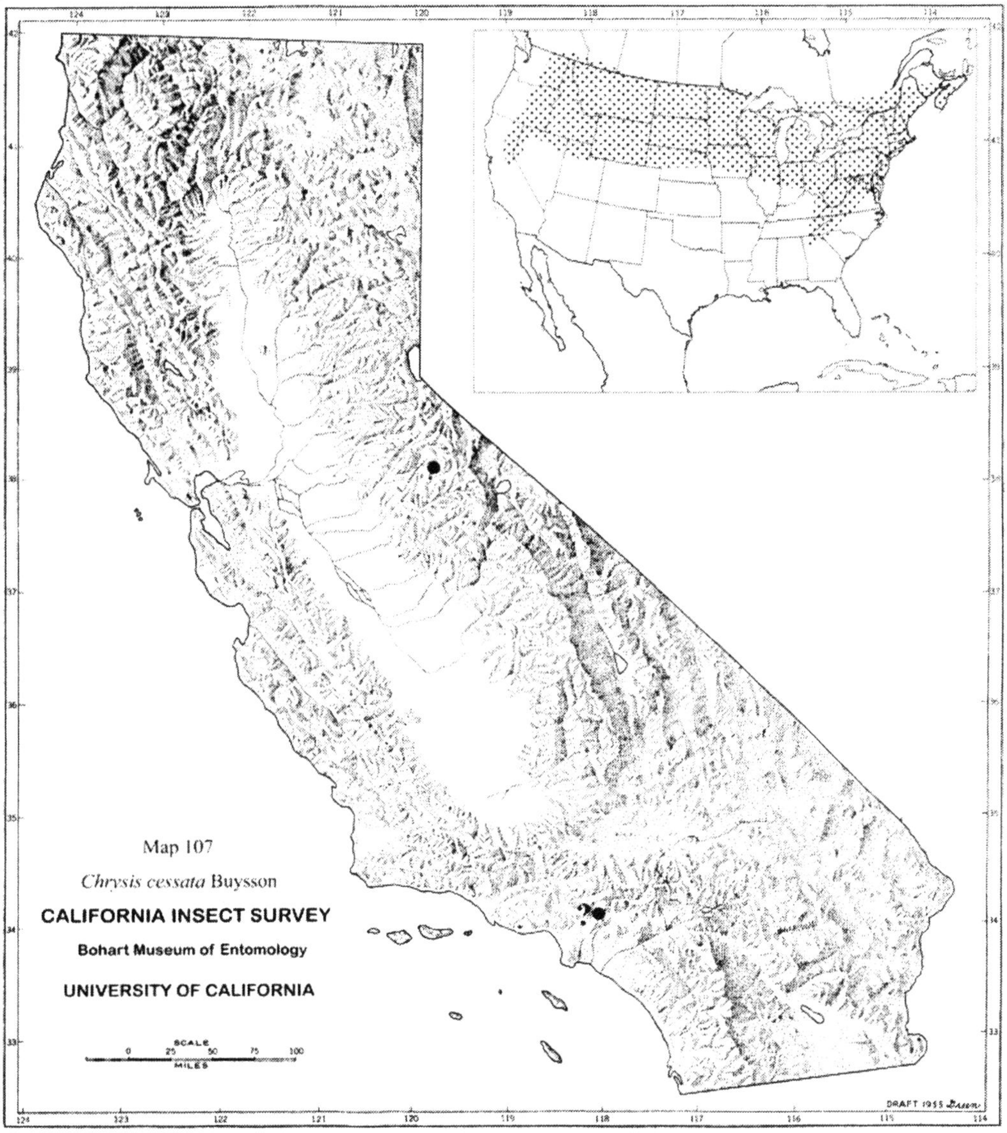

Map 107. California distribution of *Chrysis cessata*. Inset: overall distribution.

California Records. Collected May through July; **Los Angeles Co**.: Glendale; **Tuolumne Co**.: Leland Meadow.

Hosts. *Chrysis cessata* (given as *coerulans*) was recorded by Krombein (1979) as a nest parasite of *Ancistrocerus antilope* (Panzer), *A. catskill* (Saussure), *A. adiabatus*

(Saussure), *Euodynerus foraminatus* (Saussure), *E. schwartzi* (Krombein) and *Symmorphus cristatus* (Saussure) in old mud cells, twig nests, stem borings and wood borings.

Discussion. This species is closely related to *nitidula*. Although widespread, *cessata* is quite uncommon in collections. Confusion over the identity of *cessata* has been confounded by nomenclatural problems. After studying the primary types of *nitidula* Fabricius 1775 and *coerulans* Fabricius 1804 Kimsey and Bohart (1991) discovered that these names had been historically misapplied. The species treated as *coerulans* was a synonym of *nitidula*. The next available name for what had been called *nitidula* was *cessata*. *Chrysis coerulans* of authors became *nitidula* Fabricius.

Chrysis cessata can be distinguished by the bright red edging above the pit row of tergum III. This band may continue basally but does not reach the basolateral corner. In addition, flagellomere I is 2.5-3.0x as long as broad, the transverse frontal carina is well-developed, tergum II has a polished medial ridge, and the apicomedial notch of tergum III is shallower than a half circle.

Chrysis coloradica Bohart

Map 108

Chrysis pulcherrima Cresson 1865b:311. Holotype ♂; USA: Colorado (ANSP). Nec Lepeletier 1806.

Chrysis coloradica Bohart 1964:224. Replacement name for *pulcherrima* Cresson 1865.

Geographic Range (Map 108). Upper Sonoran and Canadian Zones of the western USA; 343 specimens were studied.

California Records. Collected April through July; **Alameda Co**.: Tilden Park; **Contra Costa Co**.: Mt. Diablo; **El Dorado Co**.: Kelsey; 8 mi. w Icehouse Rd.; **Kern Co**.: Mill Potrero; **Lake Co**.: Hopland Grade; **Los Angeles Co**.: La Crescenta; 11 mi. ne Azuza; Tanbark Flat; Newhall; Claremont; **Mendocino Co**.: 10 mi. e Capella; **Mono Co**.: Blanco's Corral, White Mts.; Wyman Canyon, White Mts.; **Monterey Co**.: Arroyo Seco; Los Padres Dam, Carmel Valley; **Napa Co**.: Samuel Springs (beneath Lake Berryessa); Mint Canyon; **Nevada Co**.: Sagehen Creek; **Riverside Co**.: 8 mi. w Beaumont; **San Bernardino Co**.: Cajon Junction; Oak Glen; **Santa Barbara Co**.: Santa Ynez Mts.; Santa Cruz Island; 4 mi. e Los Prietos; **Santa Clara Co**.: Mt. Hamilton; **Sierra Co**.: Sardine Lakes.; Independence Lake; **Tuolumne Co**.: Dodge Ridge; **Ventura Co**.: Sespe Canyon; Fillmore; Lockwood Creek near Stauffer Post Office; **Yolo Co**.: Putah Canyon.

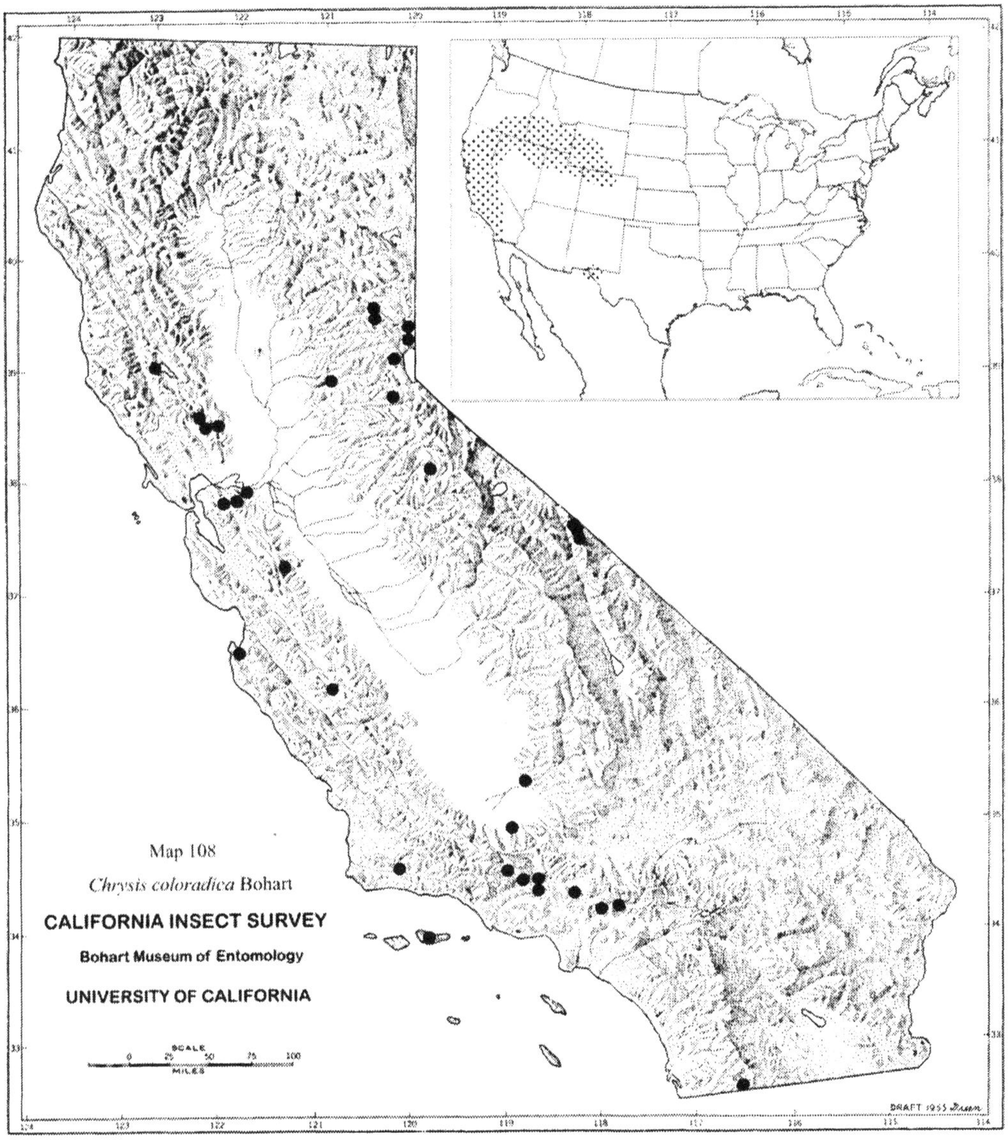

Map 108. California distribution of *Chrysis coloradica*. Inset: overall distribution.

Hosts. Bohart (1964) recorded rearing *coloradica* from ground nests of *Anthidium collectum* (Huard) (Megachilidae).

Discussion. The most distinctive feature of *coloradica* is the setal fringe on the flagellum in males, and the densely setose and unusually narrow lower frons. This species most closely resembles *xerophila*, but is more coarsely punctate overall and darker in color than *xerophila*.

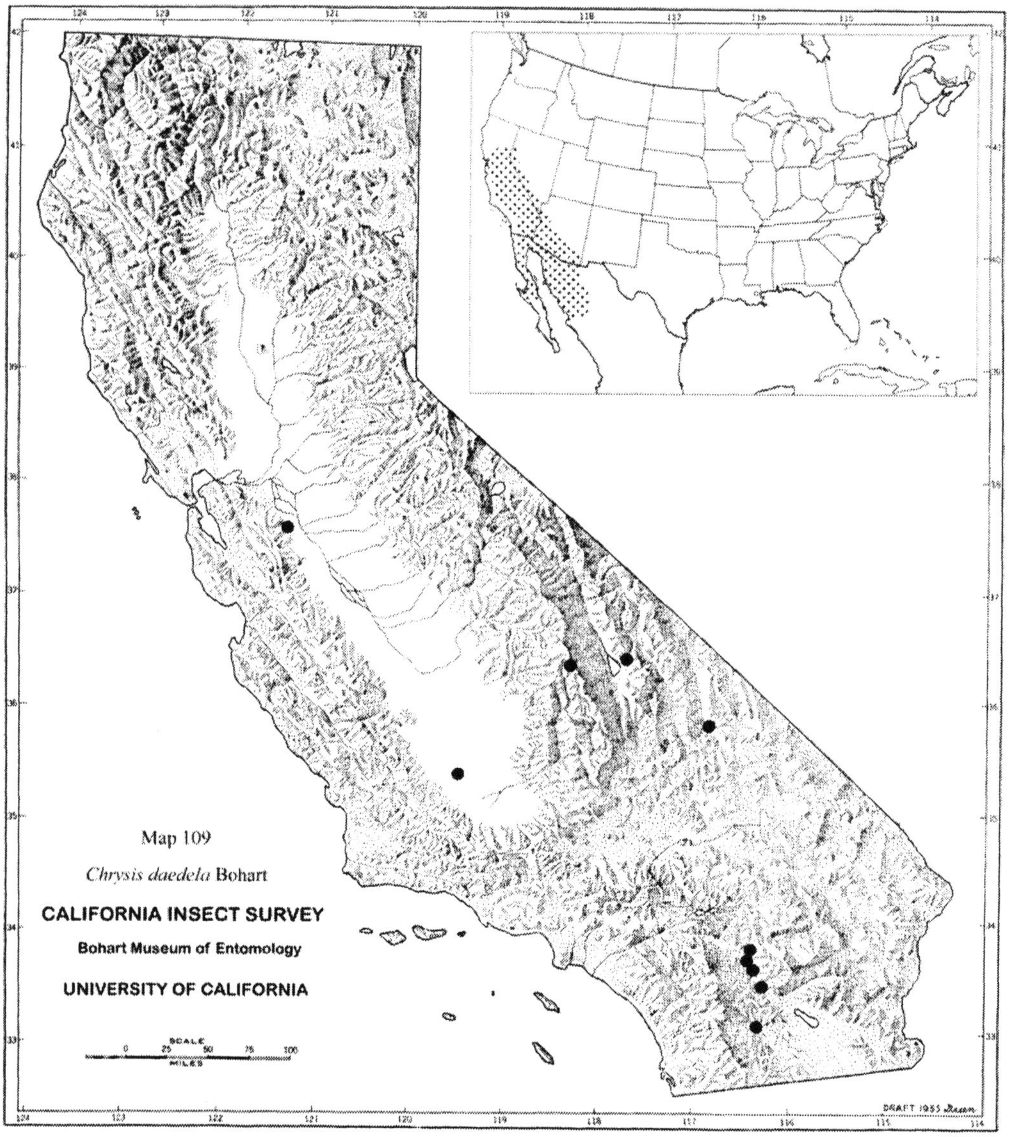

Map 109. California distribution of *Chrysis daedala*. Inset: overall distribution.

Chrysis daedala Bohart

Map 109

Chrysis daedala Bohart, in Bohart and Kimsey 1982:119. Holotype ♂; USA: California, Riverside Co., Coachella (UCDC).

Geographic Range (Map 109). Upper Sonoran Zones of the western USA (AZ, CA, NV); 18 specimens were seen.

California Records. Collected April through July; **Inyo Co**.: 12 mi. e Keeler; 13 mi. s Death Valley Junction; **Kern Co**.: 13 mi. s Shafter; **Riverside Co**.: Coachella; Thousand Palms; 1 mi. n Mecca; Willis Palms Oasis; **San Diego Co**.: Borrego; **San Joaquin Co**.: Tracy; **Tulare Co**.: Ash Mt.

Hosts. F. Parker reared this species from nests of the eumenine vespid *Leptochilus tosquineti* (Cameron) (Bohart and Kimsey 1982).

Discussion. An unusual feature of *daedala* is the presence of polished tubercles on the vertex behind the hindocelli or dorsal eye margin. The tubercles may be obsolescent in some females. This species is most similar to *charigaster* Bohart and *oreadis* Bohart, neither of which occur in California.

Chrysis derivata du Buysson

Map 110

Chrysis derivata du Buysson 1891:38. Holotype ♂; Canada: Quebec (MNHN).

Chrysis decepta Rohwer 1909:91. Holotype ♀; Colorado: Boulder (USNM).

Geographic Range (Map 110). Widespread in the USA and southern Canada; 453 specimens were studied.

California Records. Collected March through November; **Colusa Co**.: 10 mi. sw Stonyford; **Contra Costa Co**.: Antioch; Mt. Diablo; **El Dorado Co**.: Icehouse Rd.; Kelsey; **Fresno Co**.: Watts Valley; **Inyo Co**.: 2 mi. e Lone Pine; **Kern Co**.: 6 mi. w Frazier Park; **Lake Co**.: Hopland Grade; **Lassen Co**.: Hallelujah Junction. Lassen National Park; Standish; **Los Angeles Co**.: Camp Baldy; Claremont; mountains near Claremont; San Gabriel Canyon; Tanbark Flat; **Marin Co**.: Novato; **Mono Co**.: 1 mi. s Toms Place; 11 mi. n Bridgeport; **Monterey Co**.: Arroyo Seco; **Nevada Co**.: Boca; Sagehen Creek; Hobart Mills; **Placer Co**.: Folsom Lake; **Plumas Co**.: Blairsden; Johnsville; Vinton; **San Benito Co**.: 6 mi. s Idria; **San Bernardino Co**.: Mountain Home; **San Diego Co**.: Culp Canyon, Mt. Laguna; **San Luis Obispo Co**.: 10 mi. se Creston; Creston; **Santa Barbara Co**.: Santa Cruz Island; Santa Ynez Mts.; **Sierra Co**.: Kyburz Flat; **Siskiyou Co**.: Valentine Caves Lava Bed National Monument; **Sonoma Co**.: Petaluma; **Sutter Co**.: Sutter Buttes; **Trinity Co**.: Carrville; **Tulare Co**.: Wood Lake; **Tuolumne Co**.: Sequoia National Park Potwisha; Strawberry; **Yolo Co**.: 5 mi. e Guinda; Davis.

Hosts. Hosts include the eumenine vespids *Ancistrocerus catskill halophilus* Viereck, *Euodynerus foraminatus apopkensis* (Robertson), *E. f. scutellaris* (Saussure) and *Symmorphus cristatus nevadensis* (Cameron) (Parker and Bohart 1966, Krombein 1967).

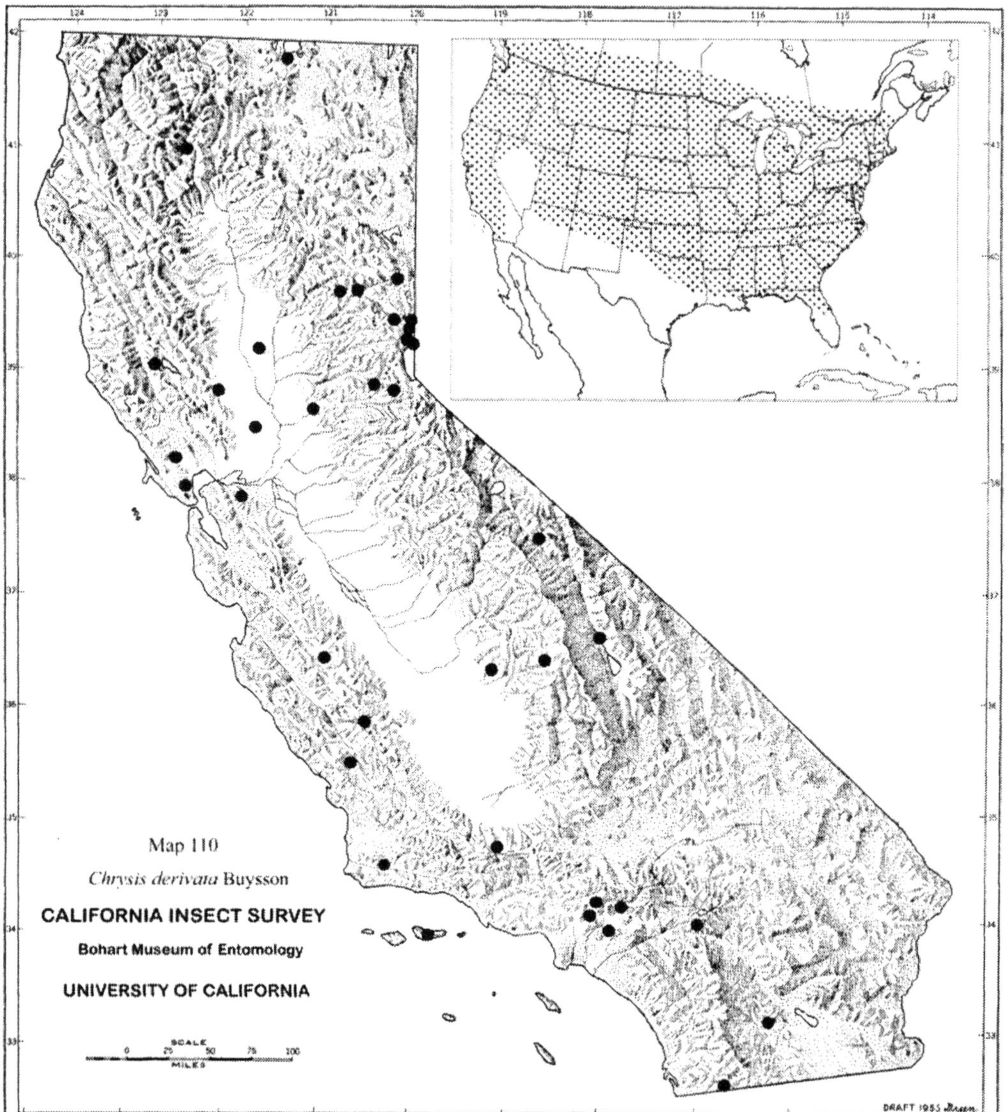

Map 110. California distribution of *Chrysis derivata*. Inset: overall distribution.

Discussion. This commonly collected species most closely resembles *Chrysis irwini*. However, the females are larger bodied (7-8 mm long), with a proportionally longer thorax (more than twice as long as broad measured by the width across the anterolateral pronotal corners). Males have tergum III markedly concave laterally, flagellomere I as broad as long (longer in *irwini*). *Chrysis irwini* is smaller than *derivata* (5-6 mm long) and the female thoracic dorsum is about as long as broad.

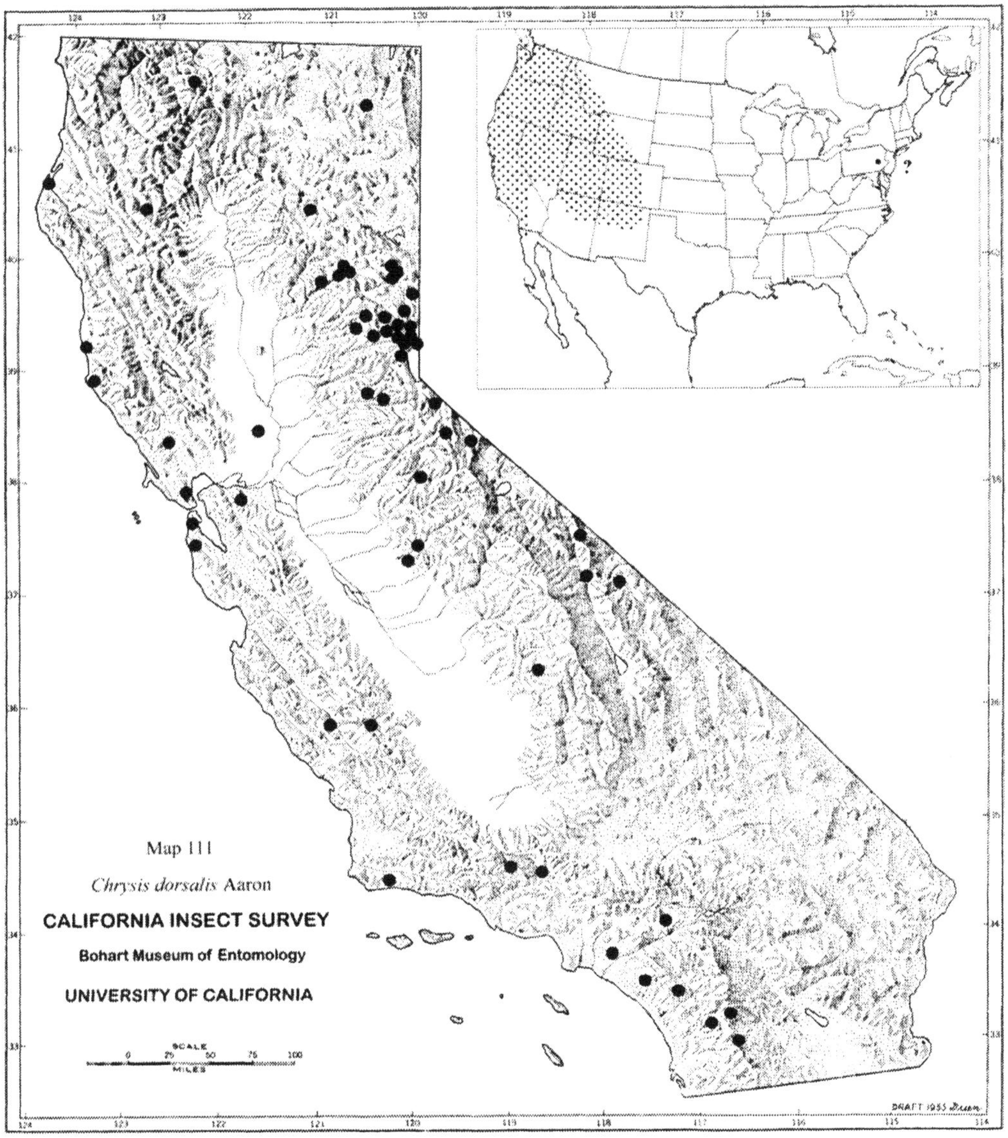

Map 111. California distribution of *Chrysis dorsalis*. Inset: overall distribution.

Chrysis dorsalis **Aaron**

Map 111

Chrysis dorsalis Aaron 1885:234. Lectotype ♀ (designated by Cresson 1928); USA: Montana (ANSP).

Chrysis alfkenella du Buysson 1904:266. Holotype ♂; USA: California, Berkeley (MNHN).

Geographic Range (Map 111). Transcontinental in the USA and Canada; 982 specimens were seen.

California Records. Collected April through September; **Alpine Co**.: Carson Pass; Hope Valley; Shot Rock; Silver Creek; **Contra Costa Co**.: Mt. Diablo; **El Dorado Co**.: 8 mi. n on Ice House Rd.; Bijon; Blodgett Forest 13 mi. e Georgetown; **Humboldt Co**.: Samoa Peninsula; **Inyo Co**.: Antelope Springs; Big Pine Creek; **Lassen Co**.: Hallelujah Junction; Mt. Lassen National Park; **Los Angeles Co**.: Tanbark Flat; **Marin Co**.: Nicasio; **Mariposa Co**.: 1 mi. e Junction 120 on Hwy 120; Mariposa; **Mendocino Co**.: 2 mi. n Point Arena; Mendocino; **Modoc Co**.: 24 mi. sw Alturas; **Mono Co**.: 11 mi. n Bridgeport; White Mts.; **Monterey Co**.: Arroyo Seco; Jolon; **Nevada Co**.: 17 mi. n Truckee; Boca; Fuller Lake; Hobart Mills; Sagehen Creek; **Orange Co**.: Brea; Peters Canyon; **Placer Co**.: Carnelian Bay, Lake Tahoe; **Plumas Co**.: 4 mi. w Quincy; Beckwourth; Buck's Lake; Greagle; Johnsville; Little Long Valley Creek; 6 mi. e Spring Garden; **Riverside Co**.: 1 mi. n Murrieta; **San Bernardino Co**.: Cajon Junction; **San Diego Co**.: 2 mi. n Warner Springs; 5 mi. e Alpine; Julian; **San Francisco Co**.: Daly City; **San Mateo Co**.: San Bruno Mts. 2 mi. se Daly City; **Santa Barbara Co**.: Goleta; **Sierra Co**.: Gold Lake; Independence Lake; Sardine Lakes; Sierra Valley; Sierraville; Yuba Pass; Kyburz Flat; **Siskiyou Co**.: Swallows; 13 mi. n Yreka; **Sonoma Co**.: Petaluma; **Trinity Co**.: Coffee Creek Ranger Station; **Tulare Co**.: Sequoia National Park Ash Mt.; **Tuolumne Co**.: Strawberry; **Ventura Co**.: Sespe Canyon; **Yolo Co**.: Davis.

Hosts. This species was recorded from a cocoon of *Photopsis orestes* (W. Fox) (Mutillidae), within a cocoon of *Anthidium collectum* Huard (Megachilidae) by Ferguson (1962).

Discussion. Most closely related to *knowltoni* and *paradisica, dorsalis* can be distinguished by the unusually narrow male frons and the medially impunctate female frons.

Chrysis florissanticola Rohwer

Map 112

Chrysis florissanticola Rohwer 1909:88. Holotype ♂; USA: Colorado, Florissant (USNM).

Chrysis florissantensis Bohart 1962:362, in error.

Geographic Range (Map 112). Canadian and Hudsonian Zones in the USA and Canada west of 100th meridian, including Alaska and northern Mexico (Chihuahua); 82 specimens were studied.

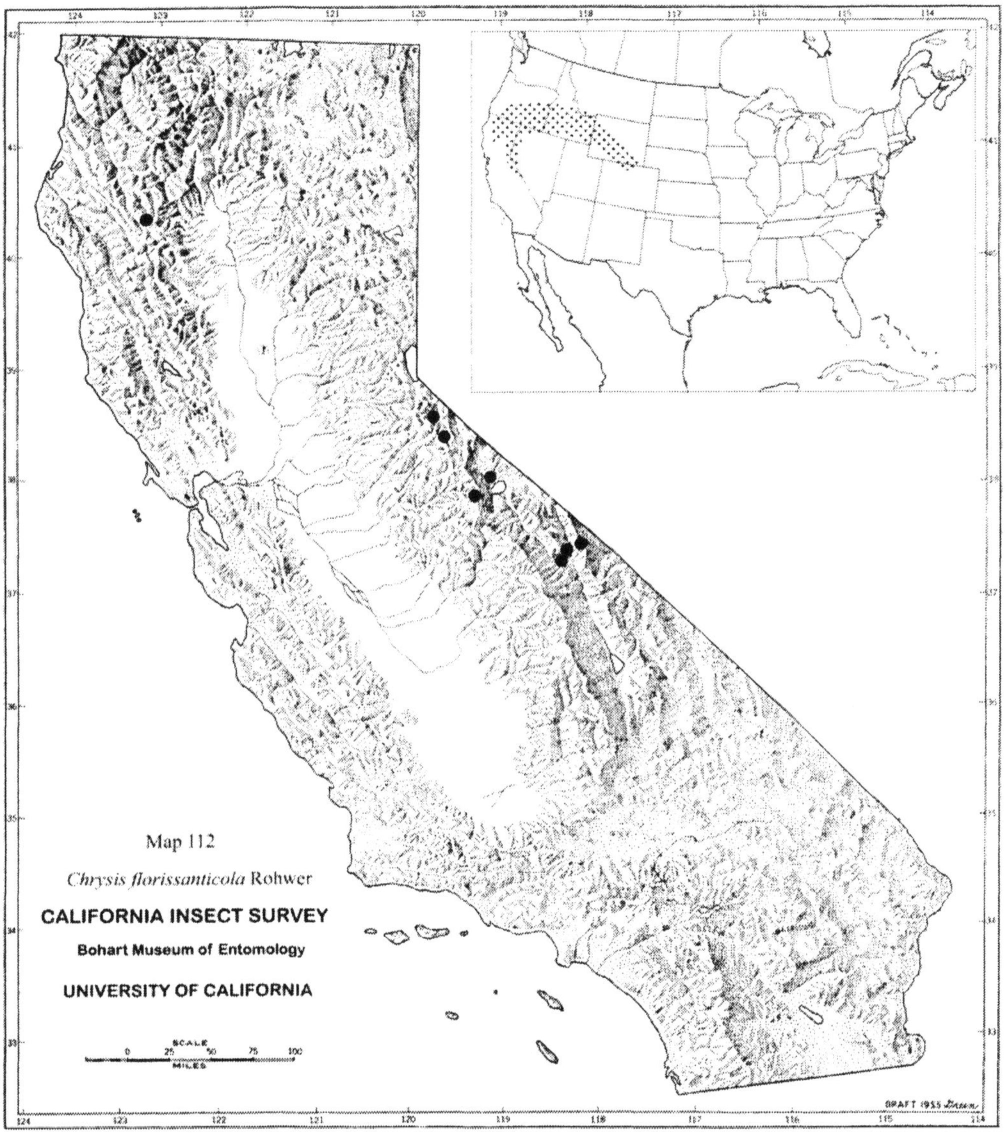

Map 112. California distribution of *Chrysis florissanticola*. Inset: overall distribution.

California Records. Collected June through August; **Alpine Co.**: Carson Pass; Ebbetts Pass; **Inyo Co.**: Mammoth Lake; Mono Pass; Ruby Lake; **Mono Co.**: Blanco's Corral, White Mts.; **Trinity Co.**: Carrville; **Tuolumne Co.**: Sonora Pass.

Hosts. Krombein (1979) reared this species from nests of *Anthidium banningense* Cockerell (Megachilidae).

Discussion. *Chrysis florissanticola* most closely resembles *coloradica*, but can be distinguished from that species by the male flagellum lacking a setal fringe, tergum II shinier due to polished interspaces between the punctures, and the pit row of tergum III weak and almost obsolescent. Specimens of *florissanticola* tend to be strongly green colored, with coppery highlights.

Chrysis inaequidens Dahlbom

Fig. 85, Map 113

Chrysis inaequidens Dahlbom 1854:334. Holotype ♂; USA: New York (MZLU).

Chrysis texana Gribodo 1879:329. Lectotype ♂ (designated by Bohart 1962); USA: Texas (MCSN).

Geographic Range (Map 113). Throughout most of the USA, and southern Canada, as well as Mexico (Baja California, Chihuahua, Chiapas, Nuevo Leon, Puebla); 229 specimens were seen.

California Records. Collected March through July; **Amador Co**.: Volcano; **Fresno Co**.: Watts Valley; **Inyo Co**.: Death Valley, Furnace Creek; **Lake Co**.: north fork Cache Creek Hwy 20; **Plumas Co**.: 1 mi. w Johnsville; **Shasta Co**.: Anderson; **Tulare Co**.: Sequoia National Park Ash Mt.; **Tuolumne Co**.: Dodge Ridge; Pine Crest; Strawberry; **Yolo Co**.: Davis.

Hosts. Hosts include the vespids *Euodynerus foraminatus scutellaris* Saussure (Parker 1962, Moore and Parker 1962), and *E. megaera* (Lepeletier), *E. foraminatus apopkensis* (Robertson), *Monobia quadridens* (Linnaeus), *Pachodynerus erynnis* (Lepeletier), *Stenodynerus pulvinatus surrufus* Krombein, *Parancistrocerus salcularis rufulus* (Bohart) (Krombein 1958, 1967).

Discussion. This is the most widespread of the six-toothed species of *Chrysis*. *C. inaequidens* can be distinguished by the strongly lobate propodeal tooth, malar space subequal in length to the subantennal distance, mesopleuron without polished knobs, and the scapal basin completely densely punctate.

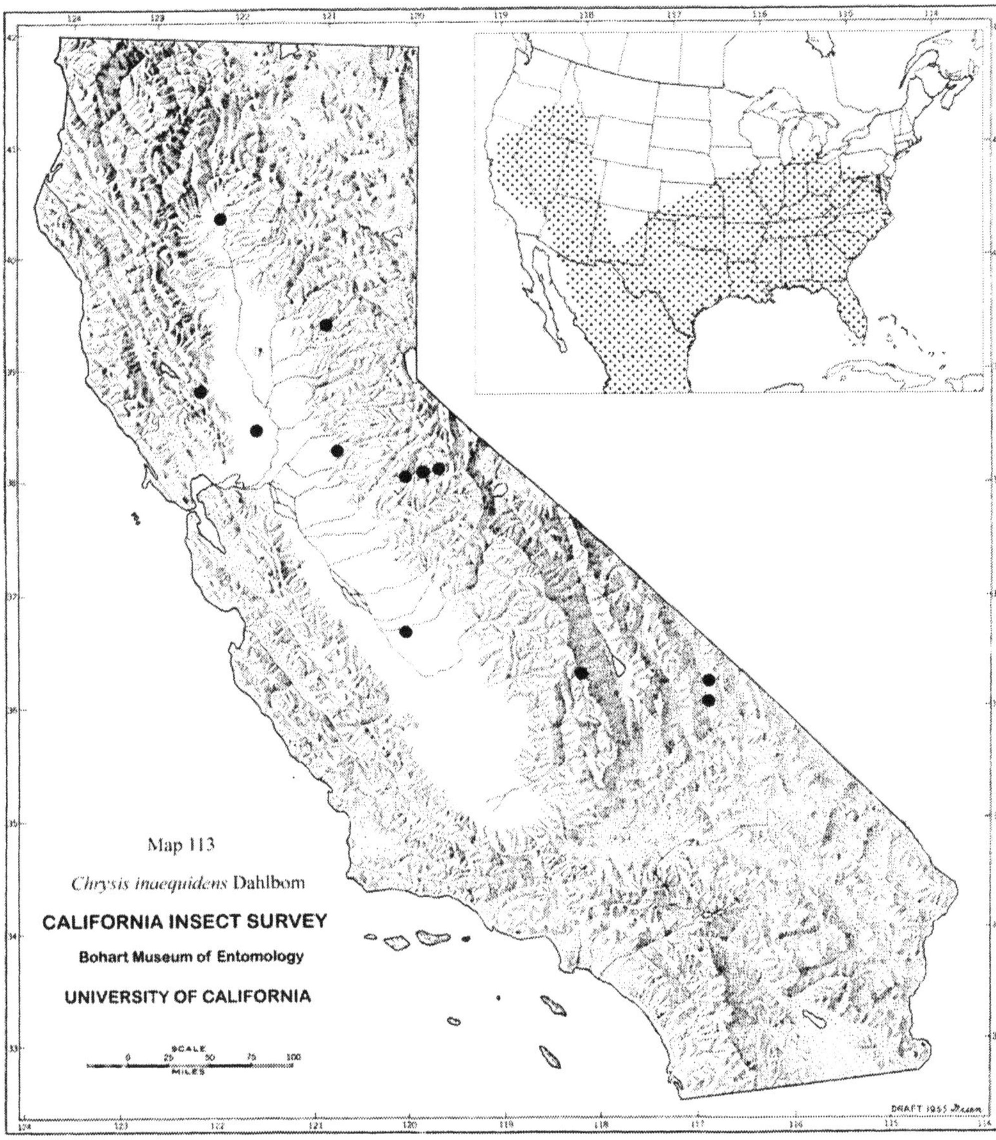

Map 113. California distribution of *Chrysis inaequidens*. Inset: overall distribution.

Chrysis inflata Aaron

Fig. 70, Map 114

Chrysis inflata Aaron 1885:237. Lectotype ♂ (designated by Cresson 1928); USA: California ("So. Cal.") (ANSP).

Chrysis nokomis Rohwer 1909:89. Lectotype ♂ (designated by Bohart, in Bohart and Kimsey 1982); USA: Colorado, Denver (USNM).

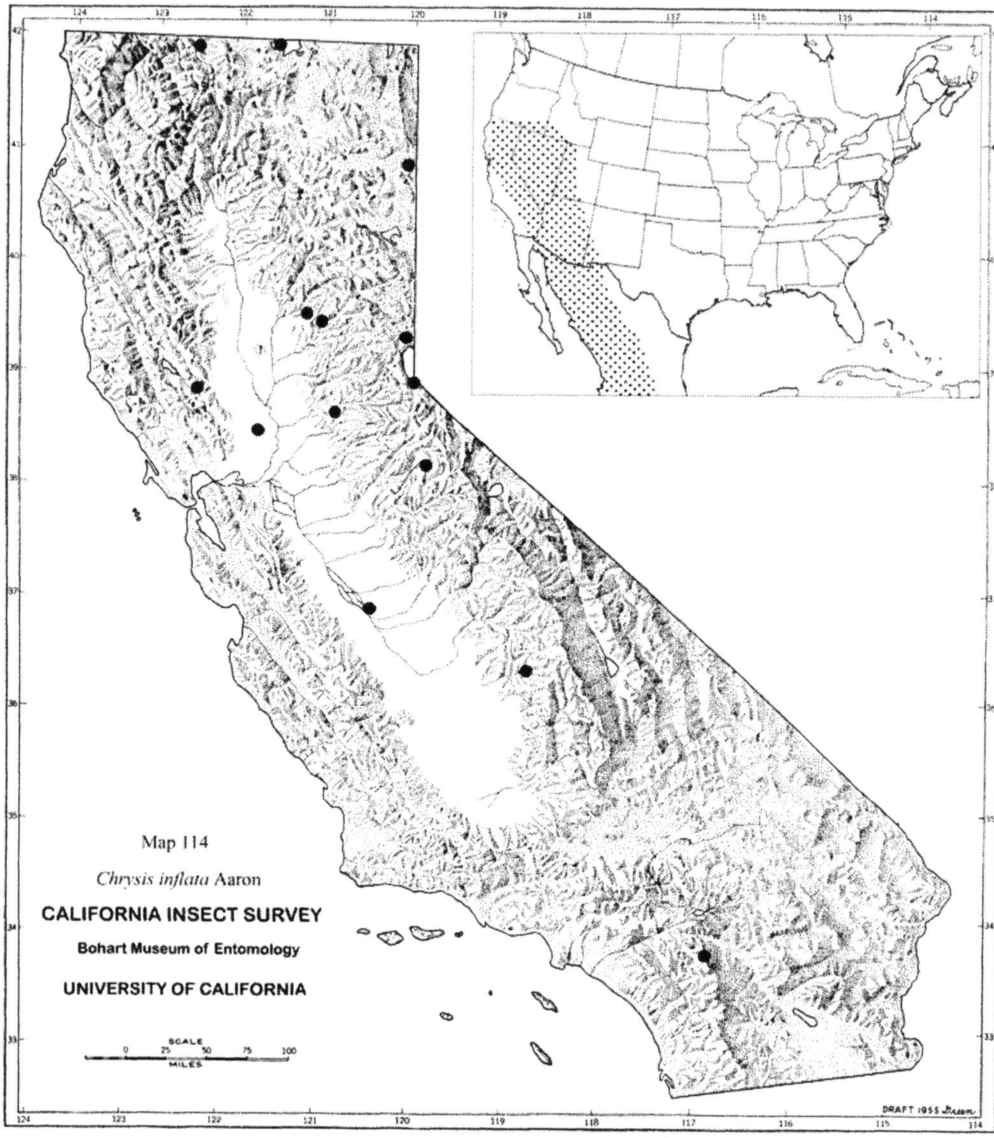

Map 114. California distribution of *Chrysis inflata*. Inset: overall distribution.

Geographic Range (Map 114). Widespread in the USA west of 100th meridian, including southern Canada (British Columbia) and northern Mexico (Baja California, Chihuahua, Coahuila, Durango, Michoacan, Nueva Leon, Zacatecas); 263 specimens were studied.

California Records. Collected March through August; **Butte Co**.: Feather Falls; **El Dorado Co**.: Lake Tahoe; Pollock Pines; **Lake Co**.: north fork Cache Creek Hwy 20; **Lassen Co**.: Madeline Plains; **Los Angeles Co**.: Camp Baldy; **Madera Co**.: San Joaquin; **Nevada Co**.: Sagehen Creek; **Plumas Co**.: Johnsville; **Riverside Co**.: San Jacinto Mts.; **Santa Barbara Co**.: Santa Cruz Island; **Siskiyou Co**.: Klamath River 8 mi. n Yreka; Tule Lake; **Tulare Co**.: Ash Mt.; **Tuolumne Co**.: Strawberry; **Yolo Co**.: Davis; **Yuba Co**.: Camptonville.

Hosts. Krombein (1979) recorded the vespids *Ancistrocerus durangoensis* Cameron, *A. lineativentris* Cameron, *A. tuberculcephalus* (Saussure) and *Euodynerus guerrero* (Saussure) as hosts. Bohart and Kimsey (1982) also listed *Parancistrocerus* sp. (Vespidae) as a host.

Discussion. An unusual feature of *inflata* is the shape of tergum III, which is strongly inflated laterally. The sides of tergum III are strongly convex above the pit row, and the bulge is strongly accentuated by the closely space apical teeth. Other diagnostic features of this species are tergum II with a polished medial longitudinal ridge, malar space longer than usual (1.8-2.1 MOD), a jutting transverse frontal carina, and the mesopleuron with a strong posterior carina.

Chrysis irwini Bohart

Map 115

Chrysis irwini Bohart 1966c:133. Holotype ♂; USA: California, Yolo Co., Davis (UCDC).

Geographic Range (Map 115). Primarily found in the Upper Sonoran to Canadian Zones of California, Oregon and Utah; 51 specimens were studied.

California Records. Collected February through August; **El Dorado Co**.: Meyers; **Kings Co**.: Hanford; **Nevada Co**.: Boca; **San Diego Co**.: Borrego; **Santa Clara Co**.: San Antonio Valley; **Sutter Co**.: Live Oak; **Tuolumne Co**.: Strawberry; **Yolo Co**.: Davis; Winters; Knights Landing.

Hosts. No host species is known, but Bohart (1966) reared *irwini* from an unidentified twig-nesting eumenine vespid.

Discussion. This is an uncommonly collected species, which closely resembles *derivata*. Distinctions between the two species are discussed under *derivata*. The male flagellomere I of *irwini* is somewhat longer than broad and the female thoracic dorsum is about twice as long as broad at the anterolateral pronotal corners. *Chrysis irwini* is smaller bodied than *derivata*, 5-6 mm long, as opposed to 7-8mm long in *derivata*.

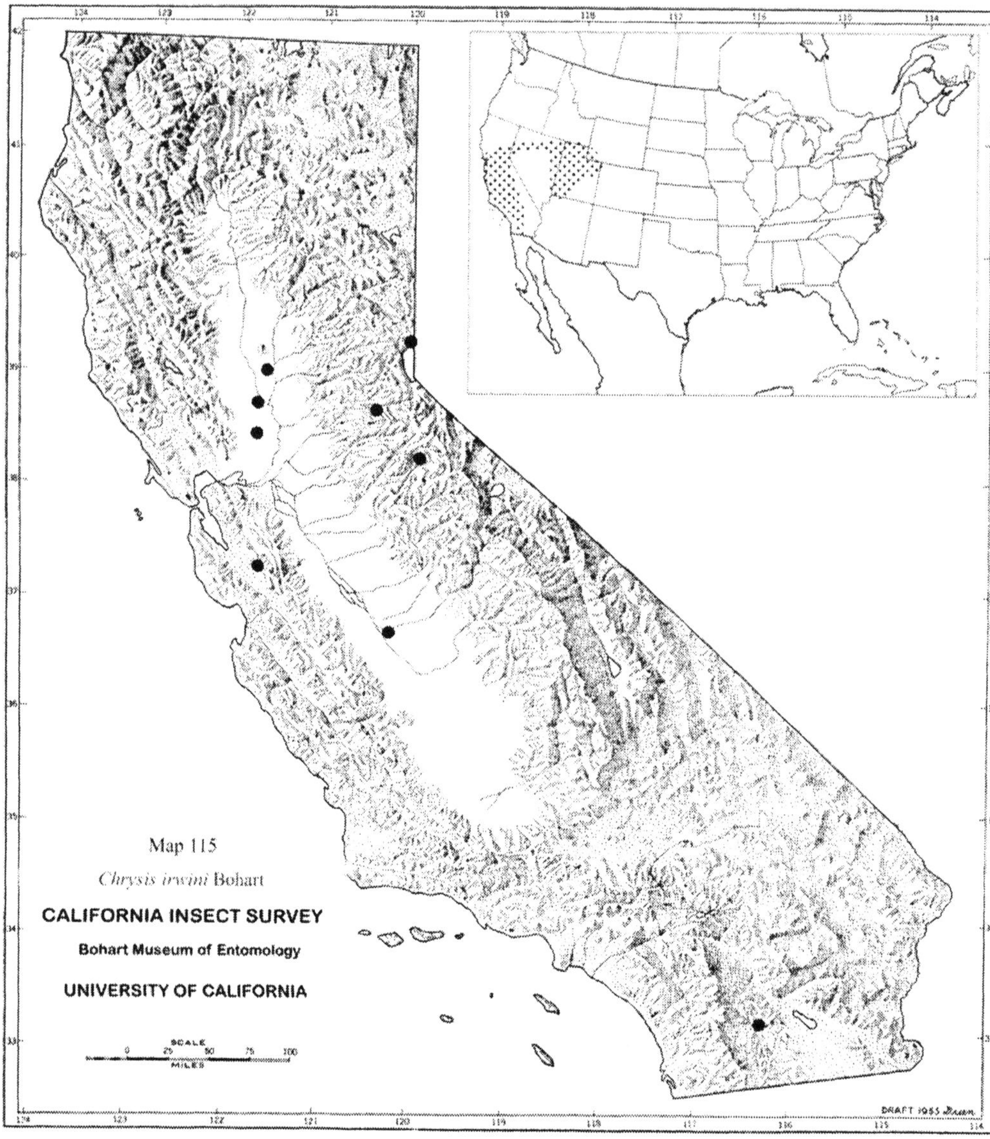

Map 115. California distribution of *Chrysis irwini*. Inset: overall distribution.

Chrysis knowltoni Bohart

Fig. 71, Map 116

Chrysis knowltoni Bohart, in Bohart and Kimsey 1982:105. Holotype ♂; USA: Utah, Roosevelt (UCDC).

Geographic Range (Map 116). Western USA; 42 specimens were seen.

California Records. Collected July and August; **Mono Co.**: Sonora Pass.

Hosts. Unknown.

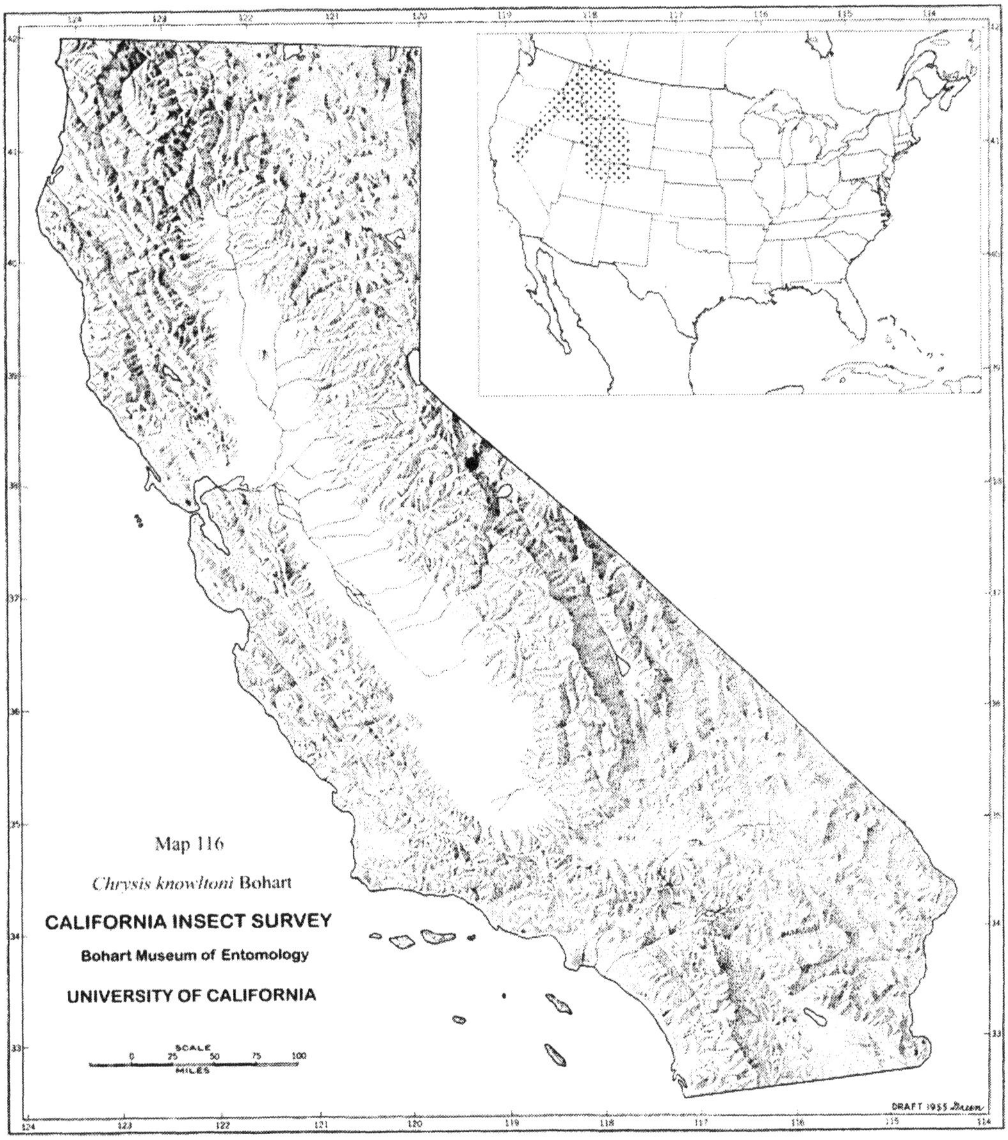

Map 116. California distribution of *Chrysis knowltoni*. Inset: overall distribution.

Discussion. This species is known only from a single locality in California. It closely resembles *dorsalis* as discussed under that species. *Chrysis irwini* is larger than *dorsalis*, ranging between 7 and 8 mm long and tergum II has the punctures across the apical third about one puncture diameter apart, with polished interspaces, giving it a shiny appearance.

Chrysis lucifera Bohart

Fig. 72, Map 117

Chrysis lucifera Bohart, in Bohart and Kimsey 1982:123. Holotype ♂; California: Los Angeles Co., Tanbark Flat UCDC).

Geographic Range (Map 117). Upper Sonoran to Canadian Zones west of the 100th meridian in the USA; 301 specimens were studied.

California Records. Collected April through September; **Alpine Co**.: Blue Lakes; **Amador Co**.: River Pines; **Contra Costa Co**.: Mt. Diablo; Antioch; **El Dorado Co**.: Chile Bar; Kelsey; Pine Hill w Rescue; **Humboldt Co**.: Orovada; **Lake Co**.: Hopland Grade; Lakeport; north fork Cache Creek Hwy 20; **Lassen Co**.: Hallelujah Junction; **Los Angeles Co**.: Newhall; Tanbark Flat; **Madera Co**.: Oakhurst; **Marin Co**.: Point Reyes Station; **Mariposa Co**.: Mariposa; **Mendocino Co**.: 10 mi. e Calpella, Hopland Grade; **Modoc Co**.: 3 mi. n Cedar Pass; 6 mi. nw Cedarville; Sugarloaf Mt.; **Mono Co**.: Mono Lake; **Monterey Co**.: 10 mi. n Parkfield; Arroyo Seco; Porterville; **Napa Co**.: 5 mi. e Pope Valley.; Butts Canyon 0.5 mi. s Napa Co. line; Monticello Dam; Pope Valley; Samuel Springs (beneath Lake Berryessa); Wooden Valley; **Nevada Co**.: Boca; Fuller Lake; Hobart Mills; Prosser Dam; Russell Valley; Sagehen Creek.; **Orange Co**.: Silverado Canyon; **Placer Co**.: 5 mi. w Foresthill; Martis Valley; **Plumas Co**.: Onion Valley; **Riverside Co**.: Mt. San Jacinto; **San Benito Co**.: 2 mi. s Idria; Clear Creek R.A.; Pinnacles National Monument; **San Diego Co**.: 2 mi. n Warner Pass; near Buckman Springs; **San Luis Obispo Co**.: 5 mi. w Nipomo; **Santa Barbara Co**.: Santa Cruz Island; Cuyler Harbor; San Miguel Island; **Santa Cruz Co**.: 9 mi. n Soquel; Loma Prieta Mts.; **Sierra Co**.: Independence Lake; Kyburz Flat; Sattley; Sierra Valley; Sierraville; Yuba Pass; **Solano Co**.: 1.5 mi. w on Mix Canyon Rd.; Monticello Dam; **Tuolumne Co**.: Strawberry; Tuolumne City; Browns Meadow; **Yolo Co**.: 2 mi. n Rumsey; Bear-Cache Creek Junction; Davis; Putah Creek Canyon.

Hosts. Unknown.

Discussion. Related to *irwini, amala, barri* and *pattoni, lucifera* can be distinguished by the short malar space (1.2 MOD in males and 2.4 MOD long in females), short apical teeth on tergum III, with the apicomedial notch considerably less than a half circle, and moderate body size, averaging 7 mm long. In males flagellomere I is subequal in length to III and twice as long as the malar space.

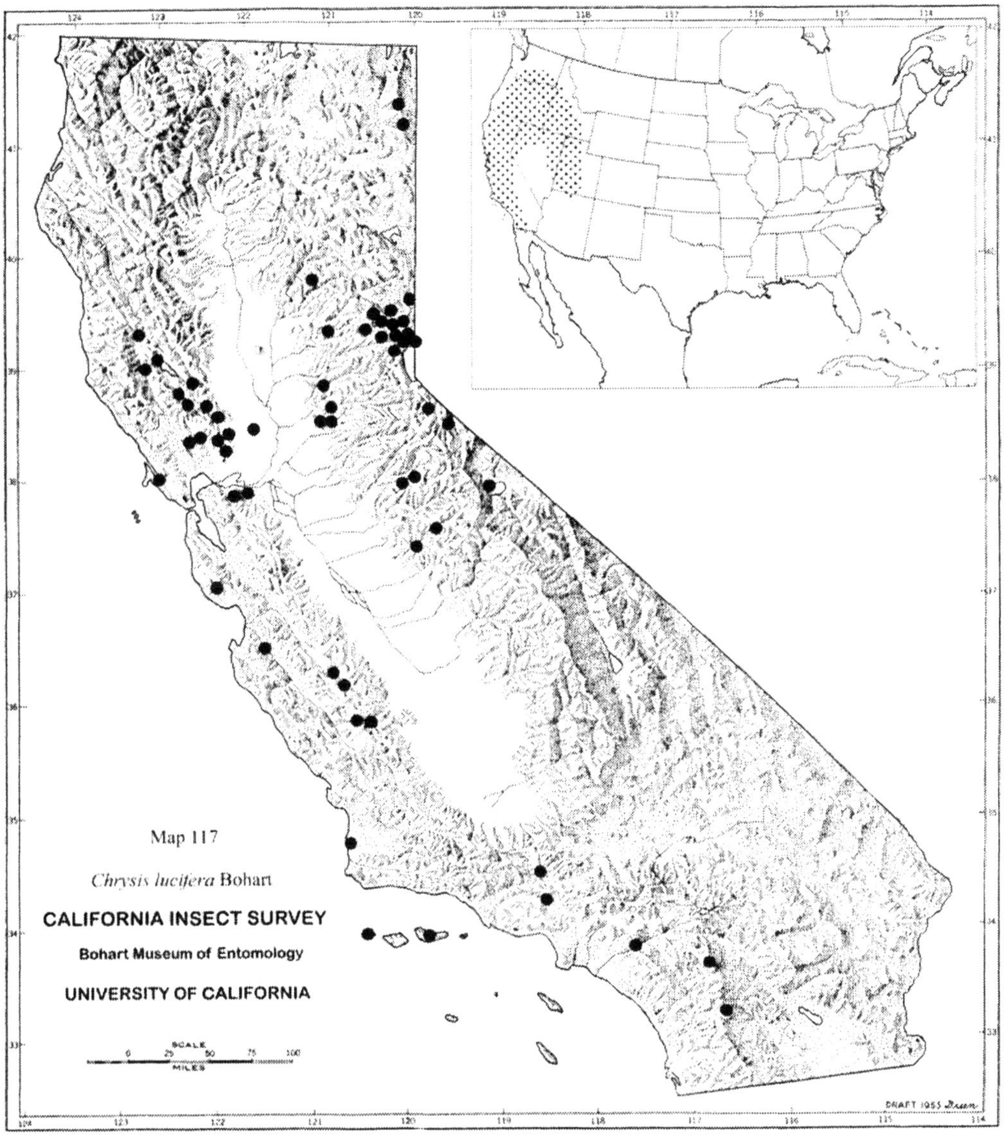

Map 117. California distribution of *Chrysis lucifera*. Inset: overall distribution.

Chrysis montana Aaron

Map 118

Chrysis montana Aaron 1885:234. Lectotype ♀ (designated by Cresson 1928); USA: Montana (ANSP).

Chrysis hirsuta Aaron 1885:235. Lectotype ♂ (designated by Cresson 1928); USA: Utah (ANSP). Nec Gerstaecker 1869.

Chrysis aaroni Mocsáry 1889:386. Replacement name for *hirsuta* Aaron.

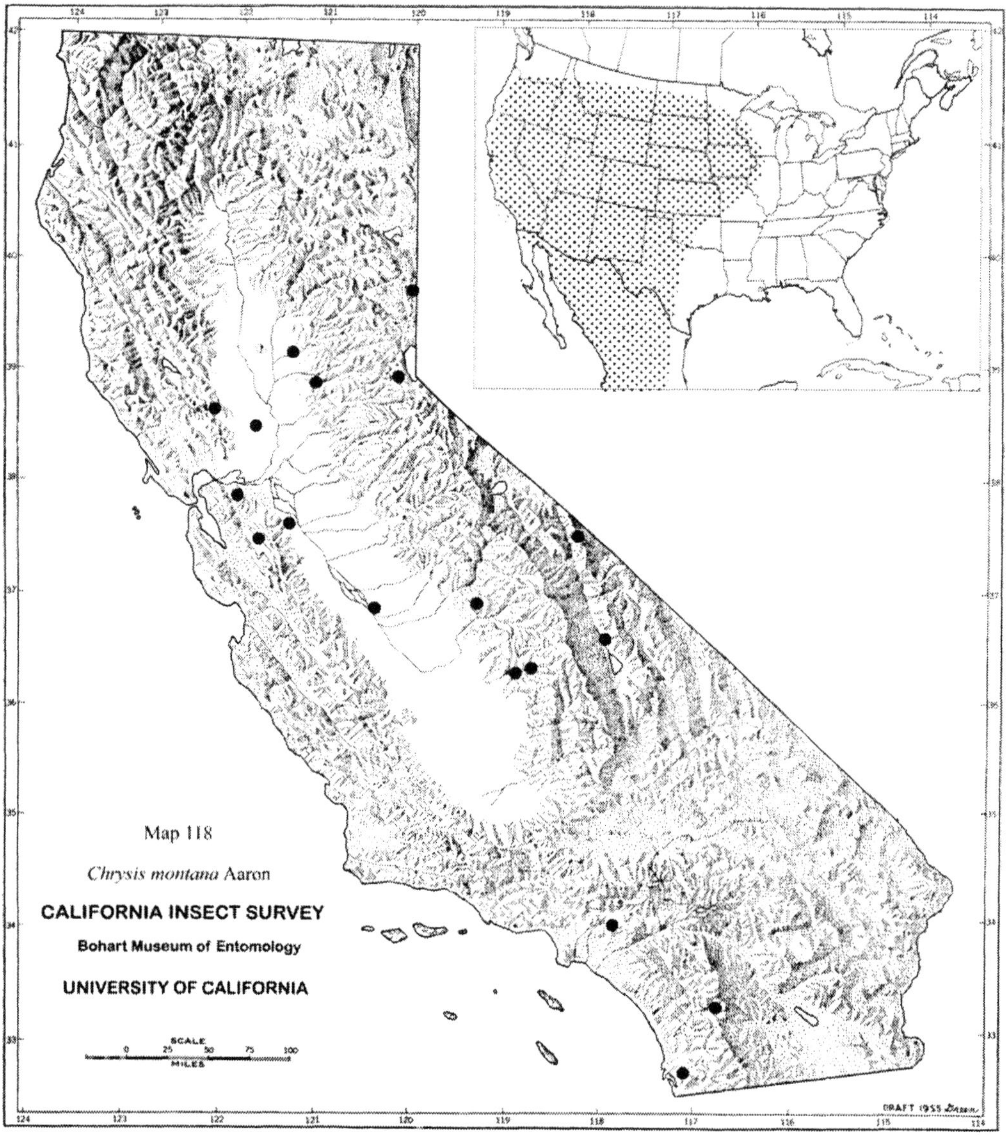

Map 118. California distribution of *Chrysis montana*. Inset: overall distribution.

Geographic Range (Map 118). Upper Sonoran and Alleghanian Zones west of the Mississippi in the USA; 78 specimens were examined.

California Records. Collected May through August; **Alameda Co**.: Arroyo Valley Park; **Contra Costa Co**.: Mt. Diablo; **El Dorado Co**.: Snowline Camp; **Fresno Co**.: Watts Valley; **Inyo Co**.: Alabama Hills; **Lassen Co**.: Hallelujah Junction; **Los Angeles Co**.: Claremont; **Mono Co**.: Campito Meadow; **Monterey Co**.: Arroyo Seco Camp; **Placer Co**.: Folsom Lake; **San Diego Co**.: 2 mi. n Warner Springs; La Mesa; **San Joaquin Co**.: Corral Hollow 8 mi. sw Tracy; **Santa Barbara Co**.: 10 mi. e Twitchell Dam, Cuyama River Canyon; **Tulare Co**.: Ash Mt.; South Fork Kaweah River; Three Rivers; **Tuolumne Co**.: Hwy. 120 Tuolumne-Stanislaus Co. lines; **Yolo Co**.: 2 mi. n Rumsey; Davis; **Yuba Co**.: 18 mi. n Marysville.

Hosts. Unknown.

Discussion. Among species related to *dorsalis, montana* has the longest, sharpest teeth on tergum III, and the female has the unusual feature of the frequent presence of a transverse frontal carina across the prominent frons. In addition, males have flagellomere I 1.8x as long as broad. Males average 5 mm and females 6 mm in length.

Chrysis moorei Bohart

Fig. 78, Map 119

Chrysis moorei Bohart, in Bohart and Kimsey 1982:131. Holotype ♂; USA: California, Nevada Co., Sagehen Creek (UCDC).

Geographic Range (Map 119). Upper Sonoran to Canadian Zones in the USA west of 100th meridian; 42 specimens were seen.

California Records. Collected May through July; **Alpine Co**.: Hope Valley; **Contra Costa Co**.: Mt. Diablo; **Los Angeles Co**.: Claremont; **Madera Co**.: Madera; **Nevada Co**.: Sagehen Creek; Hobart Mills; **San Benito Co**.: Pinnacles; **Santa Clara Co**.: Alum Rock Park; San Jose; **Siskiyou Co**.: Mt. Shasta City; **Trinity Co**.: 6 mi. sw Denny; **Tulare Co**.: Ash Mt.

Hosts. Unknown.

Discussion. Only a few American species of *Chrysis* have metallic red markings on the body. *Chrysis moorei* is one of these, having a basolateral red spot on tergum, which may extend into the pit row. The other species with metallic red abdominal markings are *nitidula* and *violacuna*. *C. moorei* is similar to *nitdula* but can be distinguished by the position of the red markings, shorter flagellomere I and less prominent mesopleural carina. It can be separated from *violacuna* by the lack of violet red markings on the anterior declivity of tergum I, and the well-developed medial ridge of the pit row.

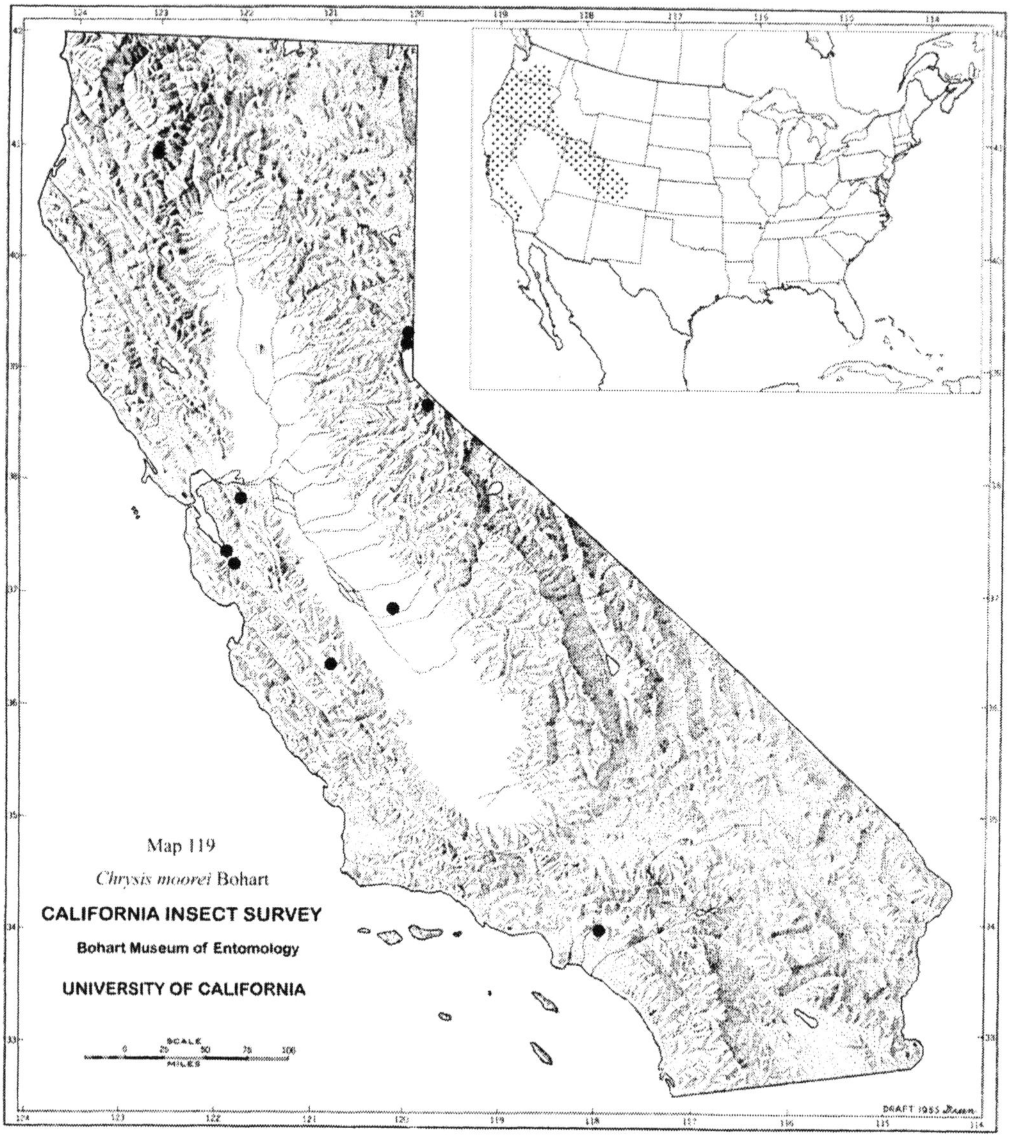

Map 119. California distribution of *Chrysis moorei*. Inset: overall distribution.

Chrysis nitidula Fabricius

Map 120

Chrysis nitidula Fabricius 1775:359. Lectotype ♀ (designated by Bohart, in Bohart and Kimsey 1991); "America" (BMNH).

Chrysis coerulans Fabricius 1804:172. Neotype ♀ (designated by Bohart, in Bohart and Kimsey 1982); USA: North Carolina ("N. C.") (ANSP).

Chrysis servillei Brullé 1846:37. Holotype ♀; USA: "Les montagnes Rocheuses" (Serville Collection, lost?).

Chrysis bella Cresson 1865b:312. Holotype ♀; USA: "Colorado Territory" (ANSP).

Chrysis nortoni Aaron 1885:237. Lectotype ♂ (designated by Cresson 1928); USA: Maine (ANSP).

Chrysis angularis Mocsáry 1889:366. Holotype ♀; Egypt (probably in error) (NHMV).

Chrysis canadensis du Buysson 1891:37. Holotype ♀; Canada: Quebec (MNHN).

Chrysis conserta du Buysson 1891:37. Holotype ♀; USA: Texas (MNHN).

Chrysis nanula Rohwer 1909:88. Holotype ♀; USA: Colorado (USNM).

Chrysis sejuncta Mocsáry 1914:51. Lectotype ♀ (designated by Bohart 1986b); USA: "Missouri" (TMB).

Chrysis praticola Mocsáry 1914:50. Lectotype ♀ (designated by Bohart 1986b); USA: Texas, Fedor (TMB).

Chrysis eurekana Linsenmaier 1994:487. Holotype female; California: Eureka (MNL). **New synonymy**.

Chrysis angustianalis Linsenmaier 1994:491. Holotype female; California: San Benito Co.: Idria (MNL). **New synonymy**.

Geographic Range (Map 120). Widespread in the Nearctic Region; 923 specimens were studied.

California Records. Collected March through September; **Alameda Co**.: Berkeley; **Alpine Co**.: Forestdale Meadow; Hope Valley; Lake Winnemucca; **Amador Co**.: Volcano; **Butte Co**.: Chico; **Contra Costa Co**.: Mt. Diablo; **El Dorado Co**.: 8 mi. n on Ice House Rd.; Pine Hill w Rescue; Pope Beach Lake Tahoe; 3 mi. s Carson Pass; Carson Pass; **Fresno Co**.: Panoche; **Humboldt Co**.: 2 mi. s Redwood Valley; Big Lagoon; Eureka; Humboldt Bay south jetty; Samoa; Arcata; Trinidad; **Inyo Co**.: Big Pine Creek; Bishop Creek; Keeler; Crooked Creek Lab., White Mts.; **Kern Co**.: Walker Pass; **Lake Co**.: Cobb Mt.; Hopland Grade; north fork Cache Creek Hwy 20; **Lassen Co**.: Bridge Creek Camp; Hallelujah Junction; Madeline

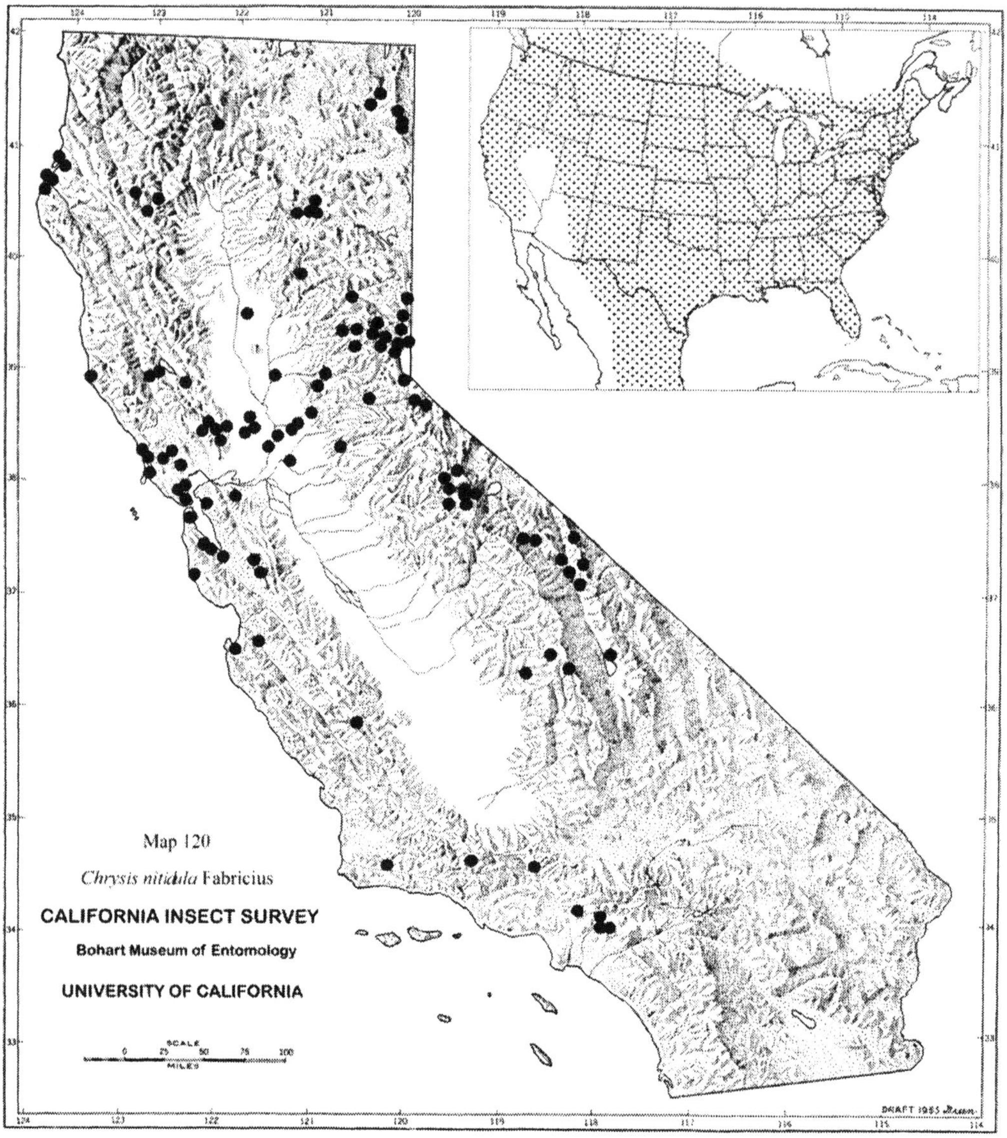

Map 120. California distribution of *Chrysis nitidula*. Inset: overall distribution.

Plains; Summit Camp; 4 mi. e Manzanita Lake; Lassen National Park; **Los Angeles Co.**: Big Dalton Canyon; Crystal Lake; Glendale; Los Angeles National Forest; mountains near Claremont; Pomona; Tanbark Flat; Walnut; **Marin Co.**: 4 mi. n San Rafael; Mt. Tamalpais; San Anselmo; Tomales Bay State Park; McClure Beach, Point Reyes National Seashore; **Mendocino Co.**: 2 mi. n Point Arena; Camp Marwedel; Point Arena; **Merced Co.**: 2 mi. n Delhi; **Modoc Co.**: 24 mi. sw Alturas; 6 mi. nw Cedarville; Cedar Pass, n Alturas; Patterson Meadow 10 mi. sw Eagleville; Sugar Loaf Mt.; **Mono Co.**: 2 mi. e Sonora Pass; Blanco's Corral, White Mts.; Hot Creek;

Leavitt Meadow; Mammoth Lakes; Mono City; Sonora Pass; **Monterey Co**.: Arroyo Seco Camp; Bixby Creek; Las Padres Dam, Carmel Valley; Porterville; **Napa Co**.: Bear Creek; Berryessa (beneath Lake Berryessa); Mt. Veeder; Samuel Springs (beneath Lk Berryessa); **Nevada Co**.: Boca; Carpenter Ridge; Fuller Lake; Jackson Lake; Russell Valley; Sagehen Creek; Hobart Mills; **Orange Co**.: Silverado Canyon; **Placer Co**.: 10 mi. n Auburn; Auburn; Brockway; Carnelian Bay, Lake Tahoe; **Plumas Co**.: Blairsden; Johnsville; Mt. Ingalls; Nelson Creek; Onion Valley; Storrie; **Sacramento Co**.: Fair Oaks; Folsom; Galt; Orangevale; Sacramento; **San Benito Co**.: 6 mi. s Idria; **San Francisco Co**.: San Francisco; **San Mateo Co**.: Pescadero; **Santa Barbara Co**.: Santa Cruz Island; Santa Ynez Mts.; **Santa Clara Co**.: Hecter Pass; Mt. Hamilton; San Jose; San Mateo; **Shasta Co**.: Lassen Park; **Sierra Co**.: 10 mi. s Sierraville; 9 mi. n Sagehen Creek; Independence Lake; Sardine Peak; Sierra Buttes; Sierra Valley; Sierraville; Weber Lake; **Siskiyou Co**.: Mount Shasta City; **Solano Co**.: Monticello Dam; Putah Canyon; **Sonoma Co**.: Bodega Bay; Glenn Ellen; Kenwood; Petaluma; **Stanislaus Co**.: La Grange; **Trinity Co**.: Big Flat Coffee Creek; Mountain Meadow Ranch Coffee Creek; Carrville; **Tulare Co**.: Ash Mt.; Mineral King; Sequoia National Park, Ash Mt. Range; **Tuolumne Co**.: Chipmunk Flat; Dodge Ridge; Leland Meadow; Sonora Pass; Tuolumne Meadows; Tioga Pass; Warner Creek, below Tioga Pass; **Ventura Co**.: Sespe Canyon; **Yolo Co**.: 2 mi. w Davis; Davis; Putah Canyon; Monticello Dam; Clarksburg; **Yuba Co**.: 18 mi. nw Marysville.

Hosts. A variety of eumenine vespid hosts have been published for this species, including *Ancistrocerus catskill halophilus* Viereck, *A. c. albophaleratus* (Saussure), *A. antilope* (Panzer), *A. spilogaster* Cameron, A. *adiabatus cytainus* (Cameron), *A. tuberculocephalus sutterianus* (Saussure), *Euodynerus foraminatus scutellaris* (Saussure), *E. f. f.* (Saussure), *E. leucomelas* (Saussure), *E. megaera* (Lepeletier), *Parancistrocerus acarophorus* Bohart, *Symmorphus cristatus* Saussure, and *S. albomarginatus* (Saussure) (Parker and Bohart 1966, Medler 1964, Krombein 1967, Kurczewski and Kurczewski 1979, Eickwort 1973, Evans 1973, Krombein 1979).

Discussion. *Chrysis nitidula* is nearly identical structurally to the palearctic species *ignita* Linnaeus, differing only in the concolorous head, thorax and abdomen (*ignita* has a metallic red abdomen). This is a widespread, commonly collected, and variable species, with many different forms as discussed by Bohart and Kimsey (1982) under the name *coerulans*. It can be distinguished from related species by the black spots on sternum II separated by at least 4 MOD, no violet red on the anterior declivity of tergum I or above the pit row on tergum III, frontal carina well-developed, tergum II two-thirds as long as well-developed medial ridge, tergum III teeth short to moderate in length and male flagellomere I twice as long as broad or shorter.

Types of the Linsenmaier species, *eurekana* and *angustianalis*, were examined and were found to be identical to *Chrysis nitidula*. As a result they are synonymized herein with that species.

As discussed under *cessata*, *nitidula* was previously known as *coerulans* but examination of the primary types revealed that the species name *nitidula* was historically misapplied to the species now known as *cessata*. Thus the correct name for the species, previously recognized as *coerulans*, is *nitidula*. Specimens identified prior to 1990 probably used the name *coerulans*.

Chrysis oraria Bohart

Map 121

Chrysis oraria Bohart 1962:370. Holotype ♂; USA: Arizona, Grand Canyon (SMEK).

Geographic Range (Map 121). Upper and Lower Sonoran Zones from southwestern New Mexico, Arizona and California, and northwestern Mexico (Baja California); 34 specimens were seen.

California Records. Collected in July; **Riverside Co**.: 4 mi. s Palm Desert; Coachella; **San Bernardino Co**.: Hwy 95; 3 mi. n Hwy 40.

Hosts. Unknown.

Discussion. A species with six teeth on tergum III, *oraria* has the strongly lobate propodeal tooth seen in *inaequidens*. The male flagellomere I is less than twice as long as broad, sternum II black spots contiguous, malar space about 1 MOD long and subequal in length to subantennal distance. Tergum III has the area beyond the pit row depressed between the medial teeth, and the teeth are not bicarinate as they are in *wasbaueri*.

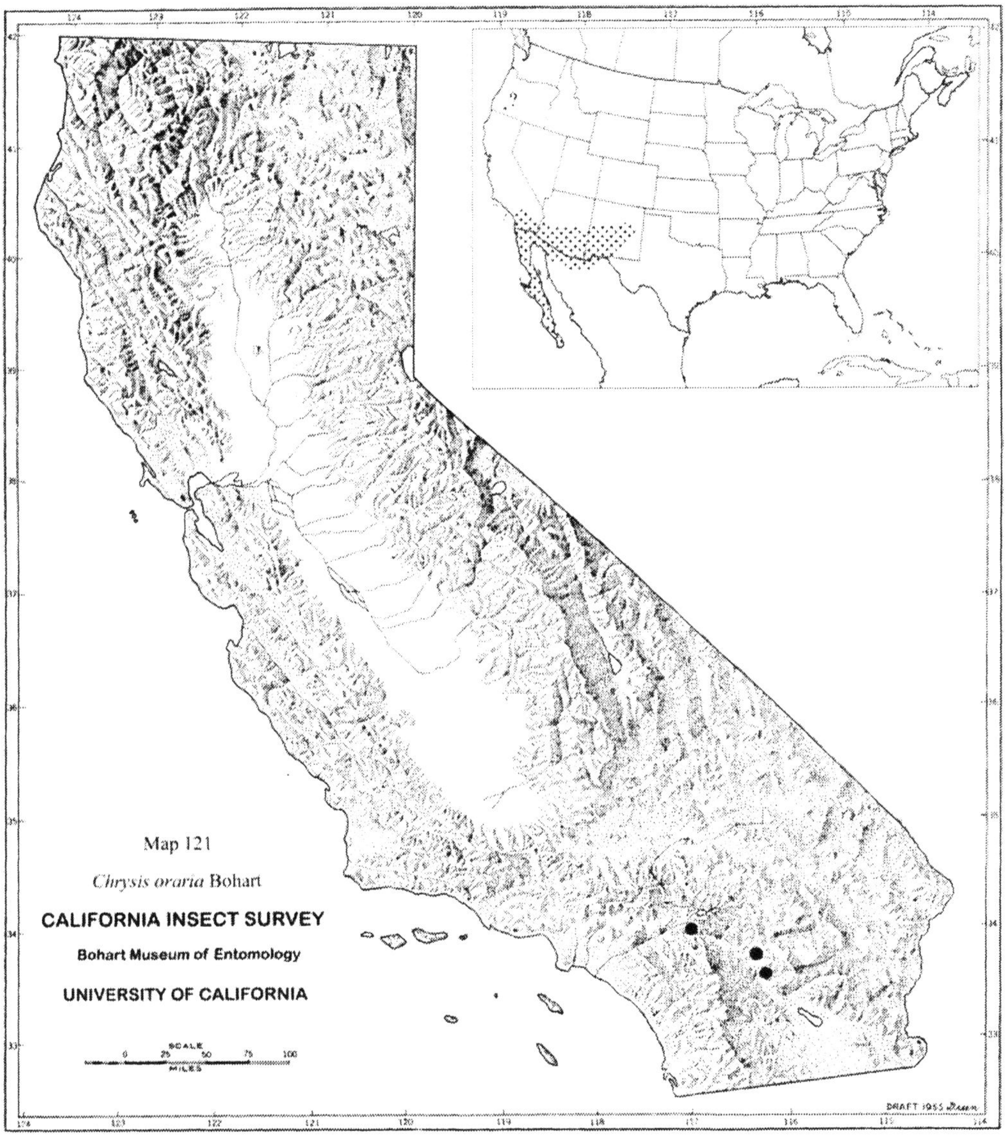

Map 121. California distribution of *Chrysis oraria*. Inset: overall distribution.

Chrysis paradisica Bohart

Fig. 73, Map 122

Chrysis paradisica Bohart, in Bohart and Kimsey 1982:106. Holotype ♂; USA: California, Nevada Co., Sagehen Creek (UCDC).

Geographic Range (Map 122). Canadian Zone of the Sierra Nevada Mountains and Transverse Ranges of California; 16 specimens were studied.

California Records. Collected May through July; **El Dorado Co.**: El Dorado; **Mono Co.**: Paradise Camp; **Nevada Co.**: Sagehen Creek; **San Bernardino Co.**: Apple Valley.

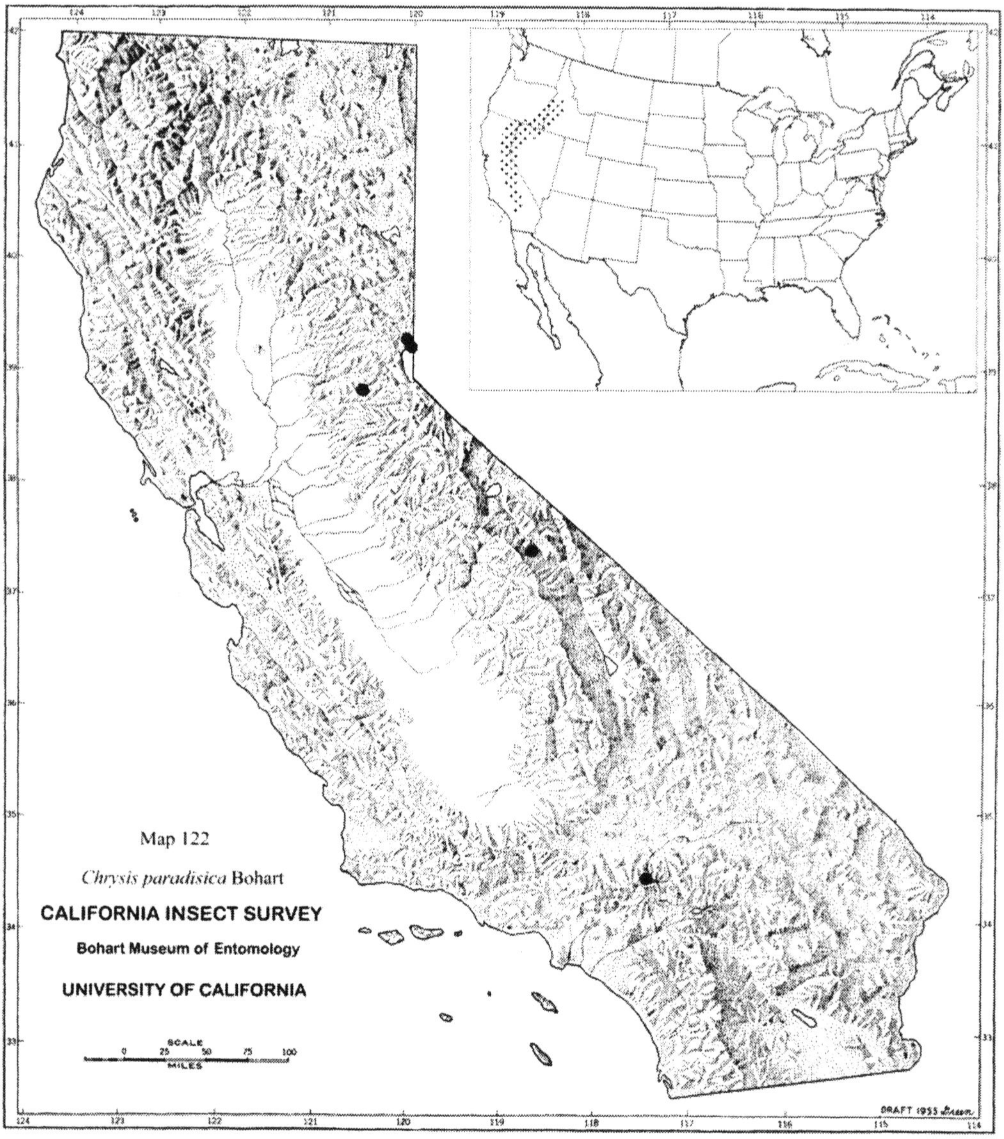

Map 122. California distribution of *Chrysis paradisica*. Inset: overall distribution.

Hosts. Unknown.

Discussion. This California endemic is known from a relatively small number of specimens. Female *paradisica* are nearly indistinguishable from those of *allectoris* (a non-Californian species) and less so *dorsalis*. Males can be readily distinguished from those of *allectoris* by genitalic characters: the paramere apex is slender and apically arcuate in *allectoris* and straight in *paradisica*. *Chrysis paradisica* is a relatively small-bodied species; ranging from 4.5-5.0 mm in length.

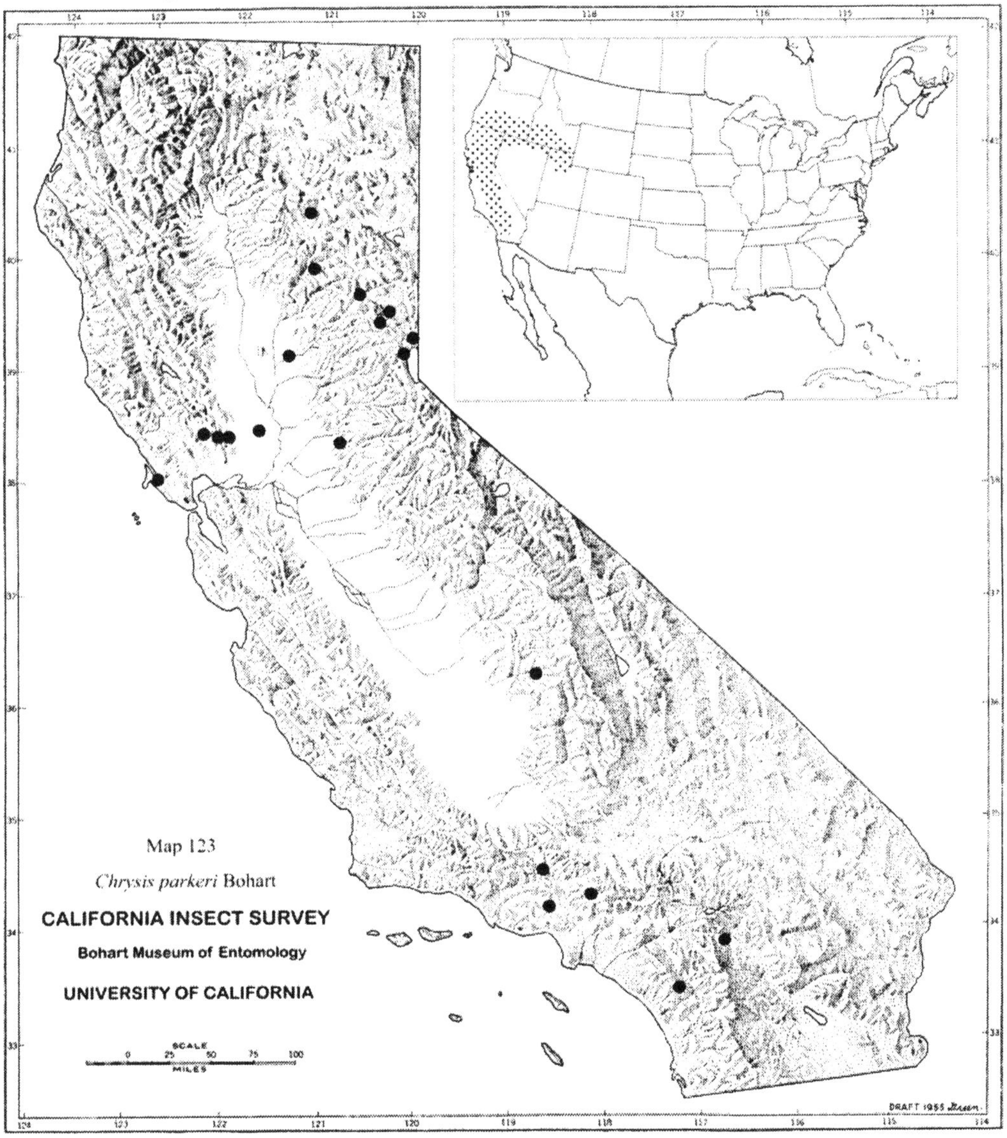

Map 123. California distribution of *Chrysis parkeri*. Inset: overall distribution.

Chrysis parkeri Moore

Fig. 83, Map 123

Chrysis parkeri Moore 1966:1130. Holotype male; USA: Nevada, Washoe Co., Verdi (UCDC).

Geographic Range (Map 123). Upper Sonoran Zone in the USA west of 100th meridian; 382 specimens were seen.

California Records. Collected April through September; **Amador Co**.: Volcano; **Lassen Co**.: Bridge Creek Camp; Summit Camp; Hallelujah Junction; **Los Angeles Co**.: Tanbark Flat; Placerita Canyon; Newhall; **Marin Co**.: Point Reyes Station; **Napa Co**.: Samuel Springs (beneath Lake Berryessa); **Nevada Co**.: Sagehen Creek; **Placer Co**.: Carnelian Bay, Lake Tahoe; **Plumas Co**.: Onion Valley; Blairsden; **Riverside Co**.: Gavilan; Banning; **Sierra Co**.: Sierra Valley; Sierraville; **Tulare Co**.: Sequoia National Park, Ash Mts.; **Yolo Co**.: Monticello Dam; Davis; Putah Canyon; **Yuba Co**.: 18 mi. nw Marysville.

Hosts. Hosts include the eumenine vespids *Leptocheilus tricolor* (Provancher), *L. rufinodus* (Cresson), *Microdynerus bakerianus* (Cameron), *Ancistrocerus simulator* Cameron and *Ancistrocerus* sp. (Bohart and Kimsey 1982).

Discussion. The large number of *parkeri* specimens seen is misleading. This is a relatively uncommon species in California. The bulk of the specimens in collections come from twig nests put out by Frank Parker in Utah.

Closely related to *nitidula, parkeri* can be distinguished by the following combination of characters: tergum II with large black spots that reach the midline, flagellomere I mostly green and 2.5-3.0x as long as broad, frons with strong transverse carina, tergum II medial longitudinal carina partial or obsolescent, tergum III pit row with 8-10 medial pits in posterior view, divided by slightly depressed medial ridge and medial notch usually less than a half circle. The female tergum III is somewhat saddled in profile, with a prepit bulge, but not laterally inflated.

Chrysis pattoni Aaron

Map 124

Chrysis pattoni Aaron 1885:235. Holotype ♀; USA: Colorado (ANSP).

Geographic Range (Map 124). Transcontinental in the USA and the western half of Canada; 182 specimens were seen.

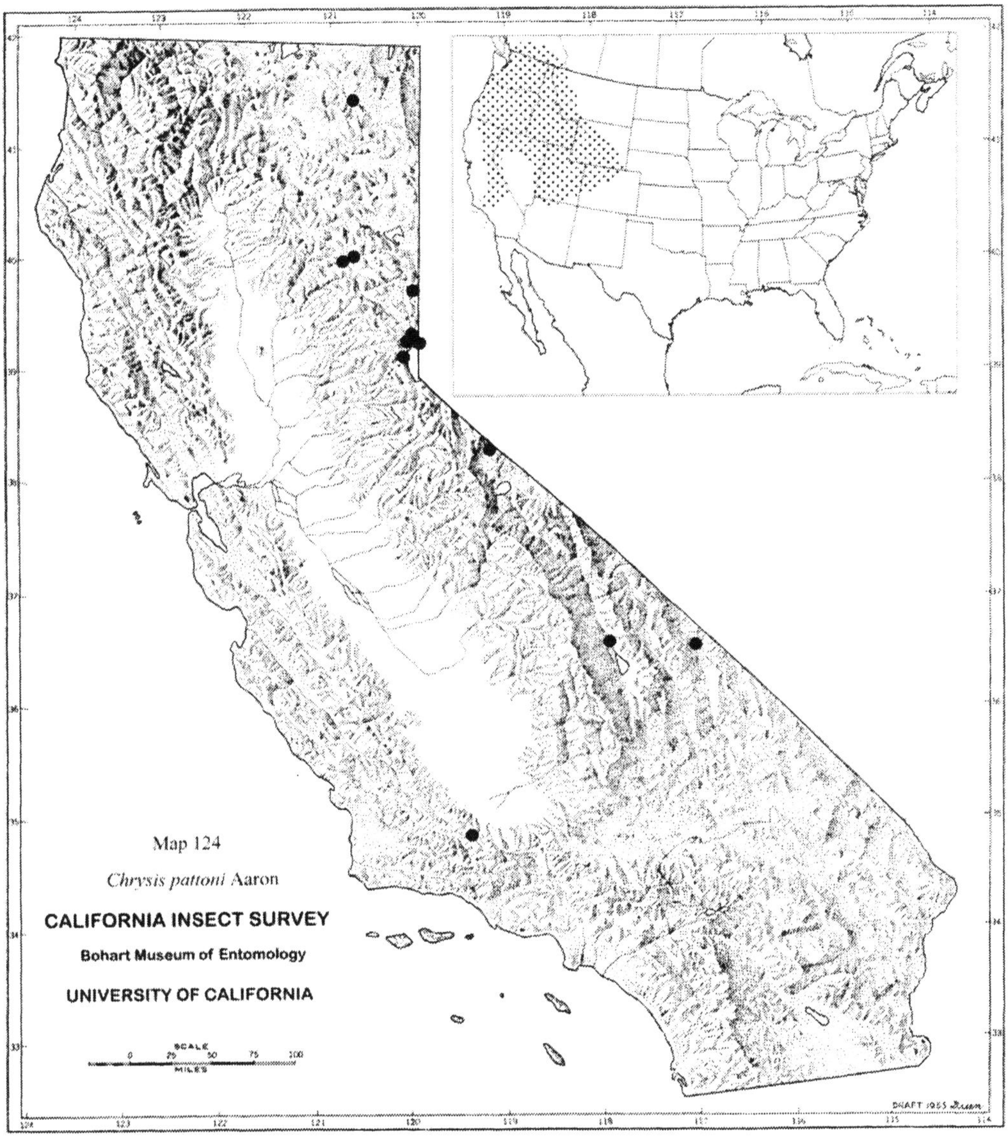

Map 124. California distribution of *Chrysis pattoni*. Inset: overall distribution.

California Records. Collected May through August; **Alpine Co**.: Lake Winnemucca; **Inyo Co**.: 3 mi. e Lone Pine; Eureka Valley Dunes; **Kern Co**.: Mt. Pinos 2 mi. w YMCA Summer Camp; **Lassen Co**.: Hallelujah Junction; **Modoc Co**.: 24 mi. sw Alturas; **Mono Co**.: Bridgeport; Topaz Lake; White Mts.; **Nevada Co**.: Boca; Hobart Mills; Sagehen Creek; **Placer Co**.: Carnelian Bay, Lake Tahoe; **Plumas Co**.: Little Long Valley Creek; 6 mi. e Spring Garden; **Santa Barbara Co**.: Santa Cruz Island.

Hosts. Parker and Bohart (1966, 1968) and Parker (1970) reared *pattoni* from *Ancistrocerus spilogaster* Cameron, *Leptochilus washo* Parker and *Microdynerus bakerianus* (Cameron) (Vespidae).

Discussion. This widespread species appears closely related to *derivata* and *irwini*. It can be distinguished by the unusually long malar space, particularly in the female (about 2 MOD long), tergum III with the lateral edge sinuous in the males and the apicomedial notch usually forming a half circle. The body color is a mixture of blue, green and some purple.

Chrysis pellucidula Aaron

Map 125

Chrysis pellucidula Aaron 1885:235. Holotype ♀; USA: "Virginia" (ANSP).

Geographic Range (Map 125). Transcontinental in the USA, southern Canada, and in the mountains of Baja California; 173 specimens were studied.

California Records. Collected May through October; **Mendocino Co**.: UC Hopland Field Station; **Riverside Co**.: Beaumont; Riverside; **Tulare Co**.: Ash Mt.; Wood Lake; **Yolo Co**.: Davis.

Hosts. Krombein (1979) reared this species from *Trypargilum collinum rubrocinctum* (Packard) and *T. tridentatum* (Packard).

Discussion. *Chrysis pellucidula* does not appear to be closely related to other California species. It is a distinctive wasp with a short, bulging face, lidded midocellus, flagellomere I twice as long as broad, and tergum III with the pit row deep having largely confluent pits and short sharp teeth.

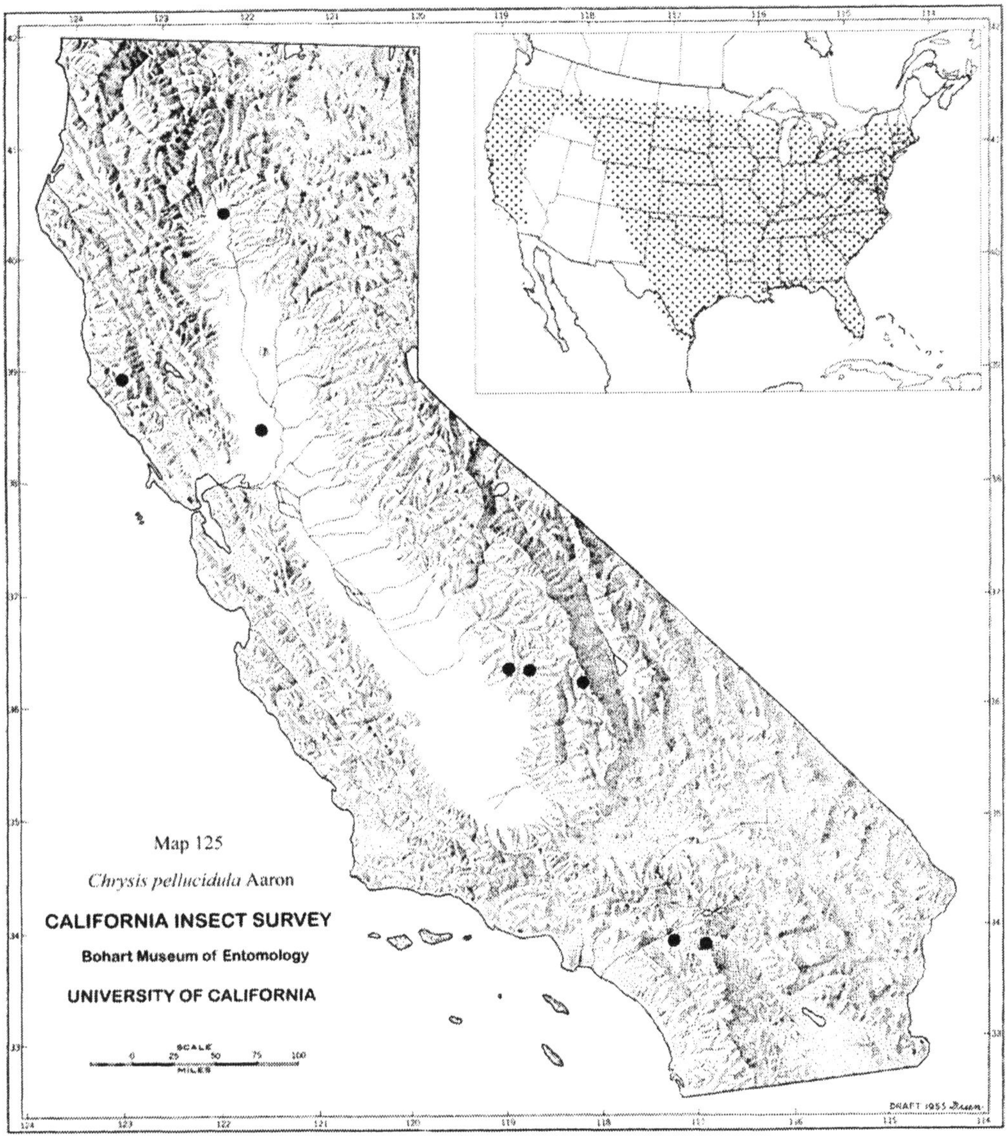

Map 125. California distribution of *Chrysis pellucidula*. Inset: overall distribution.

Chrysis pompella Bohart

Map 126

Chrysis pompella Bohart, in Bohart and Kimsey 1982:107. Holotype ♂; USA: California, Apple Valley (CAS).

Geographic Range (Map 126). Lower and Upper Sonoran Zones in southern California and Arizona; 23 specimens were examined.

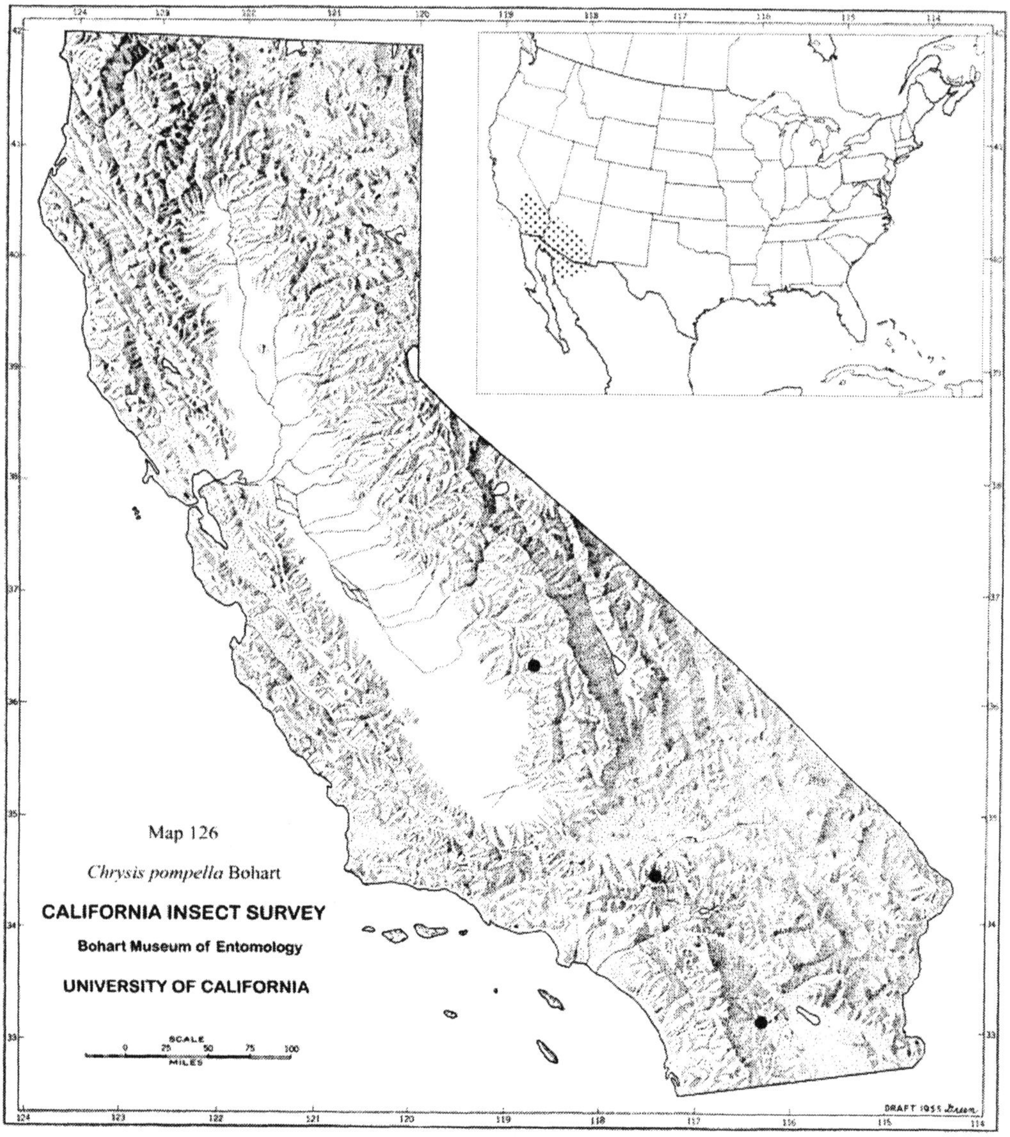

Map 126. California distribution of *Chrysis pompella*. Inset: overall distribution.

California Records. Collected in May; **San Bernardino Co**.: Apple Valley; **Los Angeles Co**.: Desert Springs; **San Diego Co**.: Borrego.

Hosts. Unknown.

Discussion. The coloration of *pompella* is similar to that of two non-Californian species, *meta* Aaron and *provancheri* Schulz. These species all have an extensively bright coppery-red abdomen, contrasting strikingly with the blue to greenish blue head and thorax. Their coloration is unusual in the Americas, but would be typical in the Palearctic Region.

Chrysis propria Aaron

Fig. 84, Map 127

Chrysis propria Aaron 1885:238. Lectotype male (designated by Cresson 1928); USA: Montana (ANSP).

Chrysis kahli Viereck 1906:194. Holotype ♂; USA: Kansas (SMEK).

Chrysis pattonella Viereck 1906:194. Holotype ♂; USA: Kansas, Hamilton Co. (SMEK).

Chrysis submontana Rohwer 1909:91. Syntype ♀♀; USA: Colorado, Rifle (USNM).

Chrysis rinconensis Mocsáry 1914:54. Holotype ♂; Mexico: Guerrero, Rincon (BMNH).

Geographic Range (Map 127). USA, west of the Mississippi River, southern Canada and northern Mexico (Chihuahua, Distrito Federal, Zacatecas, Tamaulipas, Coahuila, Guerrero, Durango); 352 specimens were studied.

California Records. Collected May through September; **Alpine Co**.: Lake Winnemucca; **El Dorado Co**.: 8 mi. on Ice House Rd.; **Inyo Co**.: Independence; **Kern Co**.: Greenhorn Ranger Station; **Lake Co**.: Hopland Grade; **Lassen Co**.: Hallelujah Junction; **Los Angeles Co**.: Claremont; Tanbark Flat; **Mendocino Co**.: 7 mi. w Eel River Ranger Station; **Modoc Co**.: .5 mi. w Willow Ranch; 3 mi. n Cedar Pass; **Monterey Co**.: Arroyo Seco; **Nevada Co**.: 3 mi. n Boca; Boca; Hobart Mills; Sagehen Creek; **Placer Co**.: Carnelian Bay, Lake Tahoe; **Plumas Co**.: Onion Valley; **San Diego Co**.: San Diego; **Santa Barbara Co**.: Upper Oso Camp, Los Padres National Forest; **Sierra Co**.: 12 mi. se Sierraville; Sierraville; Weber Lake; **Tulare Co**.: Sequoia National Park, Ash Mt.; **Tuolumne Co**.: Strawberry; **Ventura Co**.: Foster Park.

Hosts. Hosts include *Ancistrocerus* sp. and *Leptochilus rufinodus* (Cresson) (Vespidae) (Bohart and Kimsey 1982)

Discussion. The species most closely resembling *propria* is *tensa*. Both have the malar space less than 2 MOD long. In *propria* the apicomedial notch of tergum III is simply and sharply margined at its deepest point. The shape is more complex in *tensa*.

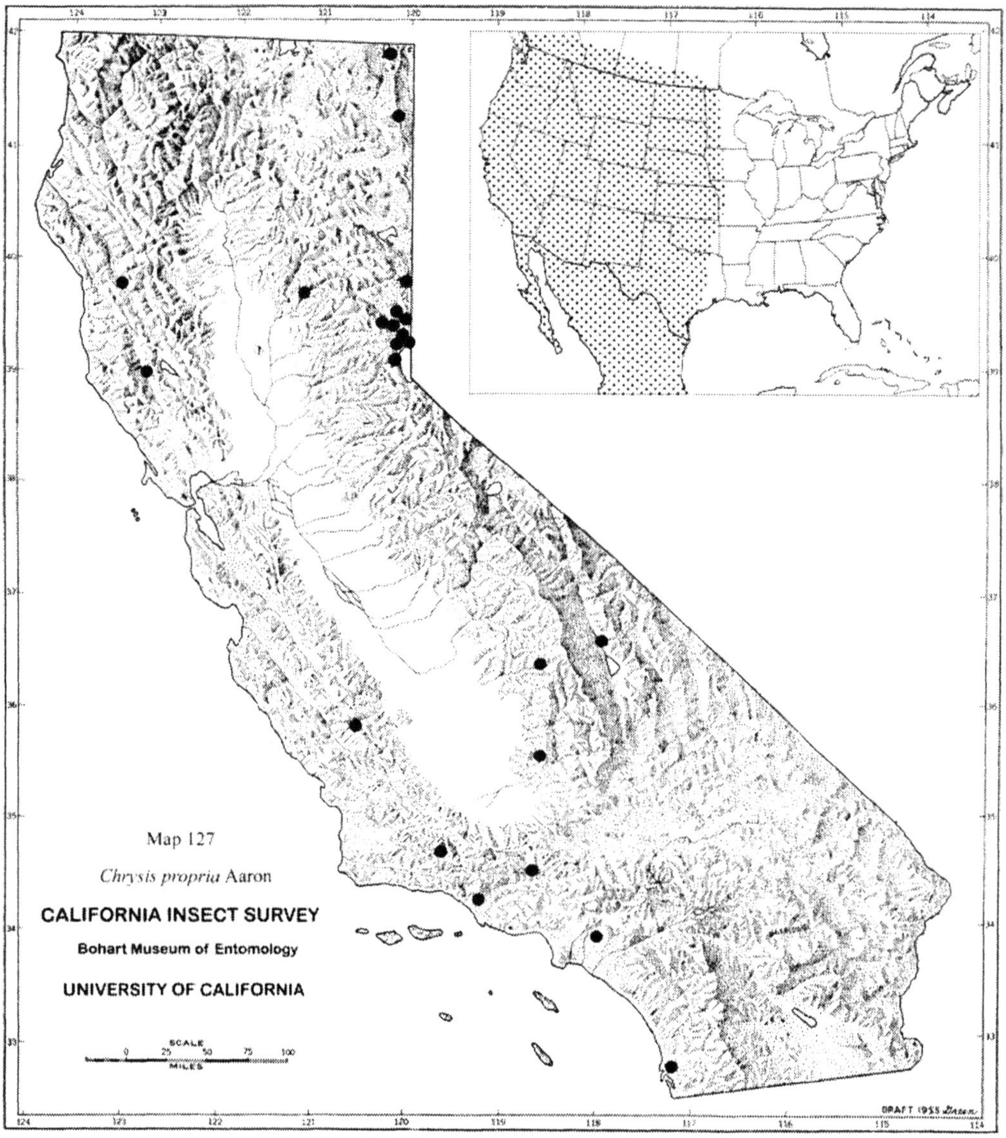

Map 127. California distribution of *Chrysis propria*. Inset: overall distribution.

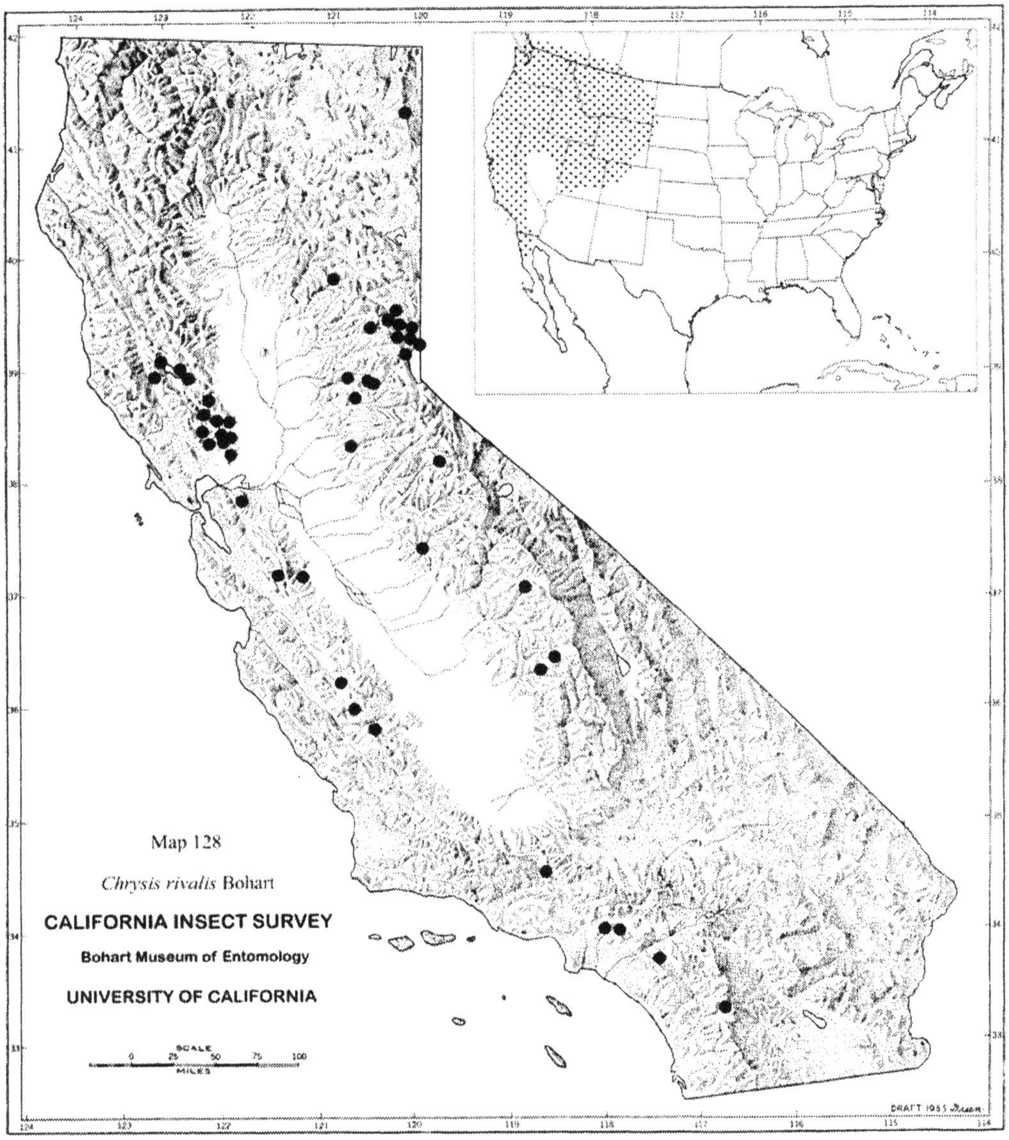

Map 128. California distribution of *Chrysis rivalis*. Inset: overall distribution.

Chrysis rivalis Bohart

Map 128

Chrysis rivalis Bohart 1964:229. Holotype ♂; USA: California, Napa Co., Samuel Springs (beneath Lake Berryessa) (UCDC).

Geographic Range (Map 128). Upper Sonoran to Transition Zones in the USA from Colorado and Wyoming west; 221 specimens were studied.

California Records. Collected May through October; **Amador Co**.: Daffodil Hill; Volcano; **Contra Costa Co**.: Mt. Diablo; **El Dorado Co**.: 10 mi. w El Dorado; Kelsey; Pine Hill near Rescue; **Fresno Co**.: Auberry; **Lake Co**.: Borax Lake; Hopland Grade; Lakeport; Laurel Beach; north fork Cache Creek Hwy 20; **Lassen Co**.: Hallelujah Junction; **Los Angeles Co**.: San Gabriel Canyon Rd. 11 mi. n Azusa; Tanbark Flat; **Mariposa Co**.: Mariposa; **Modoc Co**.: 5.5 mi. e Cedarville; **Monterey Co**.: 10 mi. n Parkfield; Arroyo Seco; **Napa Co**.: 2.5 mi. n Knoxville; Angwin; Monticello Dam; Mt. Saint Helena; Pope Valley; Samuel Springs (beneath Lake Berryessa); **Nevada Co**.: 3 mi. n Boca; Russell Valley; Sagehen Creek; Hobart Mills; **Placer Co**.: 5 mi. w Foresthill; Martis Valley; **Plumas Co**.: Butterfly Valley; **Riverside Co**.: Mt. San Jacinto; **San Benito Co**.: Clear Creek; **San Diego Co**.: 2 mi. n Warner Springs; **Santa Clara Co**.: Mt. Hamilton; **Sierra Co**.: Kyburz Flat 14 km se Sierraville; Sattley; Sierra Valley; **Solano Co**.: 3.2 mi. w Mix Canyon; Gates Canyon; Monticello Dam; 9 mi. w Winters; **Stanislaus Co**.: Del Puerto Canyon; **Tulare Co**.: Ash Mt. and Yucca Creek Sequoia National Park; **Yolo Co**.: 2 mi. n Rumsey; Putah Canyon.

Hosts. Unknown.

Discussion. Most similar to *coloradica* and *xerophila, rivalis* can be distinguished by the overall nearly contiguous small to medium-sized punctation, lower frons moderately silvery and slightly broader than lengths of flagellomeres I-II, sternum II with large nearly contiguous black spots, and tergum III punctation finer than on scutum. This is a small-bodied species ranging from 6-7 mm long.

Chrysis schusteri Bohart

Fig. 82, Map 129

Chrysis schusteri Bohart, in Bohart and Kimsey 1982:133. Holotype ♂; USA: California, Santa Barbara Co., Santa Cruz Island, Canada del Medio (UCDC).

Chrysis antiochicola Linsenmaier 1994:486. Holotype ♀; California: Antioch (MNL). **New synonymy**.

Geographic Range (Map 129). Coastal California and offshore islands; 42 specimens were studied.

California Records. Collected March through October; **Marin Co**.: Bolinas; McClure Beach; Point Reyes National Seashore; **Mendocino Co**.: Point Arena; **Santa Barbara Co**.: Bay Point Slope; San Miguel Island; Santa Cruz Island; **Sonoma Co**.: Bodega Head.

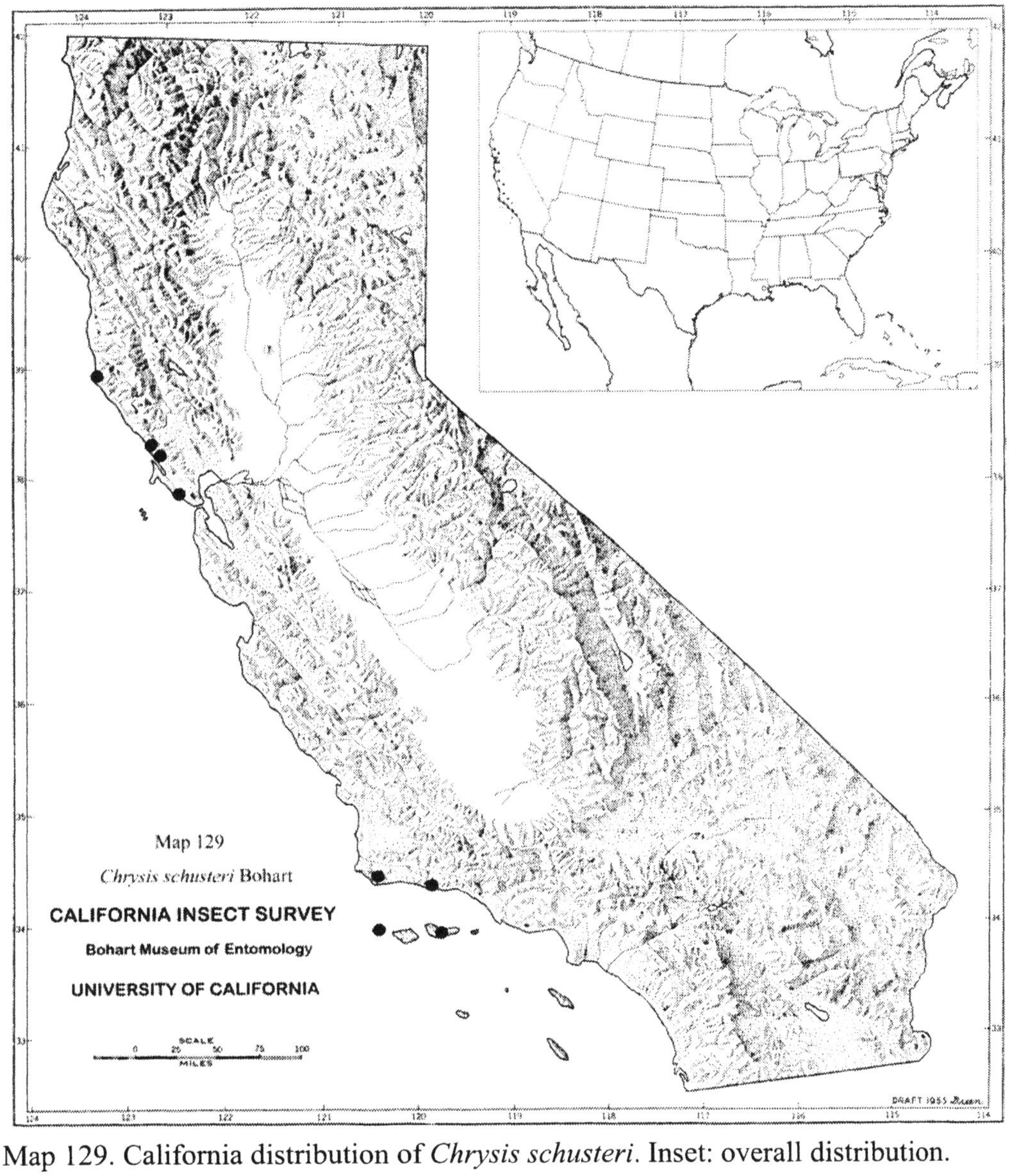

Map 129. California distribution of *Chrysis schusteri*. Inset: overall distribution.

Hosts. Unknown.

Discussion. The distribution of *schusteri* is unusual, as the species is largely restricted to the outer Coastal Range of the California coast and particularly on the Channel Islands. This is one of the smaller-bodied members of the *nitidula* group, ranging from 6-7 mm long. *C. schusteri* is a closely punctate species, with an evanescent frontal carina, a broad frons, and short eye height. Examination of the holotype of *antiochicola* revealed that it was identical with the type of *schusteri*.

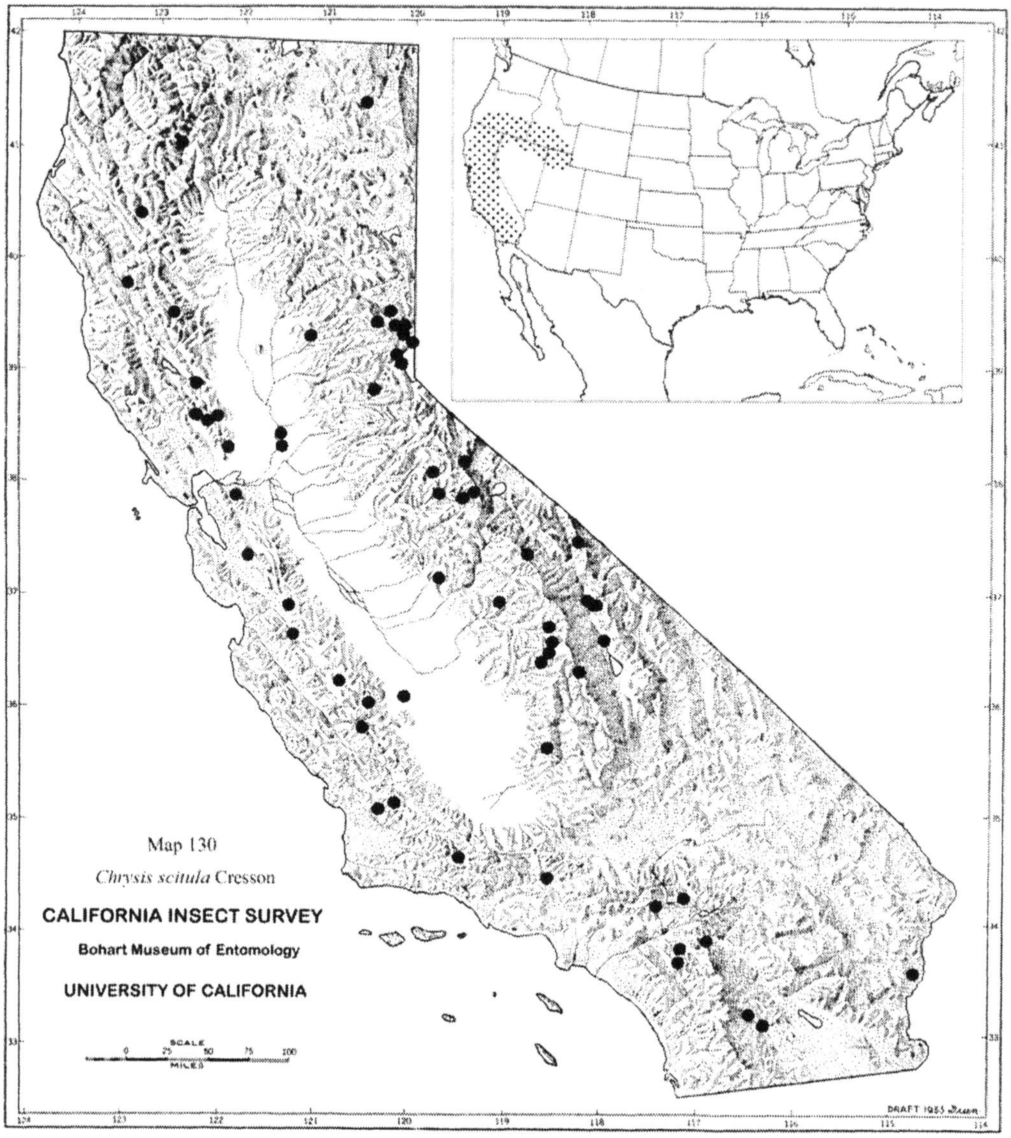

Map 130. California distribution of *Chrysis scitula*. Inset: overall distribution.

Chrysis scitula Cresson

Fig. 74, Map 130

Chrysis scitula Cresson 1865:308. Holotype ♀; USA: "Colorado Territory" (ANSP).

Chrysis californica Gribodo 1879:336. Holotype ♀; USA: California (MCSN).

Geographic Range (Map 130). Lower Sonoran to Canadian Zones, widespread in the USA and Mexican mountains; 399 specimens were studied.

California Records. Collected April through August; **Contra Costa Co**.: Mt. Diablo; **El Dorado Co**.: 15 mi. e Georgetown; Meyers; **Fresno Co**.: 12 mi. w Coalinga; Jacalitos Canyon; Watts Valley; **Glenn Co**.: Plaskett Meadow; **Inyo Co**.: 8 mi. n Independence; Big Pine Creek; **Kern Co**.: Glennville; **Lake Co**.: north fork Cache Creek, Hwy 20; **Los Angeles Co**.: Tanbark Flat; **Madera Co**.: Knowles; **Mendocino Co**.: Eel River Camp; **Modoc Co**.: 14 mi. n Alturas; **Mono Co**.: Big McGee Lake; Sonora Pass; White Mts.; **Monterey Co**.: Arroyo Seco Camp; Carmel Valley; **Napa Co**.: 5 mi. se Pope Valley; Angwin; **Nevada Co**.: Sagehen Creek; Hobart Mills; **Placer Co**.: Carnelian Bay, Lake Tahoe; **Riverside Co**.: Herkey Creek; San Jacinto Mts.; Blythe; **Sacramento Co**.: 5 mi. sw Ryde; Sacramento; **San Benito Co**.: 40 mi. e Paicines; Bear Gulch Area, Pinnacles National Monument; **San Bernardino Co**.: Yucaipa; **San Diego Co**.: Borrego Springs; near Buckman Springs; Phelan; **San Luis Obispo Co**.: 10 mi. w Simmler; Pozo; **Santa Clara Co**.: 1 mi. n Coyote; **Sierra Co**.: Independence Lake; Sardine Lakes; Sattley; Sierra Valley **Solano Co**.: 3.2 mi. w Mix Canyon; **Trinity Co**.: Carrville; Hayfork Ranger Station; **Tulare Co**.: Kings Canyon; Potwisha Sequoia National Park; Ash Mt. Range; Mineral King; White Chief Creek; **Tuolumne Co**.: Blue Canyon; Sonora Pass; Strawberry; **Ventura Co**.: Mt. Pinion; **Yolo Co**.: Cache Creek Canyon; **Yuba Co**.: 18 mi. nw Marysville.

Hosts. Unknown.

Discussion. Similar to *dorsalis, scitula* can be distinguished by the considerably reduced teeth on tergum III, which may be hardly recognizable as such, the unusually narrow frons that is rarely broader than the eye in front view, and the large polished area in the scapal basin.

Chrysis serrata Taylor

Fig. 75, Map 131

Chrysis serrata Taylor 1924:329. Holotype ♀; USA: Washington, Wenass Valley (CAS).

Geographic Range (Map 131). Upper Sonoran and Canadian Zones, west of the Mississippi River, and in Michigan; 273 specimens were examined.

California Records. Collected May through August; **Calaveras Co**.: Mokolumne Hill; **El Dorado Co**.: 6 mi. e on Ice House Rd.; Chile Bar; Emigrant Gap; **Inyo Co**.: Independence; Lone Pine Canyon; **Lassen Co**.: Bridge Creek Camp; Hallelujah Junction; **Mono Co**.: 1 mi. s Toms Place; Hot Creek; **Monterey Co**.: Arroyo Seco;

San Ardo; **Nevada Co.**: Boca; Hobart Mills; Sagehen Creek; **Plumas Co.**: Blairsden; Johnsville; **Riverside Co.**: Anza; **San Diego Co.**: 2 mi. n Warner Springs; **San Luis Obispo Co.**: 10 mi. w Simmler; **Sierra Co.**: Gold Lake; Independence Lake; Sierraville.

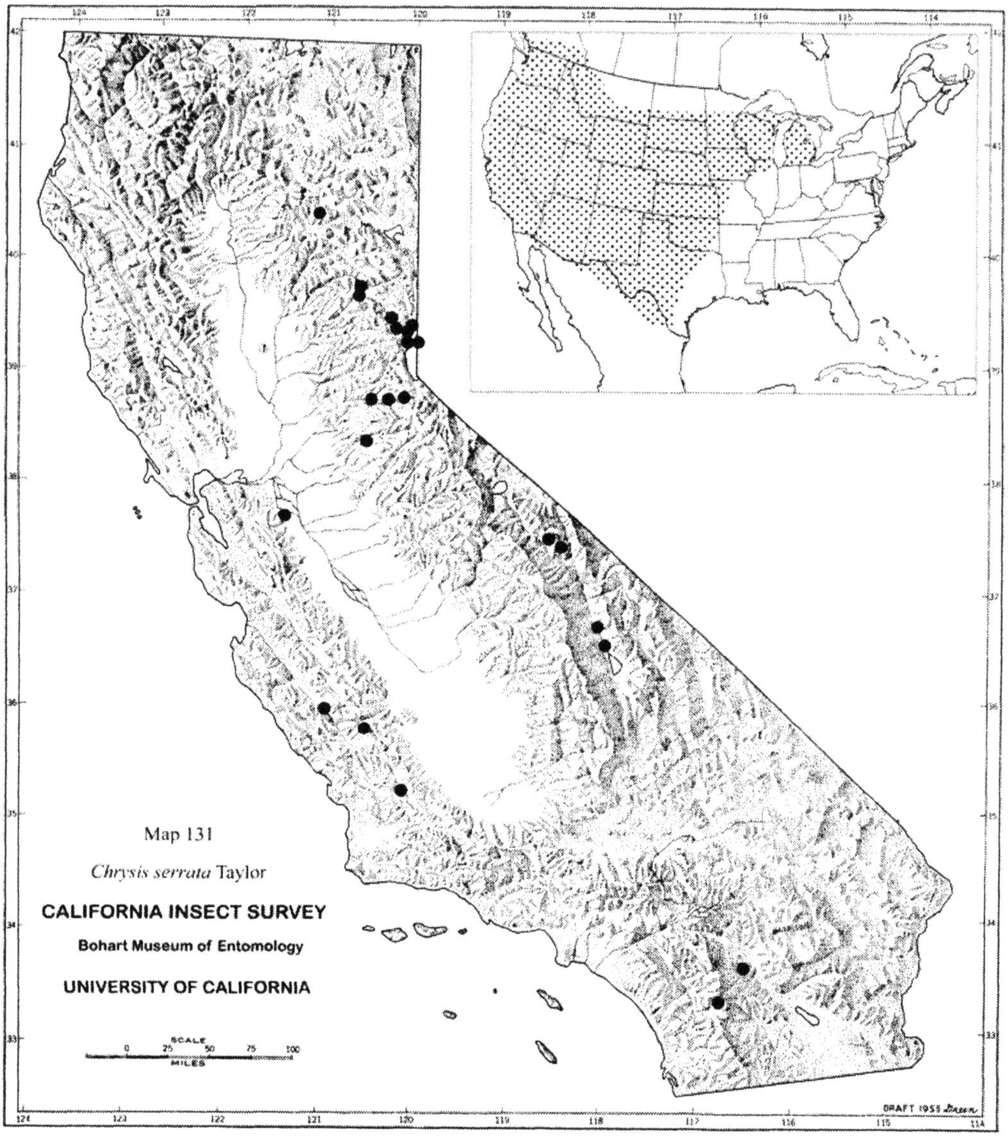

Map 131. California distribution of *Chrysis serrata*. Inset: overall distribution.

Hosts. Unknown.

Discussion. *Chrysis serrata* must be separated from the other six-toothed species by a combination of characteristics: the propodeal tooth is straight or slightly convex beneath, malar space is 1 MOD long or less and is shorter than the subantennal distance, the mesopleuron lacks polished knobs, tergum III prepit swelling is well-developed, the pit row has a narrow and sunken medial ridge, the scapal basin is covered with dense silvery setae and flagellomere I is less than twice as long as broad.

Chrysis severa Mocsáry

Map 132

Chrysis severa Mocsáry 1889:573. Holotype ♀; Mexico: Cordova (Cordoba!) (GENEVA).

Chrysis praestans du Buysson 1898a:144. Holotype ♀; West Indies: Grenada (locality in error? poss. Grenada, Mississippi) (MNHN).

Geographic Range (Map 132). Widespread in the USA, Canada, Mexico (Coahuila), and in the West Indies; 115 specimens were seen.

California Records. Collected June through August; **Alpine Co**.: Ebbetts Pass; **Inyo Co**.: Big Pine Creek; north fork Big Pine Creek below First Lake; **Lassen Co**.: Bridge Creek Camp; Hat Lake; Summit Camp; **Modoc Co**.: 6 mi. nw Cedarville; **Mono Co**.: Blanco's Corral, White Mts.; Crooked Creek Laboratory, White Mts.; Sonora Pass; **Nevada Co**.: Boca; Sagehen Creek; **Placer Co**.: Brockway; **Tulare Co**.: Mineral King.

Hosts. Unknown.

Discussion. The distinctively long malar space (2-3 MOD long) will immediately separate *severa* from other 6-tooth species. In addition, flagellomere I is more than twice as long as broad, the propodeal tooth is lobate beneath, tergum III has the area beyond the pit row not depressed between the medial teeth and the most lateral tooth forming a right or acute angle.

This species was previously known as *praestans* (as in Bohart and Kimsey 1982) in California, but examination of types by Kimsey and Bohart (1991) revealed that *praestans* was a junior synonym of *severa*.

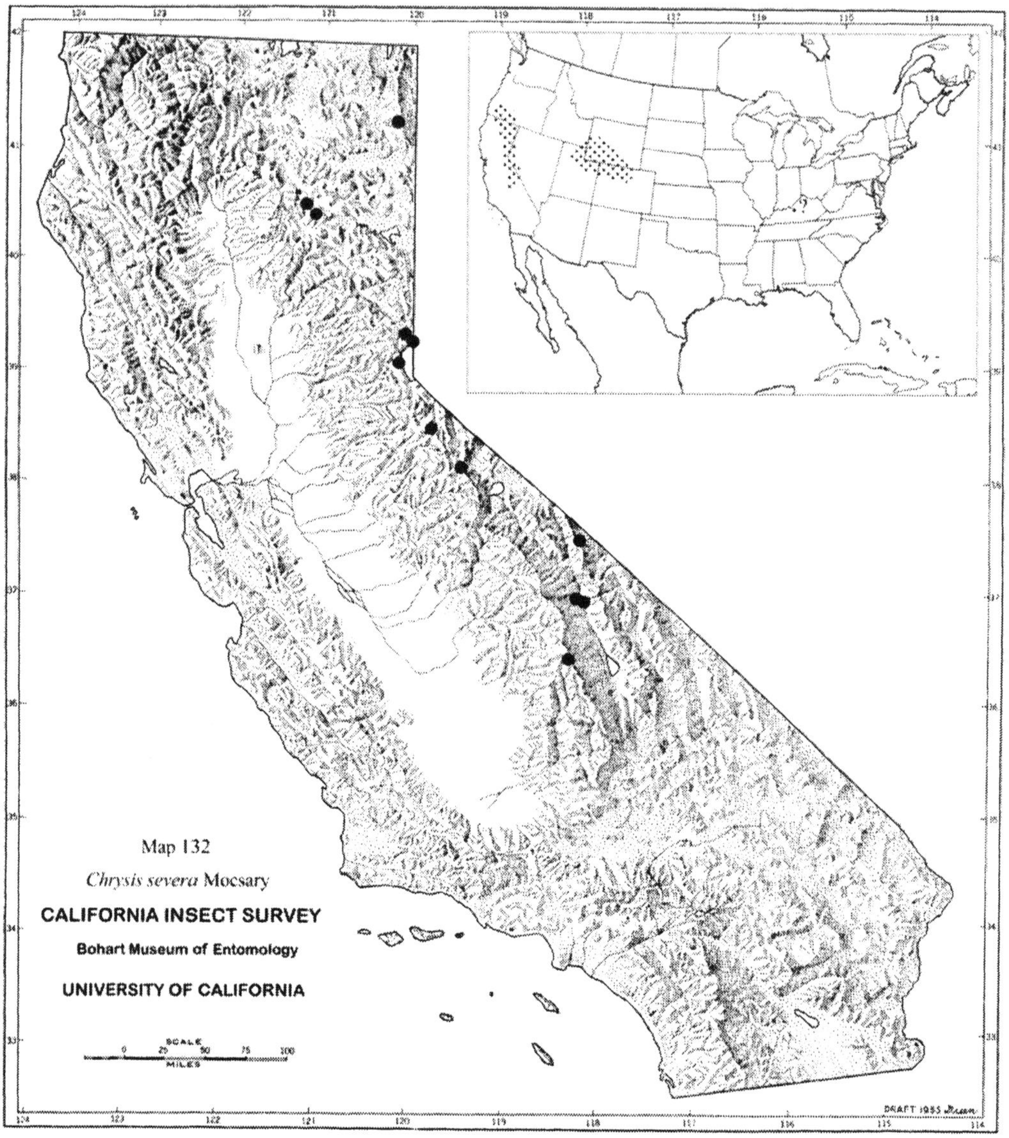

Map 132. California distribution of *Chrysis severa*. Inset: overall distribution.

Chrysis snowi Viereck

Map 133

Chrysis snowi Viereck 1906:195. Holotype ♂; USA: Kansas, Clark Co. (SMEK).

Geographic Range (Map 133). Lower Sonoran to Hudsonian Zones in the USA and Canada west of 100th meridian; 172 specimens were studied.

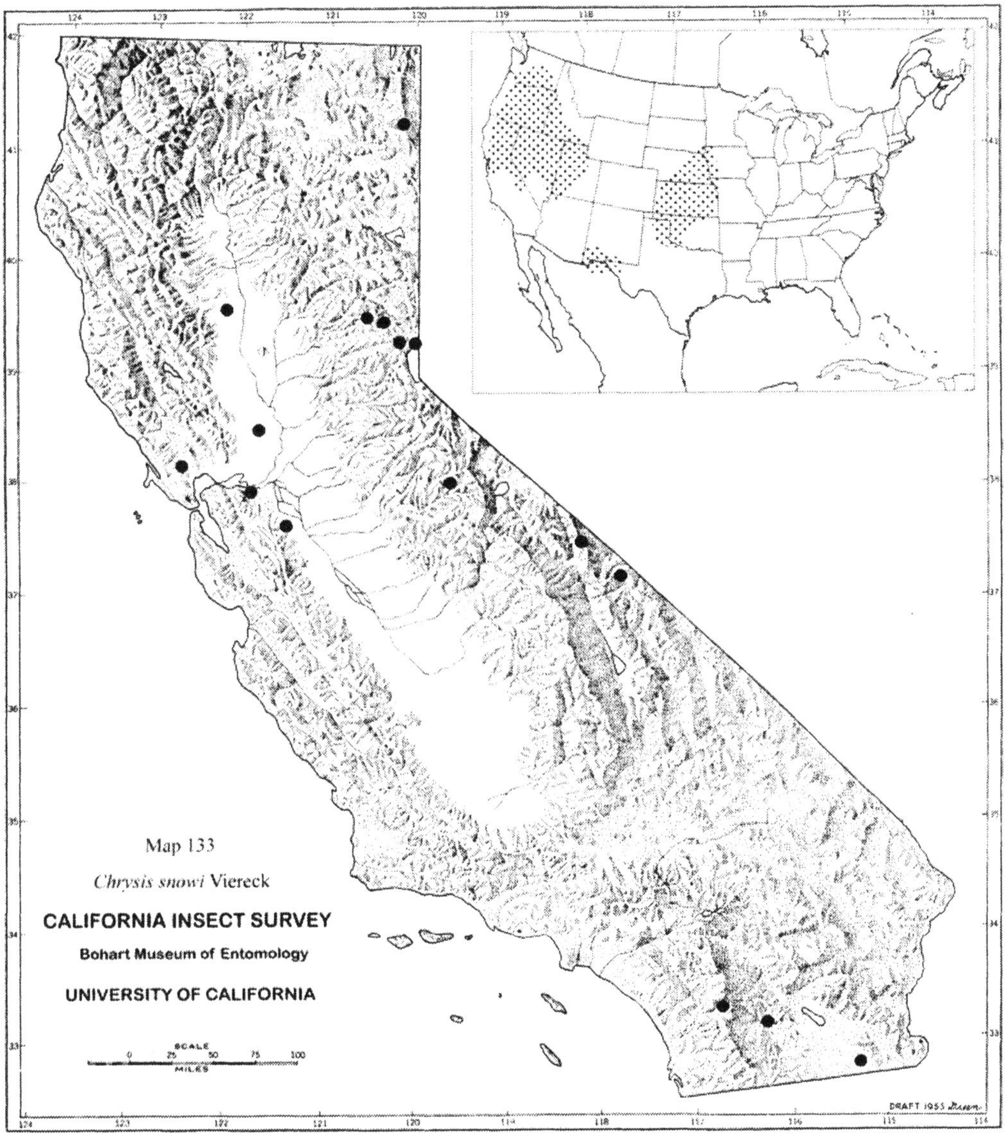

Map 133. California distribution of *Chrysis snowi*. Inset: overall distribution.

California Records. Collected April through October; **Contra Costa Co**.: Antioch; **Glenn Co**.: Artois; **Imperial Co**.: 10 mi. n Glamis; **Inyo Co**.: Deep Springs; **Modoc Co**.: 5.5 mi. e Cedarville; **Mono Co**.: Blanco's Corral, White Mts.; **Nevada Co**.: Boca; Russell Valley; **San Diego Co**.: 2 mi. n Warner Springs; **San Diego Co**.: Borrego; **San Joaquin Co**.: Tracy; **Sierra Co**.: Sattley; Yuba Pass; **Sonoma Co**.: Petaluma; **Tuolumne Co**.: Dardanelles; **Yolo Co**.: Davis.

Hosts. Unknown.

Discussion. Close to *amala* as discussed under that species, additional diagnostic characteristics of *snowi* include males with dense appressed silvery pubescence on the face, the male flagellomere I longer than II but shorter than III, the female malar space shorter than the length of flagellomere II (about 1.5 MOD long), and tergum III medial notch forming about a half circle.

Chrysis tensa Bohart

Map 134

Chrysis tensa Bohart 1964:230. Holotype ♂; USA: Arizona, 15 mi. n Yuma (UCDC).

Geographic Range (Map 134). Southwestern USA and northern Mexico; 80 specimens were studied.

California Records. Collected in June, July; **Imperial Co**.: Brawley; Imperial; **Riverside Co**.: 12 mi. n Blythe; **San Bernardino Co**.: Needles.

Hosts. Unknown.

Discussion. *Chrysis tensa* is most similar to *propria*. These species can be separated by the shape of the apicomedial notch of tergum III, which is subtriangular in *tensa* and U-shaped in *propria*. In addition, tergum III is laterally expanded and somewhat angulate or dentate in *tensa*.

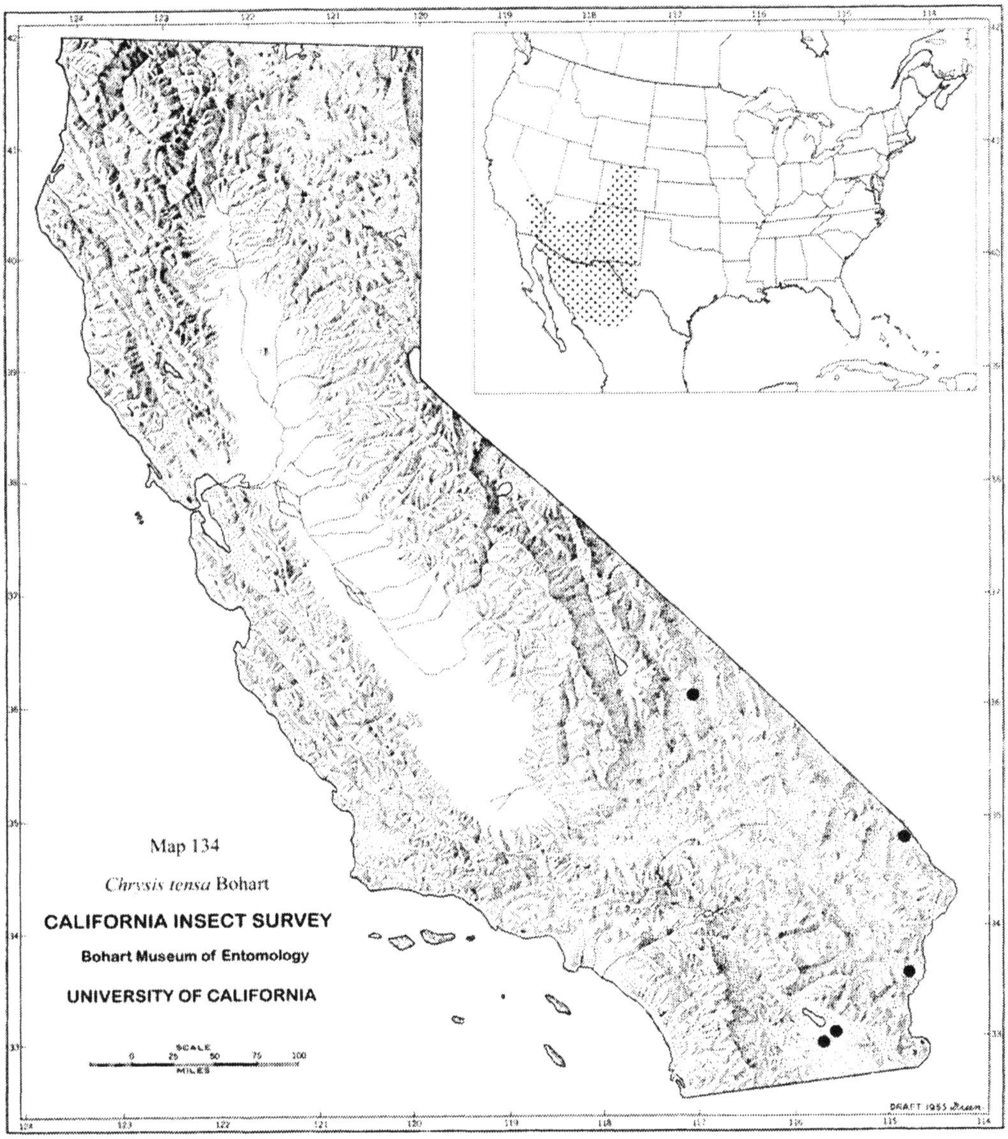

Map 134. California distribution of *Chrysis tensa*. Inset: overall distribution.

Chrysis tenuicornis Taylor

Figs. 76, 86; Map 135

Chrysis tenuicornis Taylor 1924:327. Holotype ♀; USA: Oregon, Harney Co., Steens Mts. (CAS).

Geographic Range (Map 135). Upper Sonoran Zone west of 100th meridian in the USA, northern Mexico (Chihuahua, Sinaloa, Jalisco) and southwestern Canada (British Columbia); 33 specimens were seen.

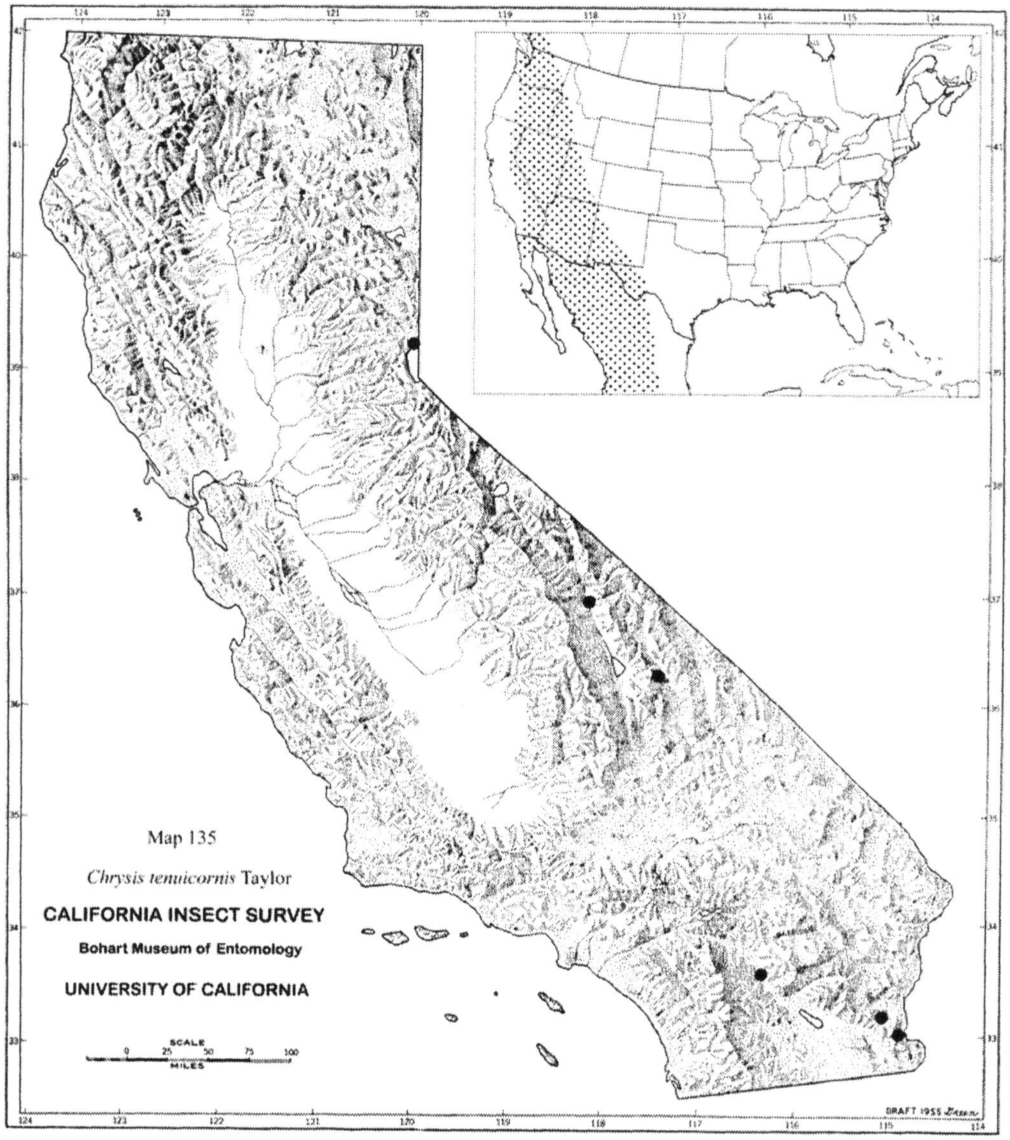

Map 135. California distribution of *Chrysis tenuicornis*. Inset:overall distribution.

California Records. Collected April through July; **Imperial Co.**: Chocolate Mts.; Black Mt.; **Inyo Co.**: 15 mi. s Big Pine; 6 mi. n Darwin; **Napa Co.**: Samuel Springs (beneath Lake Berryessa); **Nevada Co.**: Boca; Riverside Co.: 3.5 mi. s Palm Desert; **Sonoma Co.**: Boyes Hot Springs.

Hosts. Unknown.

Discussion. *Chrysis tenuicornis* has a number of distinctive features that will immediately distinguish it from other six-toothed species. Tergum III has the medial pair of teeth blunt and an oddly formed pit row. The antenna is unusually slender, with flagellomere I 6-7x as long as broad.

This species is rarely collected and is quite uncommon in collections.

Chrysis tripartita Aaron

Map 136

Chrysis tripartita Aaron 1885:238. Holotype ♀; USA: Arizona (ANSP).

Tetrachrysis indeterminabilis Bischoff 1910:476. Holotype ♀; Mexico: Sierra Mixteca (MNHU).

Geographic Range (Map 136). Widespread in the western USA and northwestern Mexico (Baja California, Chihuahua, Hidalgo, Sinaloa); 171 specimens were studied.

California Records. Collected March through September; **Kern Co**.: 1 mi. w YMCA Summer Camp Mt. Pinos; Mill Potrero; **Lake Co**.: Kelseyville; **Los Angeles Co**.: Claremont; **Placer Co**.: Folsom Lake; **Sacramento Co**.: Folsom; **San Diego Co**.: Dodge Valley, 2.7 mi. e Oak Grove; **Santa Barbara Co**.: Santa Barbara; Santa Ynez Mts.; **Shasta Co**.: 5 mi. n Hat Creek; **Siskiyou Co**.: Valentine Caves, Lava Bed National Monument; **Yolo Co**.: Clarksburg; Davis.

Hosts. Grigarick and Stange (1968) and Horning (1971) reared *tripartita* from *Anthidium collectum* Huard and *A. maculosum* Cresson (Megachilidae).

Discussion. This is a large-bodied species (8-9 mm long), with an unusually long face; the malar space is 2 MOD long. Structurally it appears most similar to *coloradica* and *xerophila*. Diagnostic features include the irregular transverse frontal carina, which nearly encloses the midocellus, distinct genal carina, tergum III with a straight lateral margin, poorly defined prepit swelling and teeth short and apically rounded, sternum II spots large and nearly contiguous, and the punctation of tergum III finer than on the scutum.

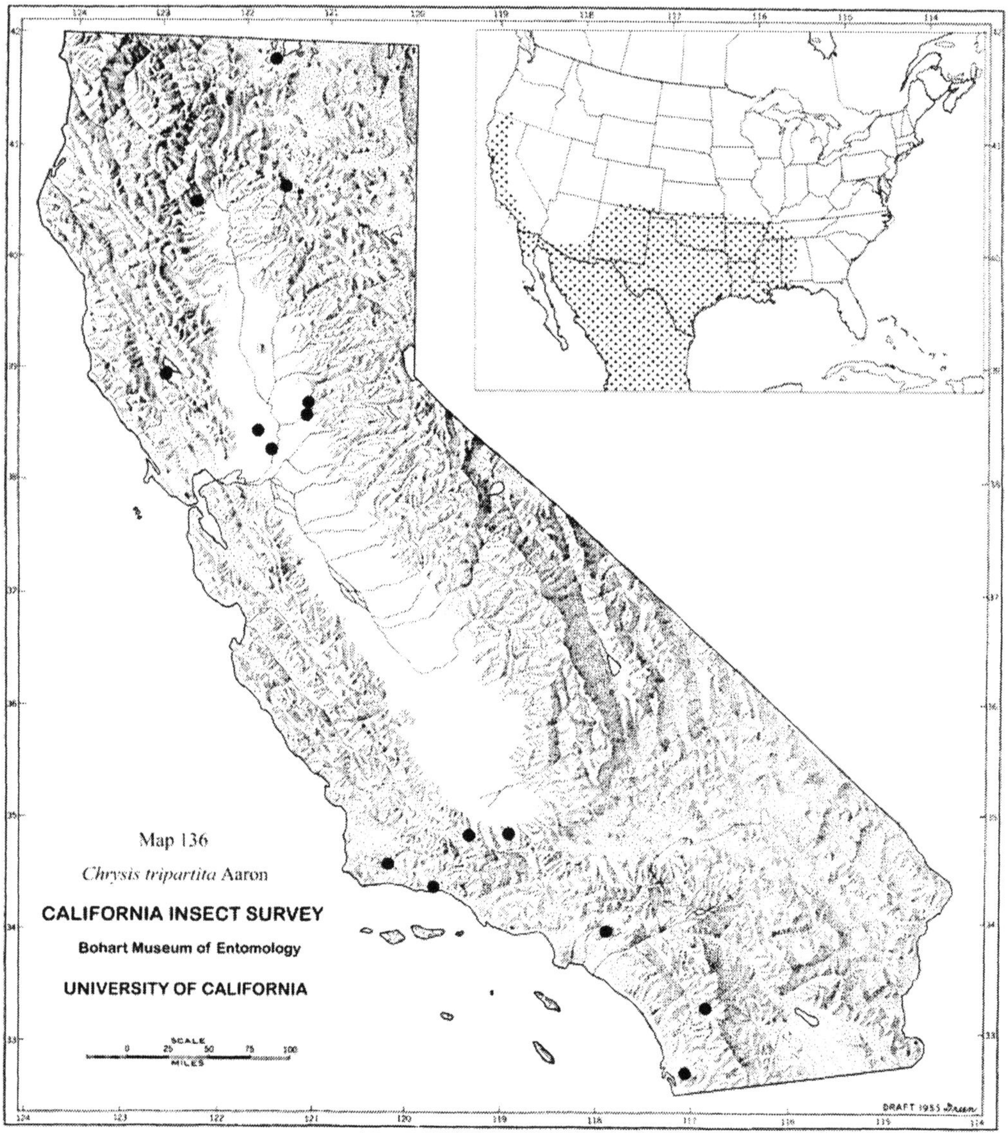

Map 136. California distribution of *Chrysis tripartita*. Inset: overall distribution.

Chrysis tularensis Bohart

Fig. 81, Map 137

Chrysis tularensis Bohart 1962:373. Holotype ♂; USA: California, Tulare Co., Woodlake (CAS).

Geographic Range (Map 137). Foothills of the San Joaquin Valley of California, three specimens were seen.

California Records. Collected April through June; **Amador Co**.: Plymouth; **Fresno Co**.: Auberry 4.1 mi. e Auberry Rd.; Wonder Valley; **Monterey Co**.: Arroyo Seco; **Tulare Co**.: Wood Lake.

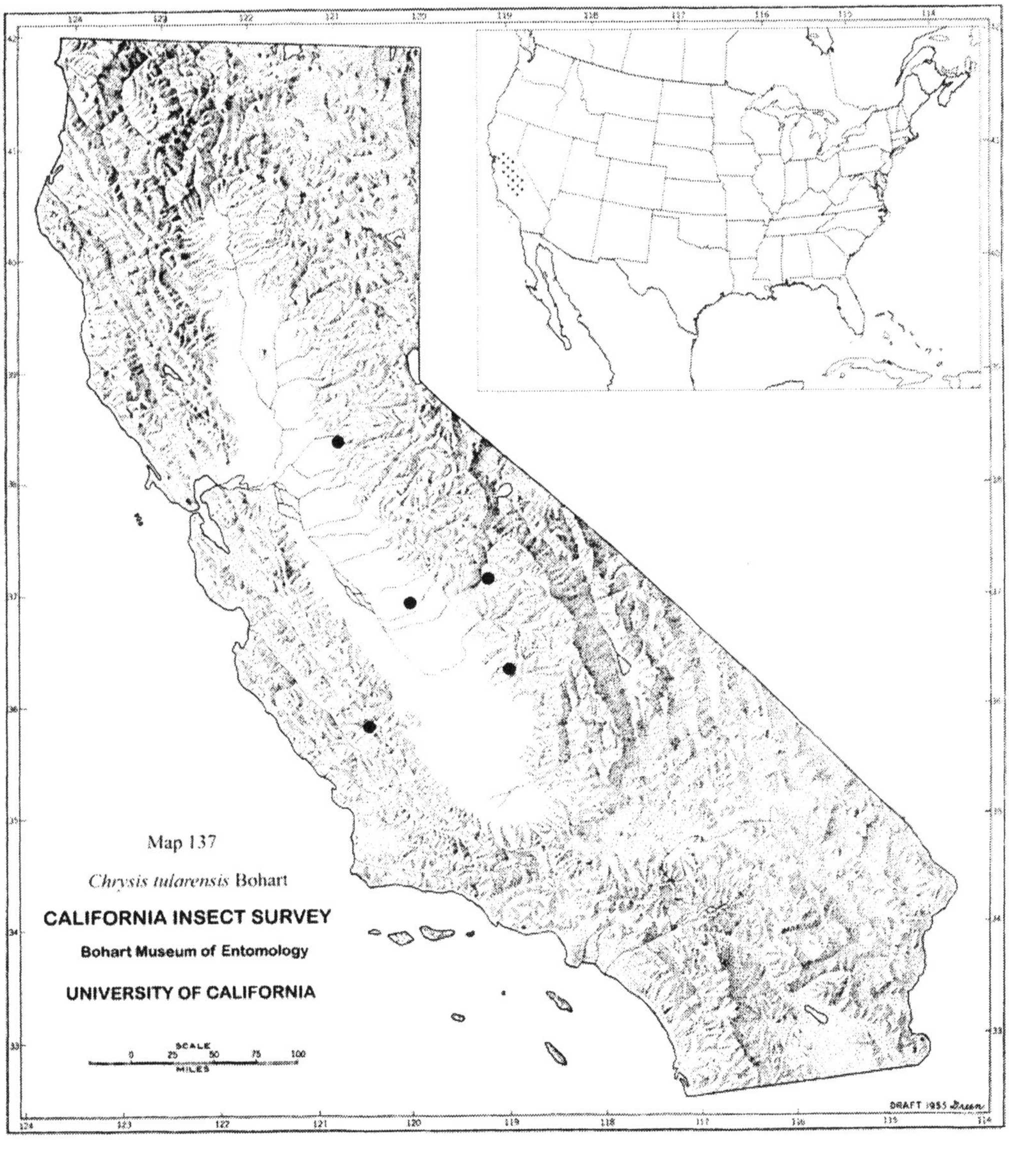

Map 137. California distribution of *Chrysis tularensis*. Inset: overall distribution.

Hosts. Unknown.

Discussion. This six-toothed California endemic is known only from a small series of specimens. It is most similar to the non-Californian species *clara*. It can be distinguished from that and other such species by the following combination of features: flagellomere I less than twice as long as broad, transverse frontal carina obsolescent, propodeal tooth nearly straight beneath, and tergum III with the area beyond the pit row not depressed between the medial teeth, medial teeth as long as wide, apicomedial notch not bicarinate, and the most lateral tooth obtuse.

Chrysis vagabunda Bohart

Map 138

Chrysis vagabunda Bohart 1964:225. Holotype ♂; USA: California: Yolo Co., Woodland (UCDC).

Geographic Range (Map 138). Upper Sonoran Zone of the western USA; 107 specimens were seen.

California Records. Collected May through September; **Contra Costa Co**.: Antioch; **Glenn Co**.: Artois; **Los Angeles Co**.: Tanbark Flat; **Monterey Co**.: Arroyo Seco Camp; San Ardo; **Riverside Co**.: Norco; **Santa Barbara Co**.: 4 mi. e Los Prietos; Santa Cruz Island; **Santa Clara Co**.: Alum Rock Park; **Stanislaus Co**.: Newman; **Yolo Co**.: Davis; Putah Canyon; Woodland.

Hosts. Krombein (1979) reported *vagabunda* from nests of *Anthidium collectum* Huard (Megachilidae).

Discussion. Diagnostic features of *vagabunda* include flagellomere I 1.3x as long as broad and subequal in length to II plus III, tergum III pit row deep with most pits confluent and teeth sharp and short, and male face covered with dense silvery pubescence. Body length in this species ranges from 6.5-7.0 mm.

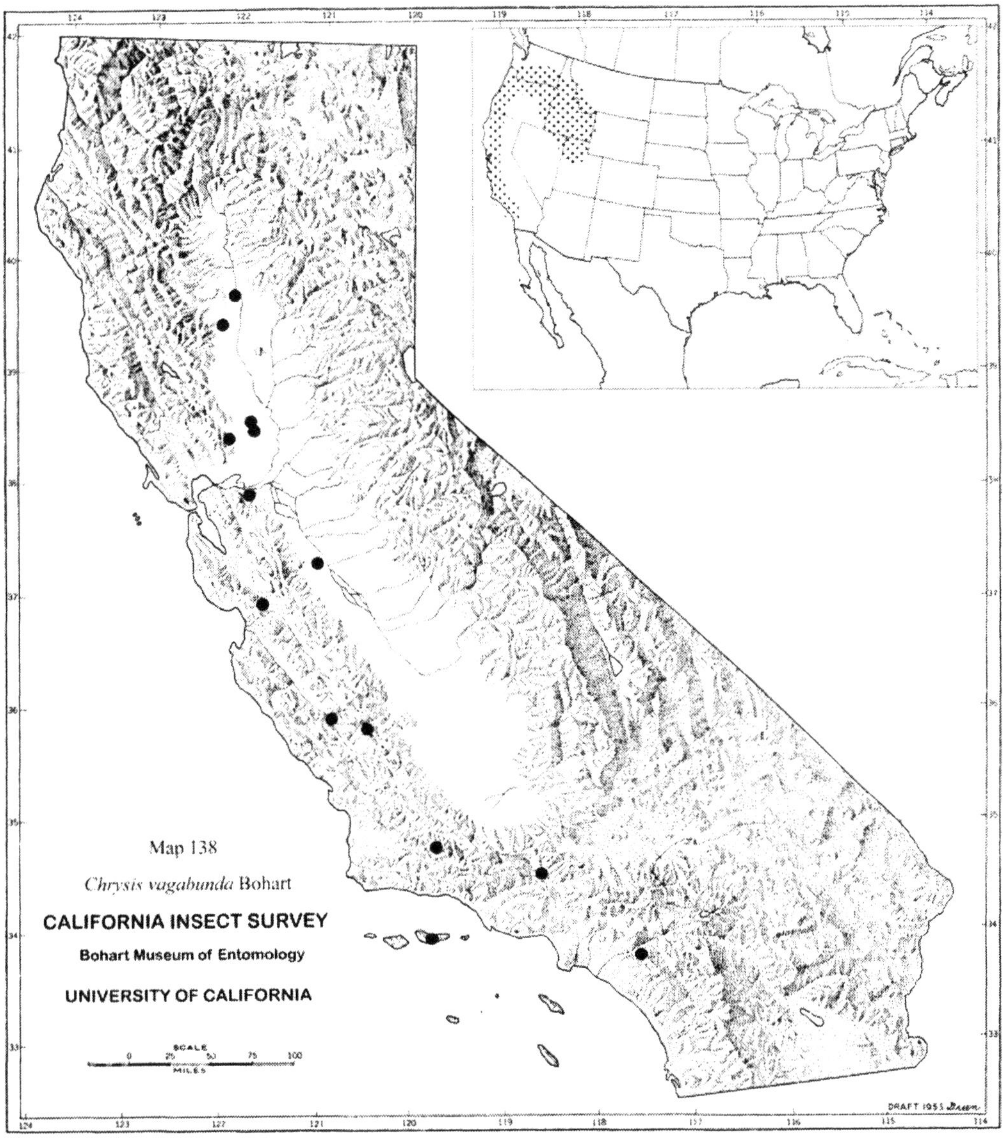

Map 138. California distribution of *Chrysis vagabunda*. Inset: overall distribution.

Chrysis venustella Bohart

Map 139

Chrysis venustella Bohart 1964:235. Holotype ♂; USA: California, Davis (UCDC).

Geographic Range (Map 139). Upper Sonoran to Canadian Zones west of the 100th meridian in the USA, and in Mexico (Durango); 134 specimens were studied.

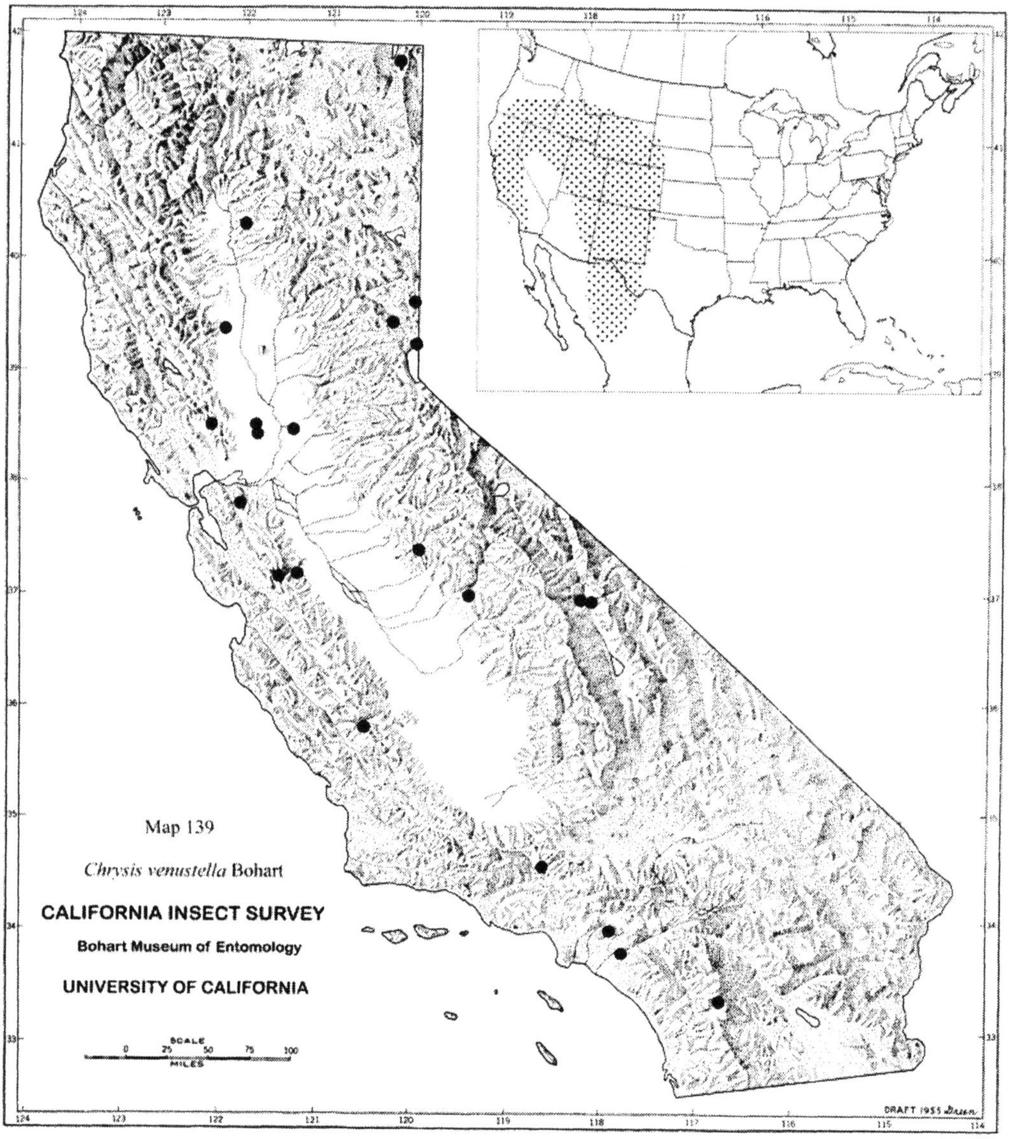

Map 139. California distribution of *Chrysis venustella*. Inset: overall distribution.

California Records. Collected May through September; **Contra Costa Co**.: 5 mi. se Walnut Creek; **Fresno Co**.: Watts Valley; **Glenn Co**.: Artois; **Inyo Co**.: 5 mi. e Big Pine; Big Pine Creek; **Lassen Co**.: Hallelujah Junction; **Los Angeles Co**.: Claremont; Tanbark Flat; **Mariposa** Co.: Mariposa; **Modoc Co**.: 0.5 mi. w Willow Ranch; **Monterey Co**.: Arroyo Seco; **Napa Co**.: Samuel Springs (beneath Lake

Berryessa); **Nevada Co.**: Boca; **Orange Co.**: Peters Canyon; **Sacramento Co.**: Carmichael; **San Diego Co.**: 2 mi. n Warner Springs; **Santa Barbara Co.**: Santa Ynez Mts.; **Santa Clara Co.**: San Antonio Valley; **Sierra Co.**: Sierraville; **Stanislaus Co.**: Del Puerto Canyon; **Tehama Co.**: Manton; **Yolo Co.**: Davis; Woodland.

Hosts. Unknown.

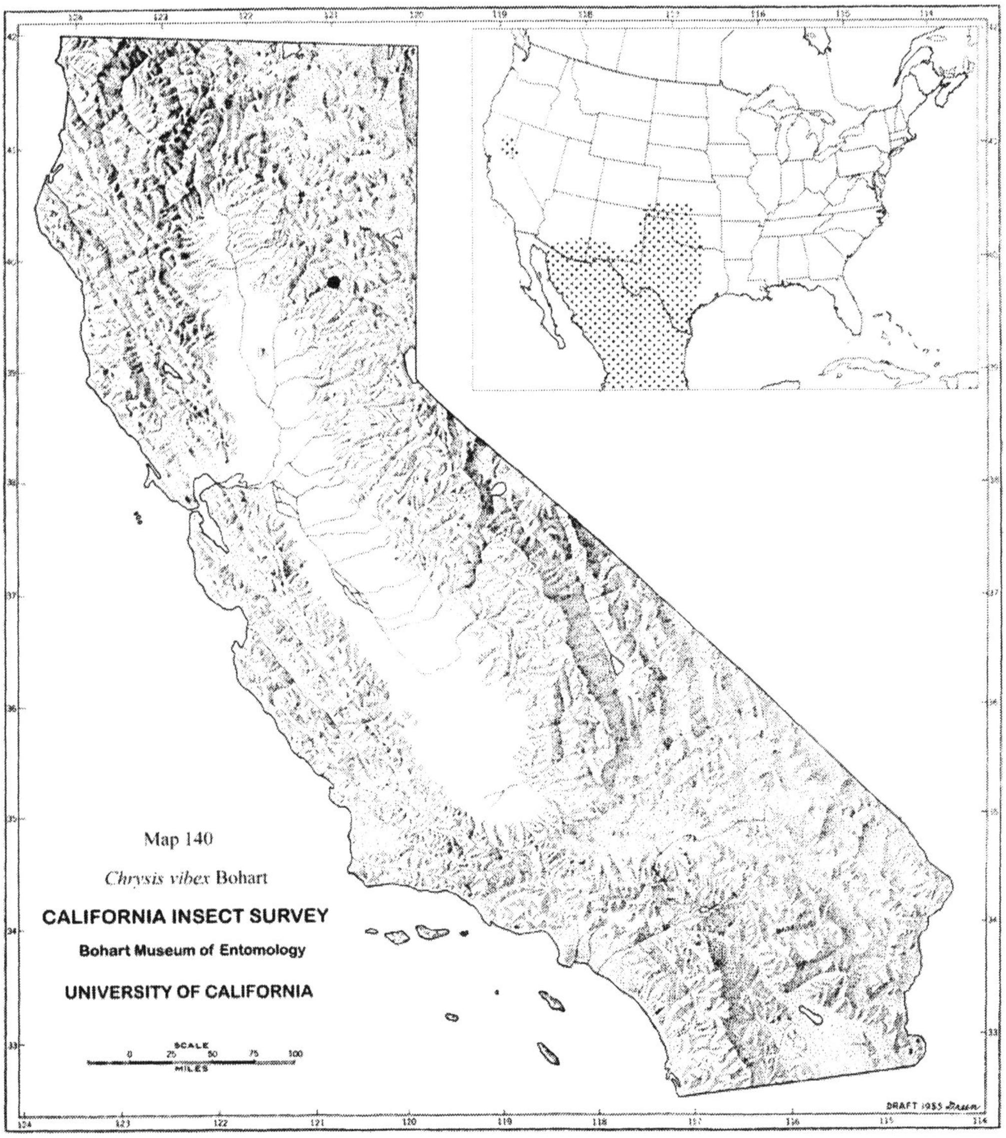

Map 140. California distribution of *Chrysis vibex*. Inset: overall distribution.

Discussion. This is another distinctive species without closely related species in California. *Chrysis venustella* can be distinguished by the unusually short malar space, which is less than 0.5 MOD long, tergum III with medial ridge of pit row sharp-edged, punctate and sunken in profile, and apical teeth closely punctate.

Chrysis vibex Bohart

Map 140

Chrysis vibex Bohart 1964:231. Holotype ♂; USA: Arizona, Portal (UCDC).

Geographic Range (Map 140). Southwestern USA west of the 100th meridian and in Mexico (Sinaloa, Nuevo Leon); 131 specimens were seen.

California Records. Collected in June, July; **Plumas Co**.: 4 mi. w Quincy.

Hosts. Unknown.

Discussion. Related to *propria* and *tensa, vibex* can be distinguished from these species by the long subantennal distance of 2 MOD and postocular area with a polished and impunctate area on the upper genal area.

Chrysis violacuna Bohart

Map 141

Chrysis violacuna Bohart (in Bohart & Kimsey) 1982:134. Holotype ♂; USA: Utah, Rich Co., Bear Lake (UCDC).

Geographic Range (Map 141). Upper Sonoran Zone of the USA west of the 100th meridian; 134 specimens were studied.

California Records. Collected April through July; **Inyo Co**.: Big Pine Creek; **Kern Co**.: Short Canyon; 6 mi. w Inyokern; **Lassen Co**.: Hallelujah Junction; **Los Angeles Co**.: Newhall; **Mono Co**.: Paradise Camp; **Riverside Co**.: Riverside.

Hosts. F. D. Parker recorded *Parancistrocerus acarigaster* (Bohart) (Vespidae) as a host for *violacuna* (Bohart and Kimsey 1982).

Discussion. This species is very similar to *nitidula*. It differs in having tergum I with red reflections on the anterior declivity of tergum I, and tergum III pit row in posterior view with three moderate-sized pits.

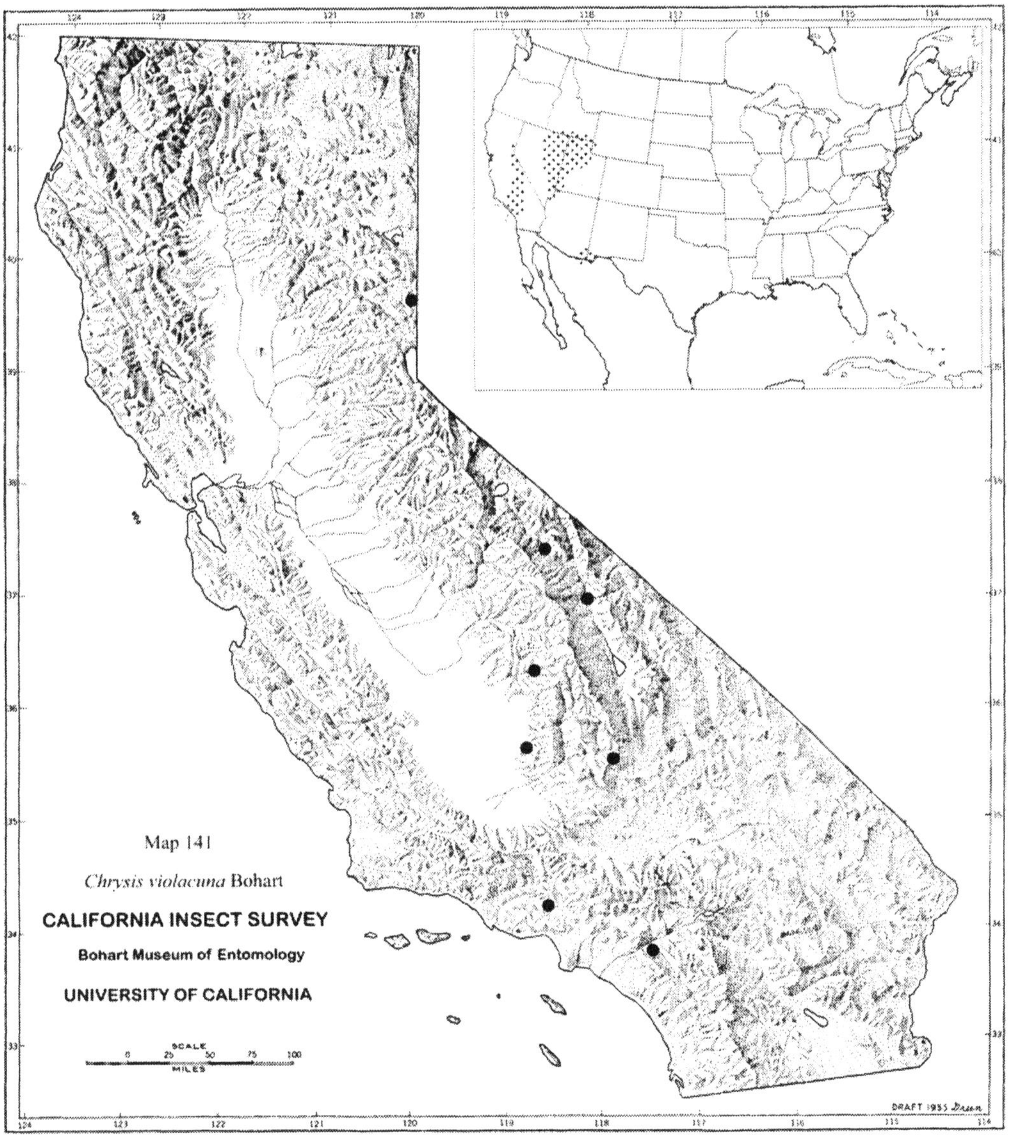

Map 141. California distribution of *Chrysis violacuna*. Inset: overall distribution.

Chrysis wasbaueri Bohart

Fig. 80, Map 142

Chrysis wasbaueri Bohart 1962:372. Holotype male; USA: Texas, Bexar Co., Ft. Sam Houston (CAS).

Geographic Range (Map 142). Transcontinental in the southern USA and in northern Mexico (Durango, Nuevo Leon); 136 specimens were examined.

California Records. Collected June through July; **Inyo Co.**: Big Pine Creek; **Kern Co.**: Kernville.

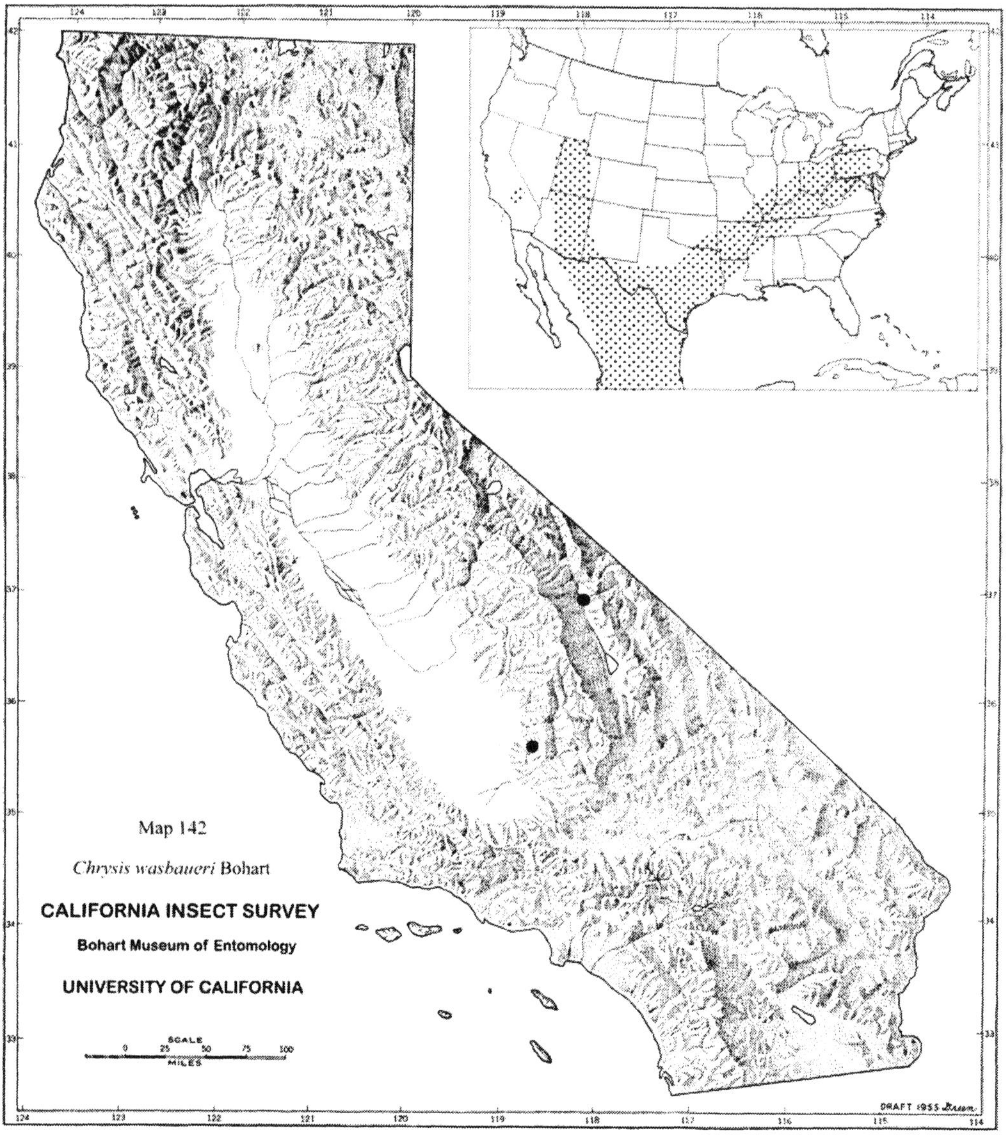

Map 142. California distribution of *Chrysis wasbaueri*. Inset: overall distribution.

Hosts. Unknown.

Discussion. This is a six-toothed species similar to *oraria* except that the teeth on tergum III are bicarinate. In addition, the propodeal tooth is strongly lobate beneath, tergum III has the distal area depressed between the medial teeth, the teeth are longer than wide basally, and the sternum II black spots are large and contiguous, or nearly so. Male flagellomere I is less than twice as long as broad.

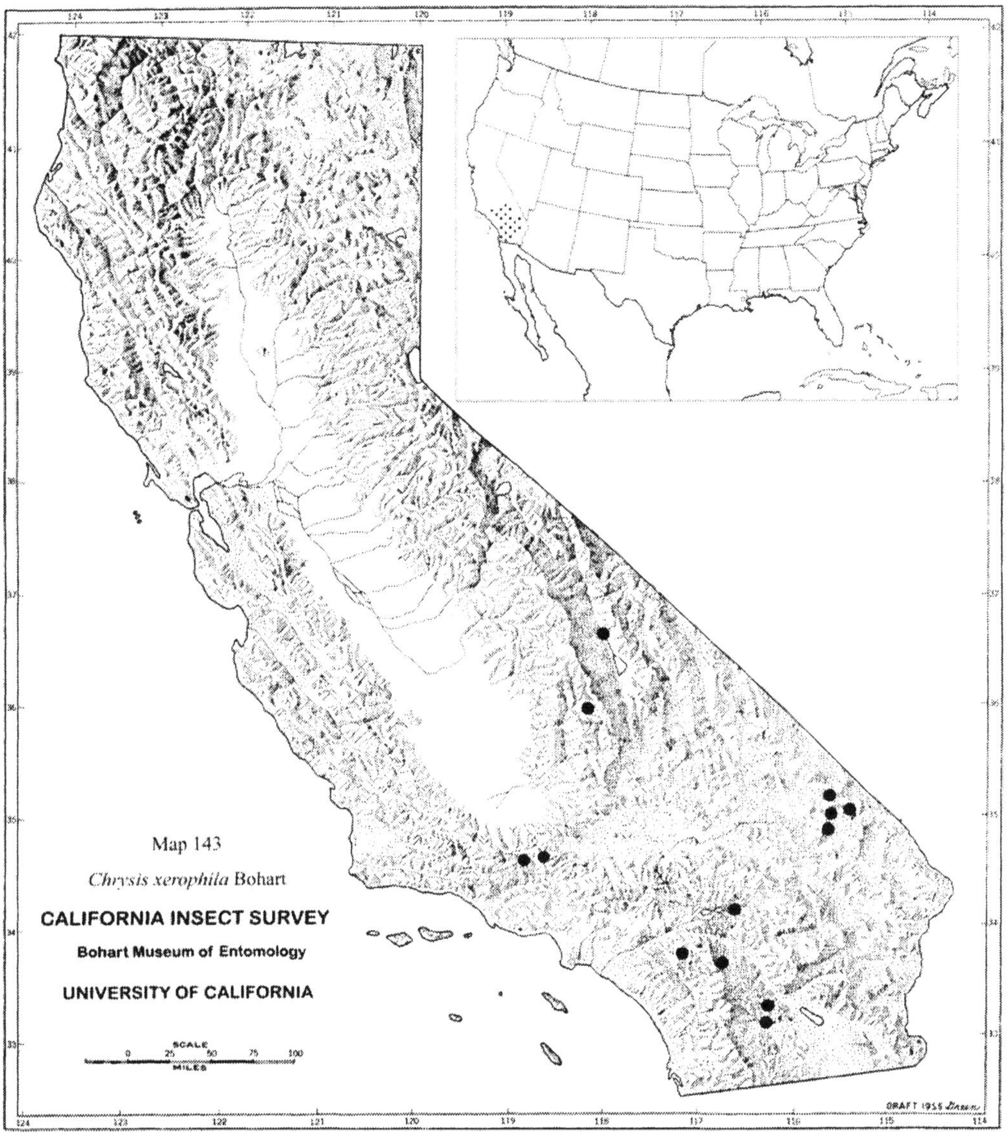

Map 143. California distribution of *Chrysis xerophila*. Inset: overall distribution.

Chrysis xerophila Bohart

Map 143

Chrysis xerophila Bohart 1964:226. Holotype ♂; USA: California, San Diego Co., Borrego Valley (UCDC).

Geographic Range (Map 143). Upper and Lower Sonoran Zones of southern California; 35 specimens were studied.

California Records. Collected in April, May; **Los Angeles Co.**: 5 mi. s Gorman; **Riverside Co.**: Pinon Flat, San Jacinto Mts.; **San Bernardino Co.**: 5.8 mi. n nw Granite Pass; Granite Mts.; Mid Hills; 9 mi. s se Cima; Morongo Valley; Providence Mts.; Upper Morongo Valley; **San Diego Co.**: Borrego Springs; Borrego Valley dunes and Palm Canyon; **Tulare Co.**: Roads End; **Ventura Co.**: Hungry Valley.

Hosts. Unknown.

Discussion. *Chrysis xerophila* is a rare, localized species, which resembles *coloradica*. It differs in the male by having a simple flagellum without a setal fringe and a broader frons without silvery pubescence on the lower part. In addition, *xerophila* has the black spots on sternum II obsolescent and the body color is greener than that of *coloradica*. Body length is on average 7 mm.

Genus *Chrysura* Dahlbom

Generic Diagnosis. Face relatively flat without medial zone of crossridging but sometimes with medial microreticulation, no transverse frontal carina; malar space usually 2 MOD long or longer; mandible with subapical tooth; midocellus not lidded; basal flagellomeres II-V of male often bulging ventrally; pronotum not longer and usually shorter than scutellum, lateral depression shallow and barely indicated; mesopleuron with scrobal sulcus and episternal sulcus; propodeal angle subtriangular; forewing discoidal cell complete; tergum II longitudinal medial ridge sometimes indicated but not sharp; tergum III pit row not deeply sunken and often quite weak, apical rim evenly rounded, truncate or indented medially; male terminalia: sternum VIII subtriangular; gonocoxa broad or narrowly tapering apically, inner margin often angulate subapically; cuspis long, digitus slender and shorter than cuspis; aedeagus slender and acute apically.

Hosts. All known hosts of *Chrysura* species are bees in the family Megachilidae. Most host records are of the widespread and essentially holarctic genus *Osmia* Panzer (Linsenmaier 1959a, Krombein 1967). Bohart and Kimsey (1982) recorded additional twig-nesting host genera for *sagmatis*, including *Ashmeadiella, Chelostoma, Hoplitis, Anthocopa,* and *Proteriades.*

Discussion. This is the second largest chrysidid genus worldwide, with species in the Afrotropical, Palearctic and Nearctic Regions. There are 26 North American species, 17 of which occur in California; all belong to the *radians* species group (Kimsey and Bohart 1991). The genus was revised by Bohart (in Bohart and Kimsey 1982). *Chrysura* is recognized by the relatively flat, almost granular appearing face without a medial zone of cross-ridging, definitive brow or transverse frontal carina. In addition, the malar space is usually long, 2.5 MOD long or longer, flagellomere I is more than twice as long as broad in both sexes and tergum III lacks definitive teeth or notches. Species are often very similar in appearance and are difficult to identify.

Key to the California Species of *Chrysura*

1 Malar space less than 2 MOD long; propodeal angles barely projecting, at most forming obtuse to right angle.. *inusitata* (Aaron)

- Malar space more than 2 MOD long; propodeal angles strongly projecting, forming acute angle...2

2 Interocellar distance between hindocelli equal to or slightly less than ocellocular distance..3

- Interocellar distance considerably less than ocellocular distance.....................5

3 Tergum III broadly and distinctly indented posteromedially (Fig. 100), pit row in both sexes preceded by strong transverse swelling and distinct saddle viewed in profile (Fig. 89).. *sonorensis* (Cameron)

- Tergum III posterior margin slightly concave, straight or convex, with or without slight to moderately developed swelling and saddle preceding pit row viewed in profile...4

4 Tergum II medially about 2.5x as long as flagellomere I and 0.6x as long as broad or less; female tergum III apical margin straight or slightly concave (Fig. 94); body usually longer than 3.5 mm...................................... *sagmatis* Bohart

- Tergum II medially about 3x as long as flagellomere I, and 0.6x as long or longer than broad; female tergum III apical margin evenly convex (Fig. 93); unusually small species, averaging 3.5 mm long.................................. *crescentis* Horning

5 Hindcoxal and midtibial pubescence black; relatively large species, averaging 10 mm long... *tota* Aaron

- Hindcoxal and midtibial pubescence pale, size variable.............................6

6 Tergum III apical margin sharply cornered in apicolateral view in male (Fig. 98), angled and toothlike in female (Fig. 99), apicomedially indented in both sexes ....*cobaltina* (Aaron)

- Tergum III apical margin not sharply cornered, sometimes broadly obtuse or evenly rounded (as in Fig. 92), apicomedially variable..7

7 Tergum III apical margin strongly convex, pit row unusually weakly indicated (Fig. 92)...*boharti* Horning

- Tergum III apical margin straight or only slightly convex, pit row distinct..........8

8 Pronotum with coarse, contiguous punctures, many punctures as large as hindocellus; large-bodied species, averaging 8 mm long in males and 10 mm long in females...*smaragidcolor* (Walker)

- Pronotum with moderate-sized punctures, separated by integument, punctures considerably smaller than hindocellus; smaller bodied species, most males less than 8 mm and females less than 10 mm long..9

9 Subantennal distance (blue or green area) more than 1.5x as long as interantennal distance..*pacifica* (Say)

- Subantennal distance (blue or green area) less than or equal to 1.5x as long as interantennal distance...*kyrae* Krombein

Chrysura boharti Horning

Fig. 92, Map 144

Chrysura boharti Horning 1971:29. Holotype ♂; USA: California, Mono Co., White Mts., Crooked Creek Lab. (UCDC).

Geographic Range (Map 144). Upper Sonoran to Hudsonian Zones of the western USA and Canada (British Columbia); 48 specimens were studied.

California Records. Collected April through July; **Glenn Co**.: Black Butte; **Marin Co**.: Novato; **Mono Co**.: Crooked Creek; **Napa Co**.: Angwin; **Nevada Co**.: Hobart Mills; **Solano Co**.: Cold Creek near Monticello Dam; Dixon.

Hosts. A specimen label record indicated that this species was taken from a colony of ground-nesting *Osmia* (Megachilidae) (Bohart and Kimsey 1982).

Discussion. The most distinctive feature of *boharti* is the weakly differentiated pit row. Individual pits are small and nearly disappear medially. Additional characteristics are the apical margin of tergum III strongly convex, with the post pit area forming a flange that is longest medially and is slightly reflexed in males.

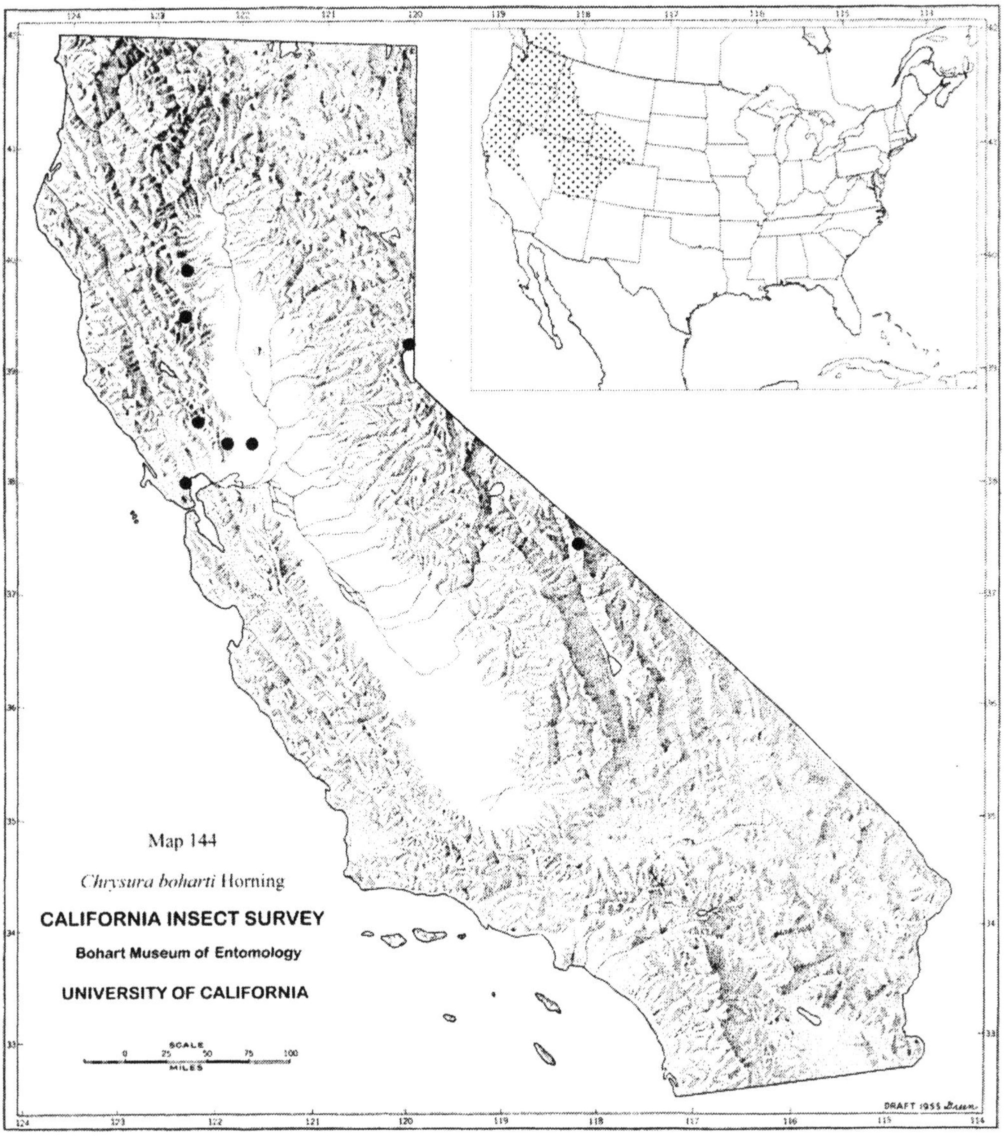

Map 144. California distribution of *Chrysura boharti*. Inset: overall distribution.

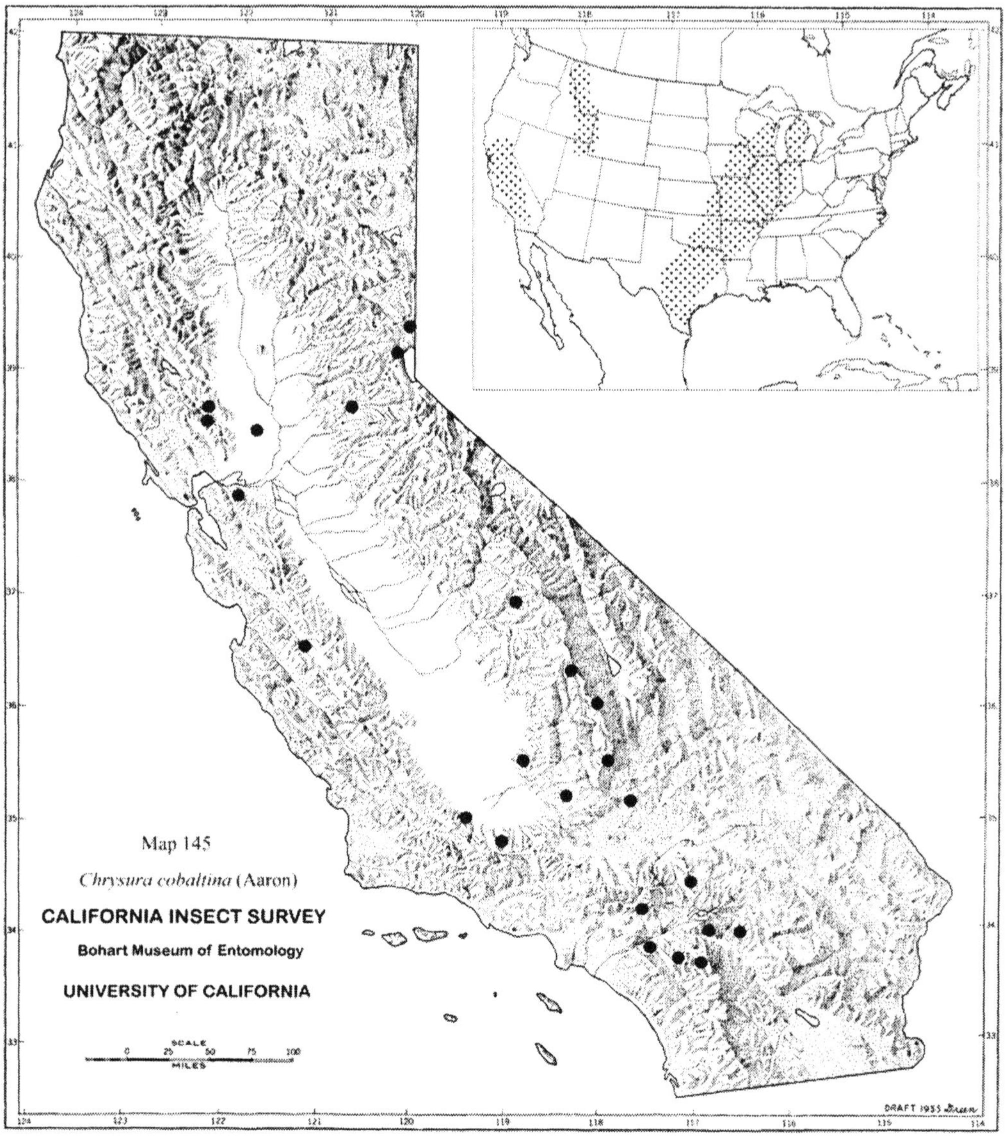

Map 145. California distribution of *Chrysura cobaltina*. Inset: overall distribution.

Chrysura cobaltina (Aaron)

Figs. 98, 99; Map 145

Chrysis cobaltina Aaron 1885:228. Holotype ♂ (not ♀); USA: Massachusetts (ANSP).

Chrysis lateridentata Aaron 1885:228. Lectotype ♀ (designated by Cresson 1928); USA: Montana (ANSP).

Geographic Range (Map 145). Upper Sonoran to Canadian Zones; transcontinental mostly between 37° and 46° latitude in the USA; also in Canada (British Columbia); 81 specimens were examined.

California Records. Collected March through July; **Contra Costa Co**.: Mt. Diablo; **El Dorado Co**.: Placerville; **Fresno Co**.: Wildman Meadow; **Inyo Co**.: 14 mi. n Little Lake; **Kern Co**.: 15 mi. w Johannesburg; Mill Potrero 18 mi. se Maricopa; Piute Mts.; Short Canyon 7 mi. w Inyokern; **Napa Co**.: 4 mi. nw Knoxville; 5 mi. se Pope Valley; **Placer Co**.: Summit; **Riverside Co**.: Idyllwild San Jacinto Mts.; White Water Canyon; **San Benito Co**.: Lime Kiln Canyon sw Paicines; **San Bernardino Co**.: 3 mi. nw Yucca Valley; 8 mi. e Lucerne Valley; Joshua Tree; **Sierra Co**.: Independence Lake Rd.; **Yolo Co**.: 4 mi. w Davis.

Hosts. Unknown.

Discussion. The configuration of the apical rim of tergum III is diagnostic in *cobaltina*. The apical margin is wavy and medially indented, with a more or less sharp corner in semilateral view in the male. This species tends to be green to blue-green in color.

Chrysura crescentis Horning

Fig. 93, Map 146

Chrysura crescentis Horning 1971:27. Holotype ♂; USA: California, Tulare Co., Johnsdale (UCDC).

Geographic Range (Map 146). Upper Sonoran Zone in southern California; 15 specimens were studied.

California Records. Collected March through September; **Kings Co**.: Kettleman City; **Los Angeles Co**.: Santa Susana Pass; Tanbark Flat; **Madera Co**.: San Joaquin Valley; **Mariposa Co**.: Miami Ranger Station; **Napa Co**.: Samuel Springs (beneath Lake Berryessa); **Riverside Co**.: Menifee Valley; Riverside; **Tulare Co**.: Ash Mt.

Hosts. Unknown.

Discussion. This is a rarely collected California endemic. Unique features of *crescentis* are the eye height slightly greater than the least interocular distance, the propodeal tooth tapering almost to a point, tergum II about as long as broad in dorsal view, and tergum III with the pit row weakly impressed, with the pits most distinct towards the middle and the apical margin strongly convex. This is a small species with the body length averaging about 3.8 mm.

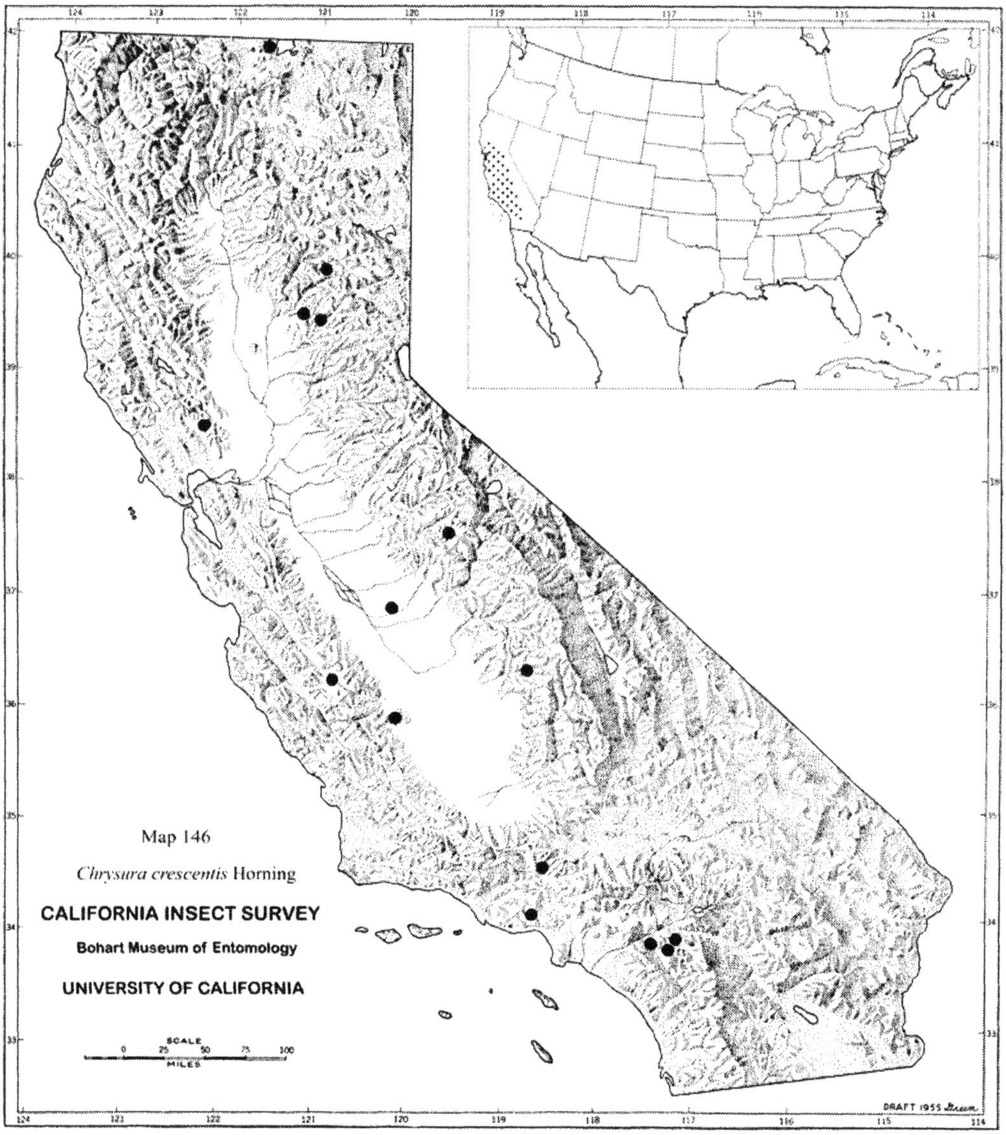

Map 146. California distribution of *Chrysura crescentis*. Inset: overall distribution.

Chrysura inusitata (Aaron)

Map 147

Chrysis inusitata Aaron 1885:227. Lectotype ♂ (designated by Cresson 1928); Canada: British Columbia, Vancouver (ANSP).

Chrysis optima Aaron 1885:227. Lectotype ♀ (designated by Cresson 1928); USA: "Cala" (ANSP).

Holochrysis interfata du Buysson 1908c:208. Holotype ♀; USA: California (MNHN).

Geographic Range (Map 147). Upper Sonoran to Canadian Zones in California, Nevada and Oregon; 222 specimens were examined.

California Records. Collected March through July; **Alameda Co**.: 1 mi. e Mission Peak; Berkeley; Murietta Caves; **Amador Co**.: Plymouth; **Calaveras Co**.: Mokelumne Hill; **Colusa Co**.: 9.7 mi. n Hwy. 20; Bear Valley; **Contra Costa Co**.: Albany; Mt. Diablo; **El Dorado Co**.: 1 mi. n Nashville; Pine Hill w Rescue; **Lake Co**.: north fork Cache Creek, Hwy. 20; **Marin Co**.: Kirby Beach; Mill Valley; Novato; **Monterey Co**.: 3 mi. se Hastings Reservation; Arroyo Seco; **Napa Co**.: 4 mi. s Pope Valley; Samuel Springs (beneath Lake Berryessa); **Placer Co**.: 10 mi. n Auburn; **Plumas Co**.: Buck's Lake; Onion Valley; **Sacramento Co**.: American River; Citrus Heights; Folsom; Sacramento; **San Mateo Co**.: Redwood City; **Santa Barbara Co**.: Santa Cruz Island; **Santa Clara Co**.: Las Animas Creek; Mt. Hamilton; **Sierra Co**.: Sattley; **Solano Co**.: 11 mi. s Dixon near Dozer; Dixon; Putah Canyon; **Sonoma Co**.: 2 mi. n Sebastopol; Geyserville; Glen Ellen; Old Adobe Rd. near Petaluma; Santa Rosa; **Stanislaus Co**.: Del Puerto Canyon; **Tulare Co**.: Ash Mt.; **Ventura Co**.: Point Magu; **Yuba Co**.: 18 mi. n Marysville.

Hosts. Unknown.

Discussion. *Chrysura inusitata* is a moderate-sized wasp, ranging from 8.0-8.5 mm in length. The species is characterized by the genal area greatly narrowed above the malar space, malar space 1.5 MOD long and shorter than the subantennal distance, and propodeal angle forming an obtuse angle. The forewing venation forms most distinctive features of *inusitata*; many veins are narrowed distally and Rs becomes obsolescent well before the wing margin.

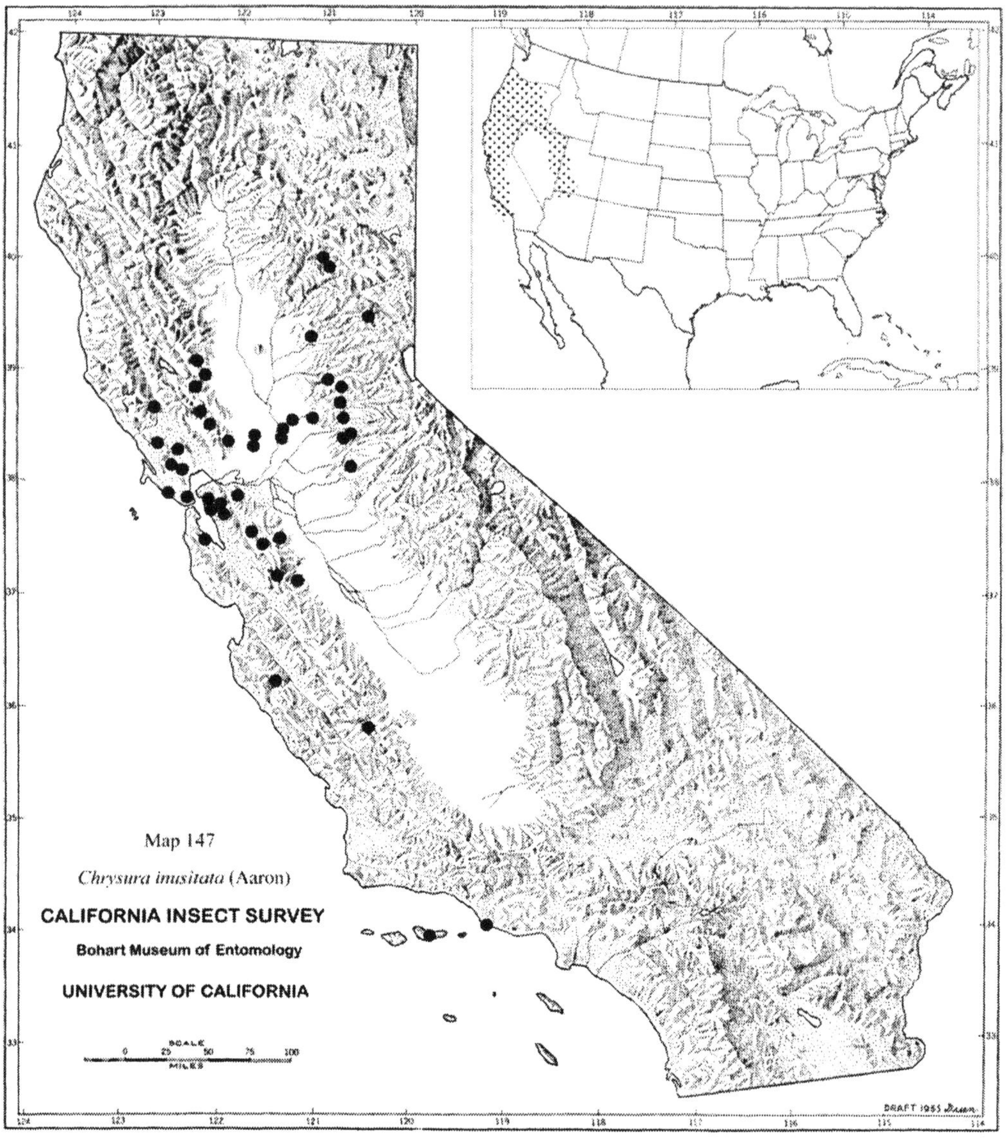

Map 147. California distribution of *Chrysura inusitata*. Inset: overall distribution.

Chrysura kyrae Krombein

Map 148

Chrysura kyrae Krombein 1963a:150. Holotype ♂; USA: Maryland, Plummers Island (USNM).

Geographic Range (Map 148). Upper Sonoran to Canadian Zones, from the northeastern USA west to California and British Columbia; 360 specimens were studied.

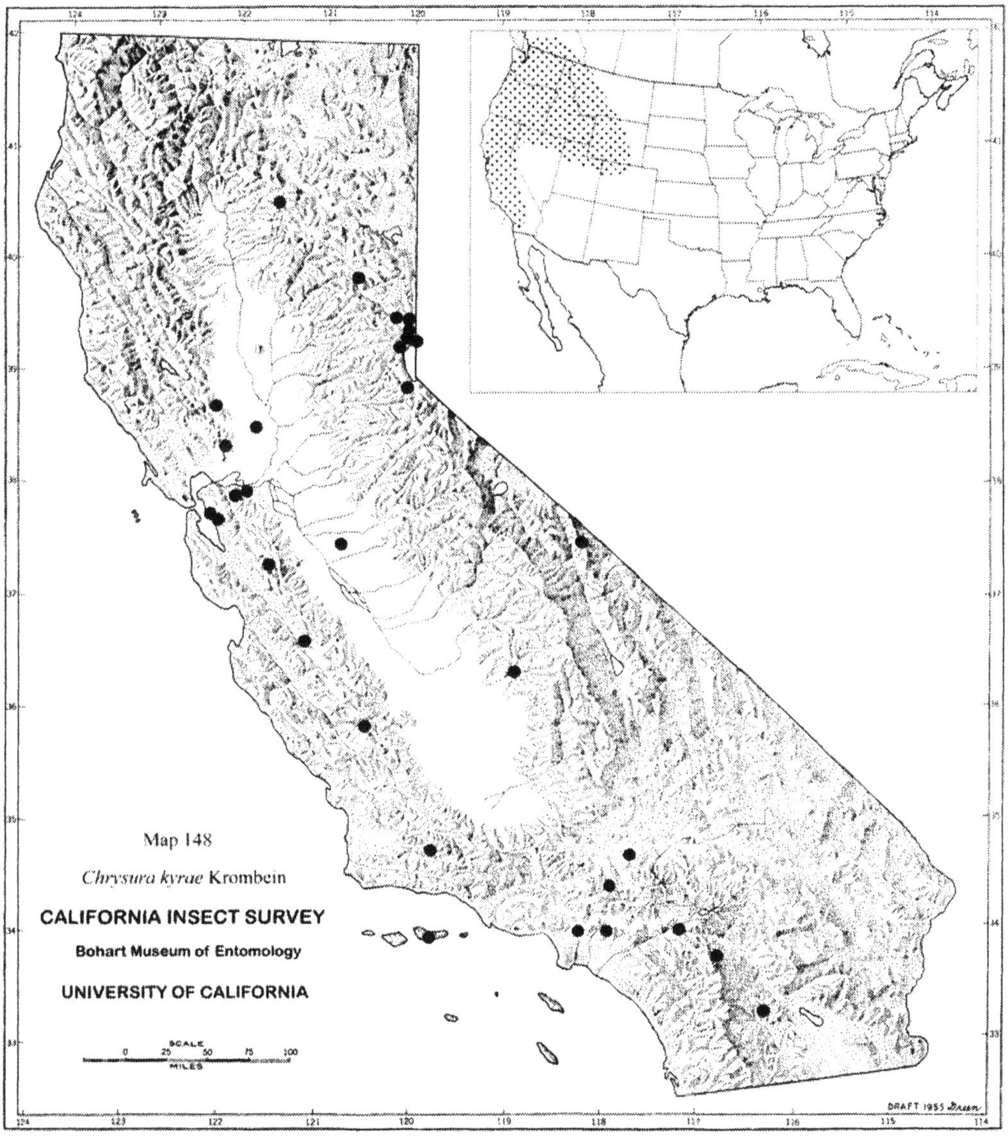

Map 148. California distribution of *Chrysura kyrae*. Inset: overall distribution.

California Records. Collected March through September; **Alameda Co.**: Berkeley; Oakland; **Contra Costa Co.**: 10 mi. s Pittsburgh; Mt. Diablo; **El Dorado Co.**: trail to Wrights Lake; **Los Angeles Co.**: Huntington Lake; 2 mi. s sw Valyermo;

Claremont; **Mono Co**.: Blanco's Corral, White Mts.; Crooked Creek; **Monterey Co**.: Arroyo Seco; **Nevada Co**.: Boca; Hobart Mills; Sagehen Creek; **Placer Co**.: Lake Tahoe; **Plumas Co**.: Johnsville; **Riverside Co**.: White Water Canyon; **San Benito Co**.: San Juan Canyon; **San Bernardino Co**.: 10 mi. sw Redlands; Tejon Pass; **San Diego Co**.: Box Canyon; **Santa Barbara Co**.: Santa Cruz Island; Upper Oso Camp Los Padres National Forest; **Santa Clara Co**.: San Antonio Valley; **Shasta Co**.: Old Station; **Sierra Co**.: 10 mi. s Sierraville; Independence Lake; **Solano Co**.: Mix Canyon; **Stanislaus Co**.: Turlock; **Tulare Co**.: Three Rivers; **Yolo Co**.: Davis; Rumsey.

Hosts. Recorded hosts are all in the family Megachilidae and include *Osmia lignaria* Say, *O. kinkaidii* Cockerell, *O. bruneri* Cockerell, *O. marginata* Michener, *O. pumila* Cresson and *Anthocopa copelandica* (Cockerell) (Bohart and Kimsey 1982, Krombein 1979, Parker and Bohart 1966).

Discussion. This species is difficult to separate from *pacifica*. The subantennal distance in *kyrae*, measured just inside the clypeal notch, is relatively short (1.5x the interantennal distance). This distance is longer in *pacifica*. In addition, the interocellar distance is about half that of the ocellocular distance in males and in many females of *kyrae*, and the apical margin of tergum III is convex and subtruncate.

Chrysura pacifica (Say)

Fig. 88, Map 149

Chrysis pacifica Say 1828:82. Type lost; USA: Indiana. Neotype ♂ (designated by Bohart, in Bohart and Kimsey 1982); USA: Indiana, Lafayette (UCDC).

Chrysis hilaris Dahlbom 1854:103. Holotype ♂; USA: New York (MZLU).

Chrysis halictula Gribodo 1874:359. Holotype ♀; USA: California (MCSN).

Chrysis resecta Gribodo 1879:334. Holotype ♀; USA: California, Mariposa (MCSN).

Geographic Range (Map 149). Transcontinental in USA, including Alaska, Canada, in the Mexican cordillera; Lower Sonoran to Hudsonian Zones; 1573 specimens were seen.

California Records. Collected April through August; **Alameda Co**.: Berkeley; Oakland; **Alpine Co**.: Carson Pass; Faith Valley; **Amador Co**.: 6 mi. w Tragedy Springs; Volcano; **Contra Costa Co**.: Mt. Diablo; **El Dorado Co**.: Lake Tahoe; Placerville; 5 mi. w Stateline; 8 and 10 mi. n on Ice House Rd.; 2km s Carson Pass; Echo Lake; **Glenn Co**.: 5 mi. w Elk Creek; **Imperial Co**.: Glamis; **Inyo Co**.: 11 mi. w Big Pine; Big Pine Creek; Mammoth; **Kern Co**.: Mill Potrero; **Lake Co**.: Hopland Grade; Kelseyville; north fork Cache Creek Hwy 20; Upper Lake; **Lassen Co**.:

Hallelujah Junction; **Los Angeles Co**.: Claremont; Camp Baldy; Crystal Lake; **Marin Co**.: Kirby Beach; McClure Beach; Mill Valley; Point Reyes near Lighthouse; Tocaloma; **Modoc Co**.: 5 mi. n Clear Lake; 6 mi. nw Cedarville; Goose Lake; **Mono**

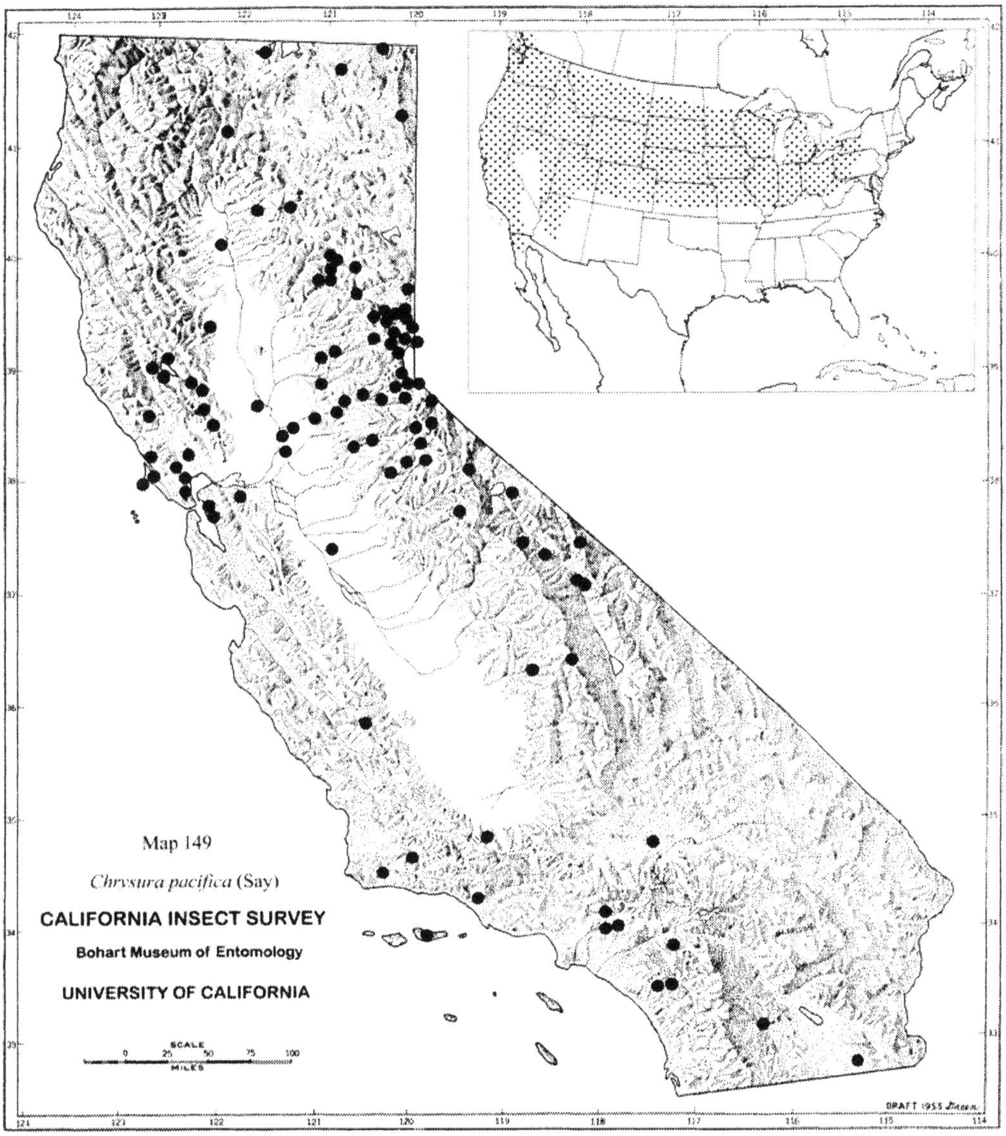

Map 149. California distribution of *Chrysura pacifica*. Inset: overall distribution.

Co.: Blanco's Corral, White Mts.; Elery Lake; Mammoth Lake; Sardine Creek; Sonora Pass; **Monterey Co**.: Arroyo Seco; **Napa Co**.: 3.2 mi. se Pope Valley; 5 mi. se Pope Valley; Samuel Springs (beneath Lake Berryessa); **Nevada Co**.: Boca; Grass Valley; Jackson Lake; Hobart Mills; Russell Valley; Sagehen Creek; **Placer Co**.: 10 mi. n Auburn; 5 mi. nw Kings Beach; 5 mi. w Foresthill; 8 mi. e Foresthill;

Brockway; Carnelian Bay, Lake Tahoe; Emigrant Gap; 4 mi. s Rocklin; 2 mi. n Donner Summit; **Plumas Co.**: 1 mi. w Johnsville; 4 mi. nw Taylorsville; 4 mi. w Quincy; Blairsden; Bucks Lake; Butterfly Valley; Greenville; Tobin; **Riverside Co.**: 1 mi. n Murrieta; Riverside; White Water; **Sacramento Co.**: American River; Elk Grove; Folsom; Sacramento; **San Bernardino Co.**: Apple Valley; **San Diego Co.**: Borrego; **Santa Barbara Co.**: Bluff Camp San Rafael Mnts.; Santa Cruz Island; Santa Ynez Mnts.; **Shasta Co.**: Old Station; Shingletown; **Sierra Co.**: 5 mi. e Weber Lake; Independence Lake; Kyburz Flat; Sardine Lakes.; Sardine Peak; Sierra Valley; Sierraville; Weber Lake; Yuba Pass; **Siskiyou Co.**: Shasta City; Valentine Caves Lava Bed National Monument; **Solano Co.**: 3.2 mi. w Mix Canyon; Monticello Dam; **Sonoma Co.**: Healdsburg; Old Adobe Rd. near Petaluma; Petaluma; Sonoma; **Stanislaus Co.**: Turlock; **Tehama Co.**: Cone Grove Park; Red Bluff; **Tulare Co.**: Ash Mt.; Mineral King; Roads End; **Tuolumne Co.**: Chipmunk Flat; Dodge Ridge; Leland Meadow; Strawberry; **Ventura Co.**: Foster Park; **Yolo Co.**: 2 mi. n Rumsey; Cache Creek Canyon; Davis; Elkhorn Ferry; Putah Canyon; Putah Creek near Monticello Dam; Rumsey.

Hosts. Hosts include megachilid bees in the species *Osmia kinkaidii* Cockerell, *O. pumila* Cresson, *O. tanneri* Sandhouse, *O. georgica* Cresson, *O. nigrifrons* Cresson, *O. coloradensis* Cresson, O. marginata Michener, *O. rostrata* Sandhouse, *O. lignaria propinqua* Cresson and *Hoplitis brachyodonta* (Cockerell) (Krombein 1979, Bohart and Kimsey 1982, Hicks 1933).

Discussion. This is one of the most abundant species of chrysidids in North America, and California specifically. Diagnostic features of *pacifica* include: tergum II broader than long in dorsal view, malar space about 5 MOD long and longer than the subantennal space, femora with pale erect pubescence, the propodeal projection forming an acute angle from the base and apically blunt, tergum III with a well-developed pit row, the margin subtruncate to convex, rarely indented, and the corner broadly rounded in semilateral view.

Chrysura sagmatis Bohart

Fig. 94, Map 150

Chrysura sagmatis Bohart, in Bohart and Kimsey 1982:154. Holotype ♂; USA: California, Yolo Co., Davis (UCDC).

Geographic Range (Map 150). Upper Sonoran Zone west of 100th meridian in the USA, and mountains of Baja California; 64 specimens were studied.

California Records. Collected March through July; **Alameda Co.**: Berkeley; **El Dorado Co.**: 6 mi. n on Ice House Rd.; **Imperial Co.**: 3 mi. n Glamis; Dave's Wash,

4 mi. sw Palo Verde; Palo Verde; **Inyo Co.**: 2 mi. e Lone Pine; Antelope Spring; Midway Well, Death Valley National Monument; **Kern Co.**: Mill Potrero; Sand Canyon 3 mi. w Brown; **Los Angeles Co.**: Claremont; **Los Angeles Co.**: La Puente; Tanbark Flat; **Marin Co.**: Kentfield; Point Reyes; **Mariposa Co.**: Mariposa; **Mono Co.**: Paradise Camp; White Mts.; **Monterey Co.**: Arroyo Seco; **Napa Co.**: 10 mi. e Shell Peak; 10 mi. se Pope Valley; 6 mi. w Monticello Dam; Monticello Dam; Pope Valley; Samuel Springs (beneath Lake Berryessa); **Nevada Co.**: Boca; Sagehen

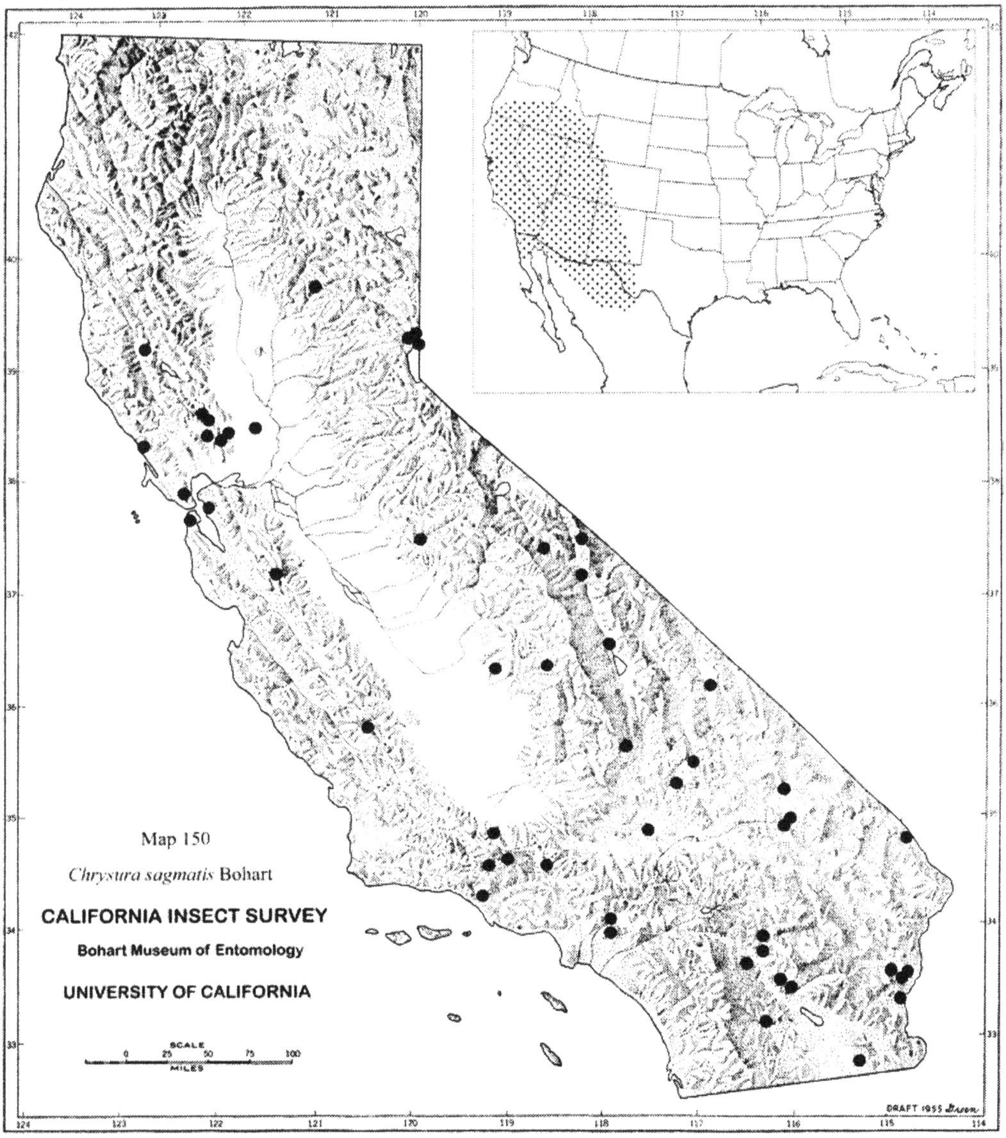

Map 150. California distribution of *Chrysura sagmatis*. Inset: overall distribution.

Creek; Hobart Mills; **Riverside Co.**: 18 mi. w Blythe; 2 mi. s Thermal; 23 mi. se Twenty-Nine Palms; Blythe; Hopkins Well; 1 mi. e Shavers Well; Thousand Palms; Deep Canyon; **San Bernardino Co.**: 12 mi. s Needles; 14 mi. s Baker; 15 mi. w Baker; 3 mi. s Kramer Junction; 6 mi. nw Adelanto; 7 mi. w Salt Wells; Kelso Dunes; Kramer Hills; 7 mi. sw Kelso; **San Diego Co.**: Borrego; **San Francisco Co.**: Lobos Creek, San Francisco; **Santa Clara Co.**: Mt. Hamilton; **Sonoma Co.**: Bodega Head; **Tulare Co.**: Wood Lake; Ash Mt. Range, Sequoia National Park; **Ventura Co.**: Foster Park; Lockwood Creek, near Stauffer Post Office; Sespe Canyon; **Yolo Co.**: Davis; Putah Canyon.

Hosts. A large number of megachilid species have been recorded as hosts for *sagmatis*, including *Ashmeadiella aridula* Cockerell, *A. breviceps* Michener, *A. bucconis denticulata* (Cresson), *A. californica* (Ashmead), *A. cubiceps* (Cresson), *A. femorata* Michener, *A. gillettei* Titus, *A. melilotti* (Cockerell), A. *opuntiae* Cockerell, *A. timberlakei* Michener, *Chelostoma phaceliae* Michener, *Hoplitis biscutellae* (Cockerell), *H. producta gracilis* (Michener), *Anthocopa copelandica* (Cockerell), *Proteriades buncocephala* (Michener), *Osmia kinkaidii* Cockerell (Thorpe 1968, Bohart and Kimsey 1982).

Discussion. Similar to *sonorensis, sagmatis* differs in having close fine punctation on tergum II giving it a dull granular appearance, tergum III in profile is slightly saddled in males and distinctly so in females, without a transverse bulge behind the saddle, and the apical margin is nearly straight to very slightly concave in females.

Chrysura smaragdicolor (Walker)

Map 151

Chrysis smaragdicolor Walker 1868:343. Holotype ♀; Canada: British Columbia (BMNH).

Geographic Range (Map 151). Upper Sonoran to Canadian Zones in the USA, west of the 100th meridian; 218 specimens were seen.

California Records. Collected June and July; **Alpine Co.**: Ebbetts Pass; **Los Angeles Co.**: Claremont; Tanbark Flat; **Mono Co.**: Pickel Meadow; Sonora Pass; **Napa Co.**: Monticello Dam; Mt. Veeder; **Nevada Co.**: Carpenter Ridge; Hobart Mills; Sagehen Creek; **Nevada Co.**: Hobart Mills; **Placer Co.**: 5 mi. w Kings Beach; Emigrant Gap; **Plumas Co.**: Onion Valley; **San Bernardino Co.**: Camp Baldy; **Sierra Co.**: 10 mi. s Sierraville; Gold Lake; Sattley; Yuba Pass; **Tulare Co.**: Ash Mt., Sequoia National Park.

Hosts. Published hosts are megachilid bees in the genus *Osmia*, including *montana* Cresson, *coloradensis* Cresson, *texana* Cresson, *californica* Cresson, *longula* Cresson and *lignaria propingua* Cresson (Krombein 1979, Bohart and Kimsey 1982).

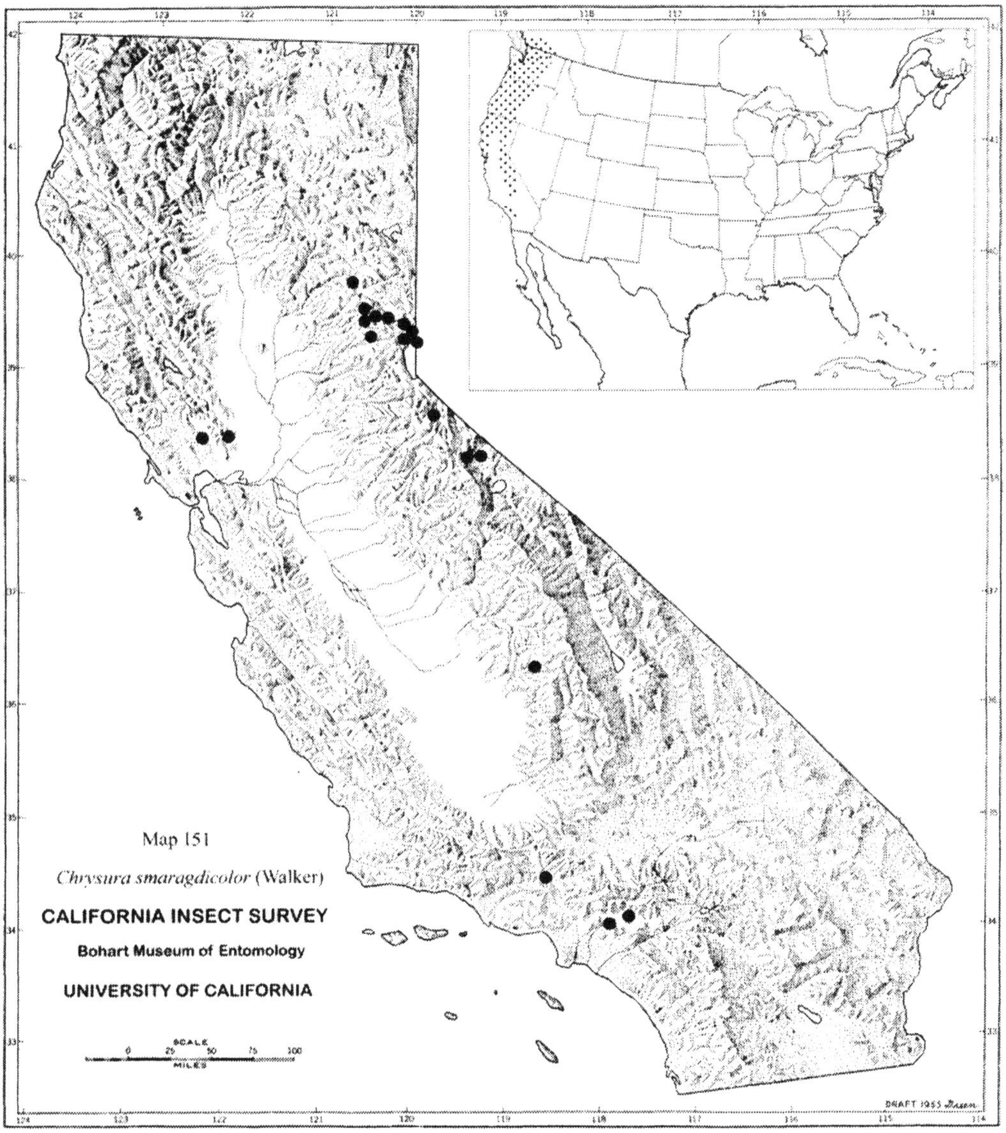

Map 151. California distribution of *Chrysura smaragdicolor*. Inset: overall distribution.

Discussion. These are moderate to large-sized chrysidids ranging from an average body length of 7.5 mm in males to 8 mm in females. Diagnostic features of *smaragdicolor* include the pronotum with many ocellus-sized punctures; tergum II punctation toward the apex much finer than toward the base and apical punctures separated by polished interspaces, giving it a shiny appearance. Tergum III has conspicuous erect black setae, particularly in females and the apical margin is subtruncate.

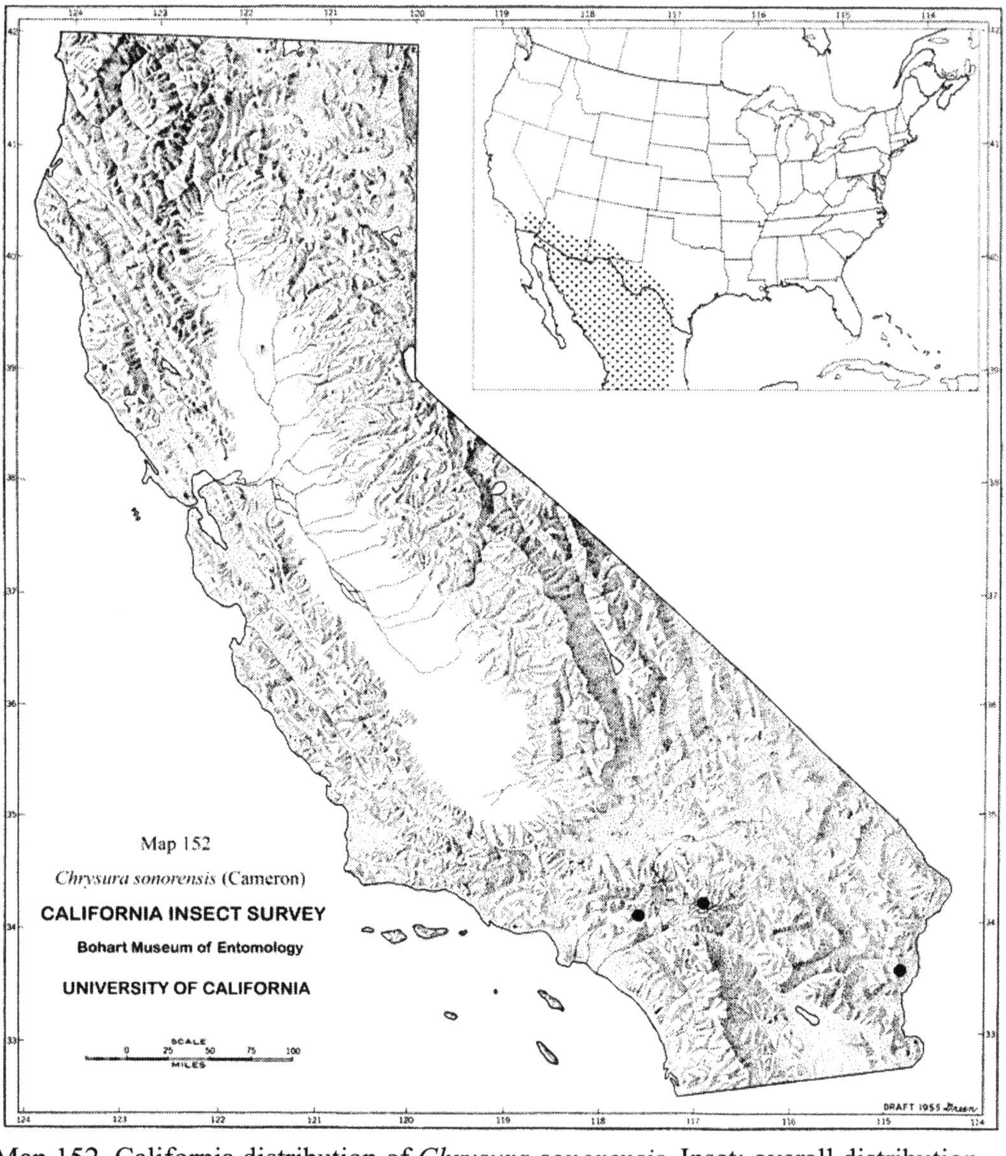

Map 152. California distribution of *Chrysura sonorensis*. Inset: overall distribution.

Chrysura sonorensis (Cameron)

Figs. 89, 100; Map 152

Chrysis sonorensis Cameron 1888:461. Holotype ♀; Mexico: northern Sonora (BMNH).

Gonochrysis sinuatocaudata Bischoff 1910:456. Holotype ♀; Mexico: Oaxaca (MNHU ?).

Geographic Range (Map 152). Upper and Lower Sonoran Zones in the southwestern USA (California, Arizona, New Mexico and Texas); 49 specimens were studied.

California Records. Collected in July; **San Bernardino Co**.: Colton Hills.

Hosts. F. D. Parker reared *sonorensis* from twig nests of *Ashmeadiella* sp. (Megachilidae) (Bohart and Kimsey 1982).

Discussion. This is an infrequently collected species known from a relatively small number of specimens. Similar to *sagmatis, sonorensis* has the brow unusually prominent and sharply rounded. *Chrysura sonorensis* also bears a superficial resemblance to *Ceratochrysis enhuyki*, but lacks the transverse frontal carina and vertex tubercles of that species. Other diagnostic features include: tergum II punctation medium-sized apically becoming coarse basally and tergum III apical margin broadly concave in both sexes, with the pit row preceded by a well-developed pre-pit swelling and saddle. Average body length is 6 mm.

Chrysura tota (Aaron)

Fig. 6, Map 153

Chrysis integra Cresson 1865b:306. Holotype ♀; USA: Colorado (ANSP). Nec Fabricius 1787.

Chrysis tota Aaron 1885:228. Replacement name for *integra* Cresson 1865.

Chrysis elongata Mocsáry 1887a:15. Replacement name for *integra* Cresson 1865.

Geographic Range (Map 153). Canadian and Hudsonian Zones in the USA and Canada, west of the 100th meridian; 91 specimens are studied.

California Records. Collected in June, July; **Mono Co**.: Blancos Corral, White Mts.; Mammoth; **El Dorado Co**.: 2 km s Carson Pass; **Sierra Co**.: Sierraville.

Hosts. *Osmia texana* Cresson (Megachilidae) has been reported as the host of *tota* (Bohart and Kimsey 1982).

Map 153. California distribution of *Chrysura tota*. Inset: overall distribution.

Discussion. The large size and conspicuous black setae on the legs, particularly the fore and midfemora, and on the notum will distinguish *tota* from other *Chrysura* species. In addition, tergum III is apically subtruncate, slightly indented apicomedially and females have obtuse apicolateral angles on tergum III. These are large-bodied *Chrysura*, averaging 9.5 mm long in males, and 10 mm long in females.

Chrysurissa Bohart

Generic Diagnosis. Medium-sized, 7-10 mm long, green to purple; flagellomere I 3.0-3.5x as long as broad, 2.5x as long as pedicel, 1.5-1.6x as long as flagellomere II or III; scapal basin moderately concave with weak to moderate microridging, brow rather sharp and often medially with remnant of transverse frontal carina; malar space 1.5-2.0 MOD; subantennal space 2.5-3.5 MOD; subgenal area absent; midocellar area undefined, midocellus unlidded; pronotum slightly shorter than scutellum, medial depression large and shallow, lateral depression deep; metanotum simple; mesopleuron with scrobal and episternal sulci and verticaulus; propodeal angle triangular, sometimes slightly convex posteriorly; tergum I broad; tergum II not sharp apicolaterally; tergum III large, about 0.8x as long as II, pit row well developed, post-pit area angled laterally, emarginate medially, shape distinctive; sternum II spots large, sometimes joined; male terminalia: sternum VIII nearly triangular, gonocoxa and cuspis stout, aedeagus simple.

Discussion. *Chrysurissa* resemble large *Chrysura*, ranging from 7-11 mm long. However, they have a well-developed brow, extensive cross-ridging in the scapal basin and the apical rim of tergum III has two well-developed basal angles. The single species in the genus is parasitic on *Pseudomasaris* species (Vespidae: Masarinae)

Chrysurissa densa (Cresson)

Figs. 53-55, Map 154

Chrysis densa Cresson 1865:307. Lectotype ♀ (designated by Cresson 1916); USA: Colorado ("Col.") (ANSP).

Geographic Range (Map 154). USA west of the 100th meridian; 495 specimens were examined.

California Records. Collected March through August; **Alpine Co**.: Carson Pass; Hope Valley; Lake Winnemucca; **Contra Costa Co**.: Mt Diablo; **Imperial Co**.: Glamis; Palo Verde; **Inyo Co**: 15 mi. e Bishop; Prospector Meadow; 7 mi. n Parchers Camp; Big Pine Creek; Darwin Falls; **Lassen Co**.: Norvell; **Los Angeles Co**.: Glendale; Tanbark Flat; **Mariposa Co**.: El Portal; **Modoc Co**.: 3 mi. ne Cedar Pass; 5 mi. n Clear Lake; **Mono Co**.: Blue Lake; Virginia Creek; Rock Creek; Mammoth Lakes; Leavitt meadows; Sheep Mt.; Sonora Pass; White Mnts., Wyman Cyn; **Napa Co**.: Samuel Springs (beneath Lake Berryessa); **Nevada Co**.: Boca; Sagehen Creek; **Placer Co**.: Brockway; Carnelian Bay, Lake Tahoe; Emigrant Gap; **Plumas Co**.: Eagle Lake; Johnsville Meadow Valley; Johnsville; 8 mi. w Crystal Lake; Rucker's Lake; **Riverside Co**.: Gavilan; Norden; Chuckawalla Mts.; Idyllwild San Jacinto Mnts; 5 mi. n Vidal Junct.; Andreas Canyon; Mt. San Jacinto; Palm Springs; Whitewater Canyon; **San Bernardino Co**.: San Antonio Creek, Yucca Valley; 2 mi. s

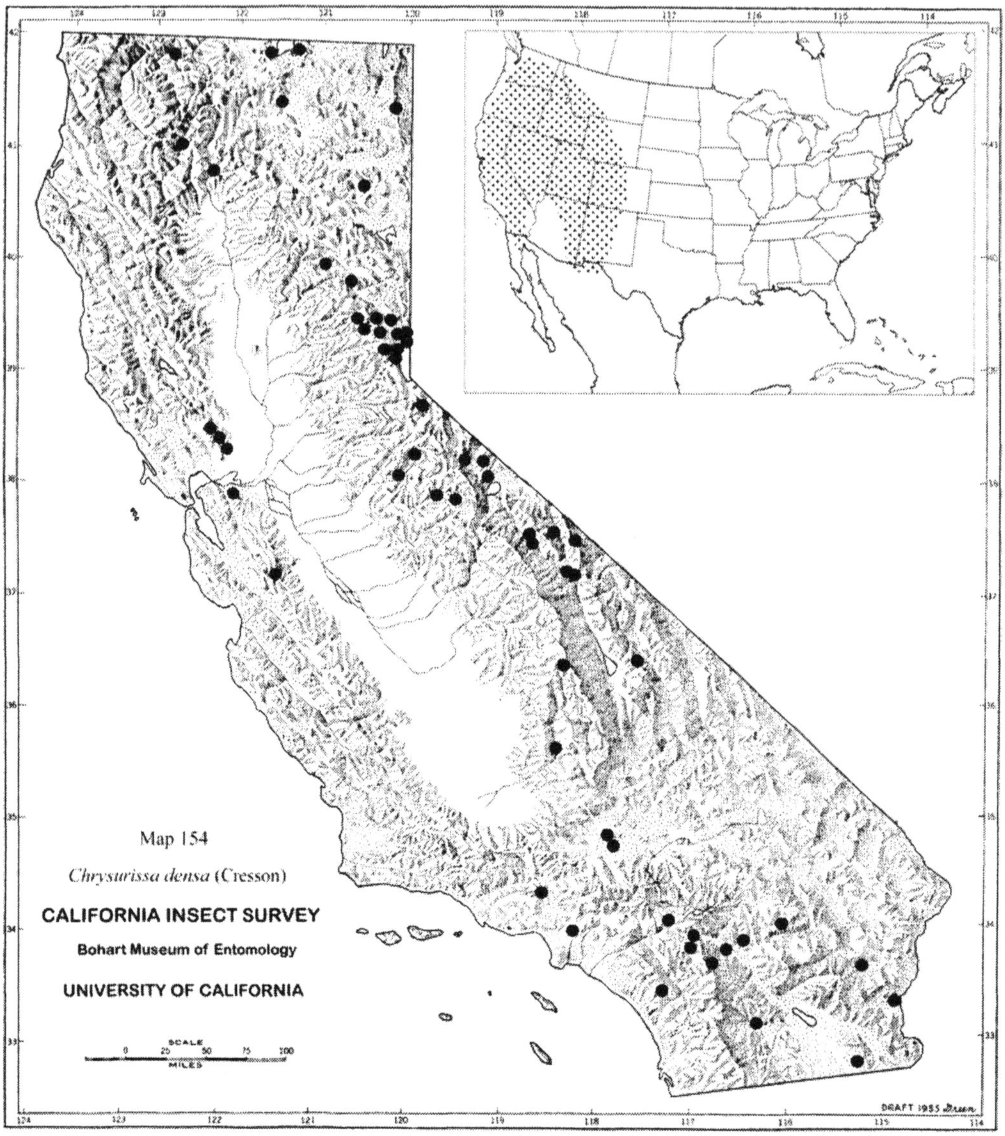

Map 154. California distribution of *Chrysurissa densa*. Inset: overall distribution.

Phelan; 2 mi. w Cajon Pass; 8 mi. n Vidal Junction; Bluff Camp San Rafael Mts.; Wrightwood; **San Diego Co.**: Borrego; **Santa Clara Co.**: 1.5 mi. n San Antone Junction; **Sierra Co.**: Smith Mill, 15 mi. se Sierraville; Gold Lake; Weber Lake; Sardine Peak; **Siskiyou Co.**: Medicine Lake; Lava Bed National Monument; Yreka; **Solano Co.**: 3 mi. w Mix Canyon; **Sonora Co.**: Tioga Pass; **Trinity Co.**: Coffee Creek; Mountain Meadows; **Tulare Co.**: 5 mi. w Kernville; Mineral King; **Tuolumne Co.**: Leland Meadows; Blue Canyon, Sonora Pass; Chipmunk Flat; Dardanelles; Pinecrest Lake; **Yolo Co.**: Putah Canyon.

Hosts. *Chrysurissa densa* has been reared from nests of species the North American masarine genus *Pseudomasaris*, including *vespoides* (Cresson), *edwardsii* (Cresson) (Hicks 1929), *zonalis* (Cresson) (Parker 1967) and *occidentalis* (Cresson) (Hungerford 1937).

Distribution. This genus occurs west of the 100th meridian.

Discussion. The apical rim of tergum III is diagnostic for *densa*, it is medially notched, with an angle at the end of the pit row. In females tergum III is elongate medially appearing subtriangular, and in males it is almost truncate, appearing subrectangular. Additional features are flagellomere I more than 3x as long as broad and the subantennal distance almost twice as long as the malar space.

Tribe Parnopini

Parnopines occur in Eurasia, Africa and North America. They are large, structurally unique chrysidids, with elongate, tubular mouthparts and the number of visible metasomal segments is sexually dimorphic (males have four and females three). All parnopines are nest parasites of bembicine sandwasps (Crabronidae). There are three genera worldwide, but only the genus *Parnopes* occurs in North America.

Genus *Parnopes* Latreille

Generic diagnosis. Tongue much longer than eye height and tubular, with greatly elongate galea and glossa; maxillary palpus with three or fewer articles; labial palpus with two articles or absent; scapal basin with or without appressed silvery setae; pronotum with transverse anterior carina broken medially and strongly projecting sublaterally; mesopleuron with large, somewhat projecting lobe above scrobe; tegula subovoid, metanotal projection large and mucronate, apically truncate and deeply notched or entire; propodeal angle broadly triangular and apically acute; forefemur without distinct ventral tooth; male terminalia: volsella reduced and membranous, not divided into digitus and cuspis; gonocoxa with articulated digitate gonostyle; aedeagus apically rounded or with apical hook or projection; sternum VIII broad and subrectangular.

Hosts. *Parnopes* have been reported from the nests of *Bembix*, *Microbembix* and *Steniolia* (Crabronidae).

Distribution. The genus *Parnopes* is found in North America, Eurasia and Africa, including Madagascar.

Discussion. The long tubular mouthparts of *Parnopes* are seen nowhere else in the Chrysididae. Kimsey (1987) reviewed the world fauna. The North American species were reviewed by Telford (1964).

Key to the California Species of *Parnopes*

1 Face in front view densely covered with silvery appressed setae, often completely obscuring integument beneath (Fig. 91)........................... *fulvicornis* Cameron

\- Face in front view with sparse or absent setae, setae not obscuring integument beneath 2

2 All terga with posterolateral angles blue or green; metanotal projection entire posteriorly.. *edwardsii* (Cresson)

\- One or more terga, with posterolateral angles whitish; metanotal projection posteriorly notched to almost entire...3

3 Foretibia extensively green or blue on outer surface; tegula strongly bicolored, medially brown with some green or blue highlights.............. *concinnus* Viereck

\- Foretibia mostly pale reddish, with little or no green to blue; tegula reddish to brown, grading externally to whitish or yellow...4

4 Metanotal projection distinctly and deeply notched posteromedially; male tergum IV much less than twice as broad as long; female tergum III with subapical foveae forming nearly right angle................................. *desertorum* Kimsey

\- Metanotal projection posteriorly truncate, weakly concave or with tiny medial notch; male tergum IV nearly twice as broad as long; female tergum III with foveae forming broadly obtuse angle (Fig. 97).................. *borregoensis* Telford

Parnopes borregoensis Telford

Fig. 97, Map 155

Parnopes borregoensis Telford 1964:9. Holotype ♂; USA: California, San Diego Co., Borrego Valley (CAS).

Geographic Range (Map 155). Upper and Lower Sonoran Zones in California, from Inyo Co. south to Baja California, Mexico; 8 specimens were seen.

California Records. Collected in June, July; **Inyo Co**.: Lone Pine Creek; Independence; **San Bernardino Co**.: Wild Horse Canyon; **San Diego Co**.: Borrego Valley.

Hosts. Unknown.

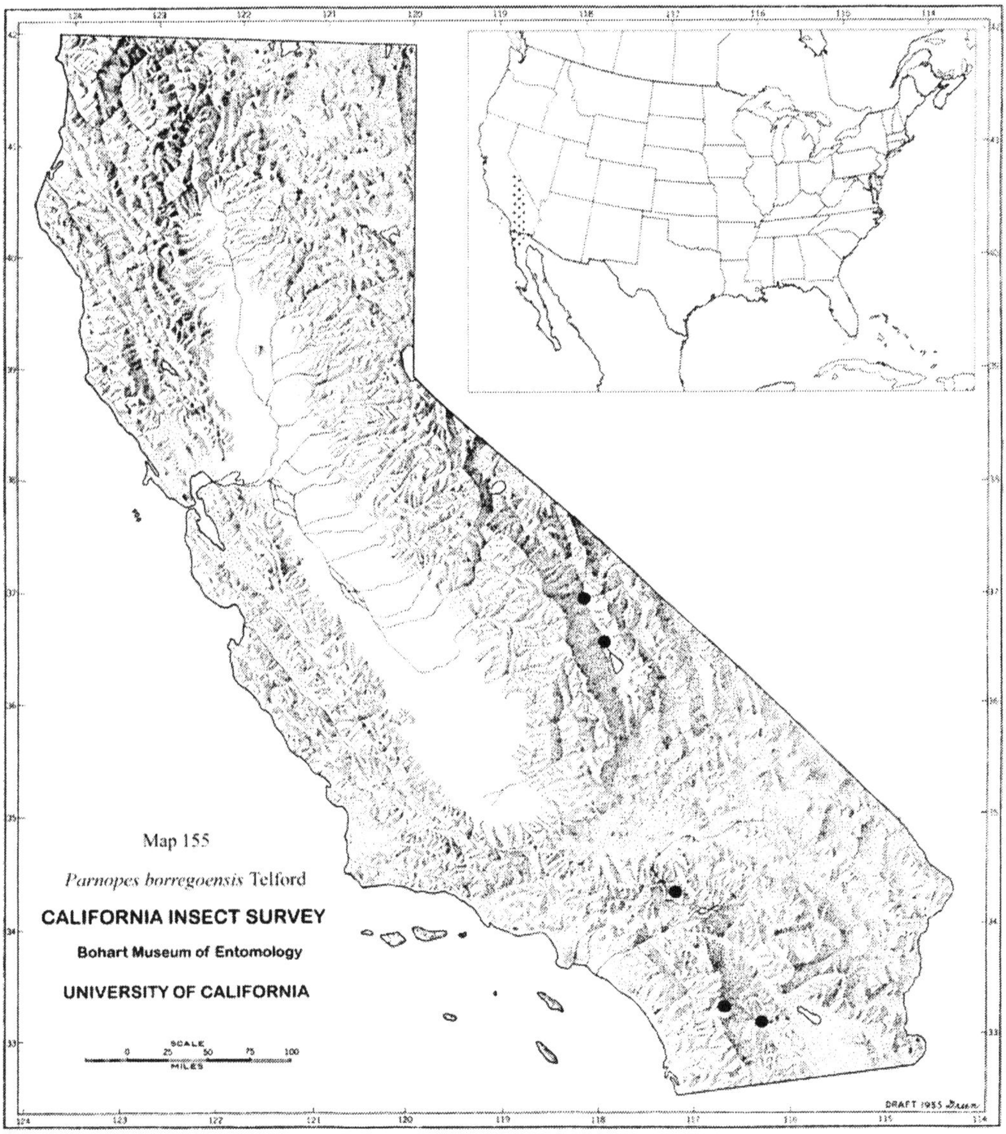

Map 155. California distribution of *Parnopes borregoensis*. Inset: overall distribution.

Discussion. This is a rare species, known from only a few specimens. It is distinguished by the posterior margin of the metanotal projection obtusely angled, the face lacks silvery pubescence, the male tergum IV is short, and the female tergum III subapical foveae form an obtuse angle.

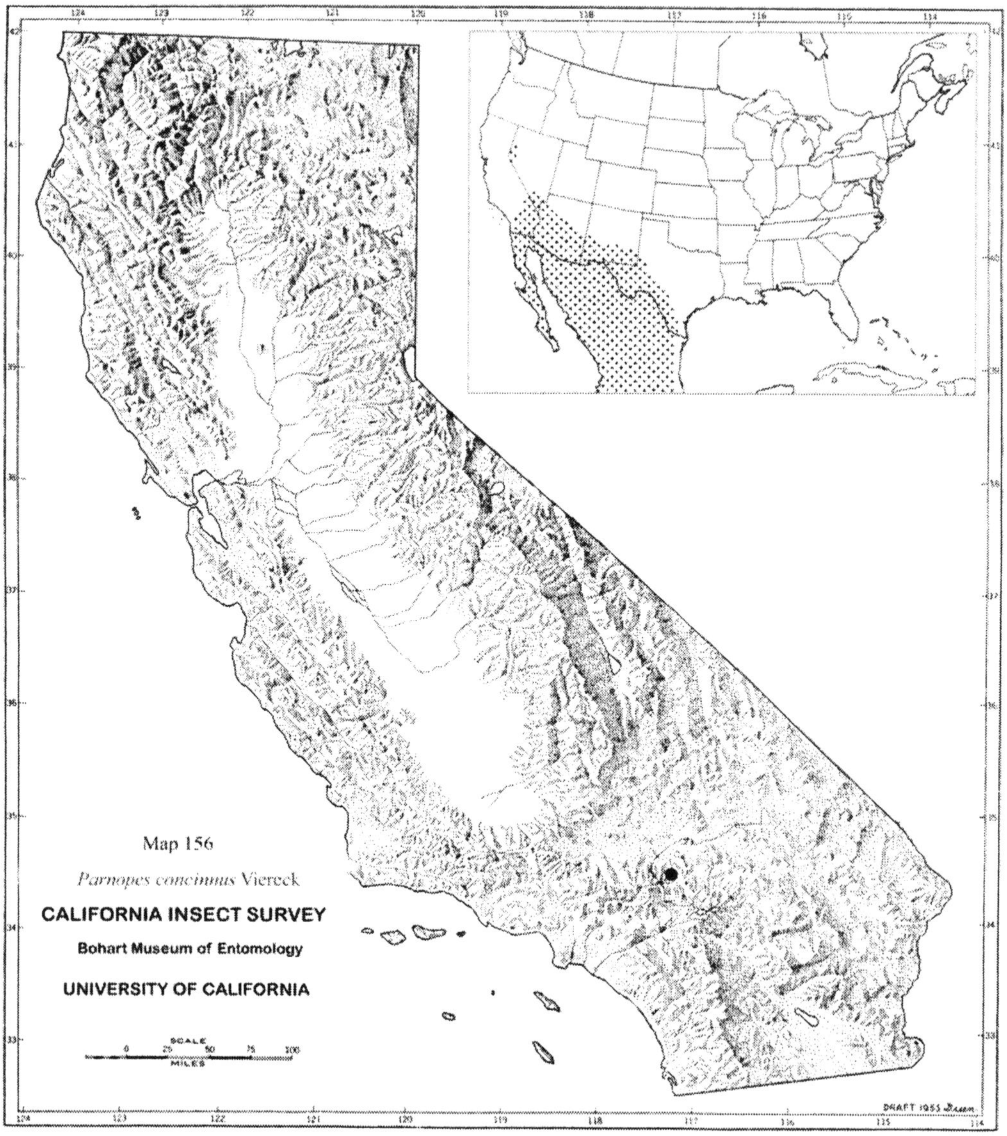

Map 156. California distribution of *Parnopes concinnus*. Inset: overall distribution.

Parnopes concinnus **Viereck**

Map 156

Parnopes concinnus Viereck 1904:248. Holotype ♀; USA: Arizona, Florence (ANSP).

Parnopes arizonensis Viereck 1904:249. Holotype ♂; USA: Arizona, Florence (ANSP).

Parnopes digueti du Buysson 1904:274. Holotype ♂; Mexico: Jalisco, Guadalajara (MNHN).

Geographic Range (Map 156). Upper and Lower Sonoran Zones west of the 100th meridian in the USA, including northwestern Mexico (Baja California); 18 specimens were examined.

California Records. Collected May through July; **Riverside Co.**: 18 mi. w Blythe; **San Bernardino Co.**: Needles.

Hosts. Unknown.

Discussion. *Parnopes concinnus* is most similar to the non-Californian species *festivus*. The foretibia is green to blue, tegula is bicolored, with some green or blue highlights, and the posterolateral corners of the terga are white. This is a moderate-sized species, ranging from 7.5-9.5 mm long, and is intermediate in size between the somewhat larger *desertorum* and *edwardsi* and smaller *fulvicornis*. It is rarely collected in California.

Parnopes desertorum Kimsey

Map 157

Parnopes festivus Cockerell 1894b:328. Holotype ♀; USA: New Mexico, Las Cruces (USNM). Nec Fabricius 1793.

Parnopes desertorum Kimsey 1987:89. Replacement name for *festivus* Cockerell 1894.

Geographic Range (Map 157). Southwestern USA, mostly west of the 100th meridian, but also in Oklahoma; 16 specimens were studied.

California Records. Collected June through August; **San Bernardino Co.**: Colton Hills; Clipper Valley.

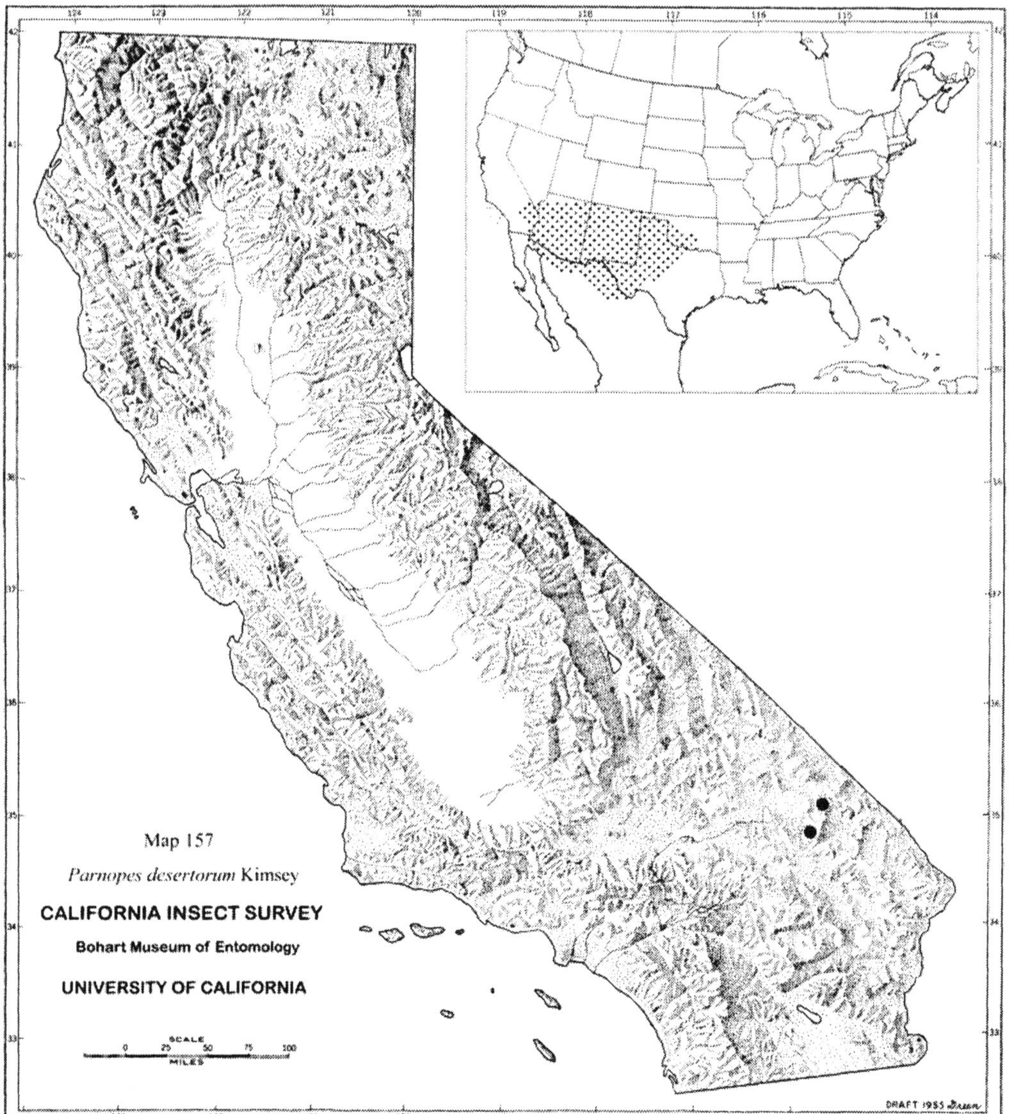

Map 157. California distribution of *Parnopes desertorum*. Inset: overall distribution.

Hosts. Unknown.

Discussion. Similar to *concinnus, desertorum* can be distinguished from that species by the reddish foretibia and tegula. This species ranges from 8.0-9.5 mm long. It is not often collected in California.

Parnopes edwardsii (Cresson)

Map 158

Euchroeus edwardsii Cresson 1879:4. Holotype ♀; USA: California (ANSP).

Parnopes hageni Viereck 1904:246. Holotype ♀; USA: Washington, Yakima (MCZ).

Parnopes henshawi Viereck 1904:247. Holotype ♀; USA: Oregon, Umatilla (MCZ).

Geographic Range (Map 158). Upper Sonoran to Canadian Zones, Rocky Mountains west including British Columbia and Baja California, Mexico; 298 specimens were studied.

California Records. Collected June through August; **Alameda Co**.: Emeryville; **Alpine Co**.: Forestdale Meadow; **Contra Costa Co**.: Antioch; **El Dorado Co**.: 6 mi. n Ice House Rd.; **Humboldt Co**.: Paradise Valley; **Inyo Co**.: 7 mi. n Parchers Camp; **Lassen Co**.: Hallelujah Junction; 1 mi. s Patterson Ranger Station, Lassen National Park; Standish; Bridge Creek Camp; **Los Angeles Co**.: Tanbark Flat; Duarte; Empalme Plantation; **Marin Co**.: 6 mi. w Inverness; **Merced Co**.: Dos Palos; **Mono Co**.: Bridgeport; White Mts.; Cottonwood Creek; Sardine Lakes.; Tom's Place; Pickel Meadow; Sonora Pass; **Monterey Co**.: Arroyo Seco; **Nevada Co**.: Big Bend near Cisco; Sagehen Creek; Truckee; Boca; **Placer Co**.: Carnelian Bay, Lake Tahoe; **Plumas Co**.: Vinton; Butte Lake; 2 mi. n Taylorsville; Johnsville; Antelope Valley; **Riverside Co**.: Anza; Banning; Riverside; **Sacramento Co**.: Sacramento; **San Bernardino Co**.: Mill Creek; Rialto; Big Bear Valley; Colton; **San Diego Co**.: 2 mi. n Warner Springs; 9 mi. s Warner Springs; 9 mi. ne Ramona; **Santa Barbara Co**.: Goleta; **Santa Clara Co**.: San Jose; **Shasta Co**.: Cassel; Hat Creek Post Office; **Sierra Co**.: Sattley; Sierra Valley; Independence Lake; Gold Lake; Kyburz Flat 14 km se Sierraville; Sierra Buttes; **Sonoma Co**.: Petaluma; **Trinity Co**.: Coffee Creek Ranger Station; **Tulare Co**.: Visalia; Ash Mt., Sequoia National Park; Indian Basin, Kings Canyon National Park; **Tuolumne Co**.: Mather; Leland Meadow; **Yolo Co**.: Davis; 5 mi. ne Woodland; Putah Creek.

Hosts. Reported hosts include *Bembix americana comata* Parker, *B. a. spinolae* Lepeletier, *B. pallidipicta* F. Smith (as *pruinosa* W. Fox), *B. amoena* Handlirsch, and *Steniolia obliqua* (Cresson) and *Microbembix* (Evans 1966, Evans and Gillaspy 1964, G. Bohart and MacSwain 1940, Krombein 1979).

Discussion. Next to *fulvicornis* this is the most commonly collected *Parnopes* species in California. It can be distinguished by the large size (body length 8.0-10.5 mm), lack of white markings on the abdomen, bright green, blue or purple coloration, and posteriorly truncate metanotal process.

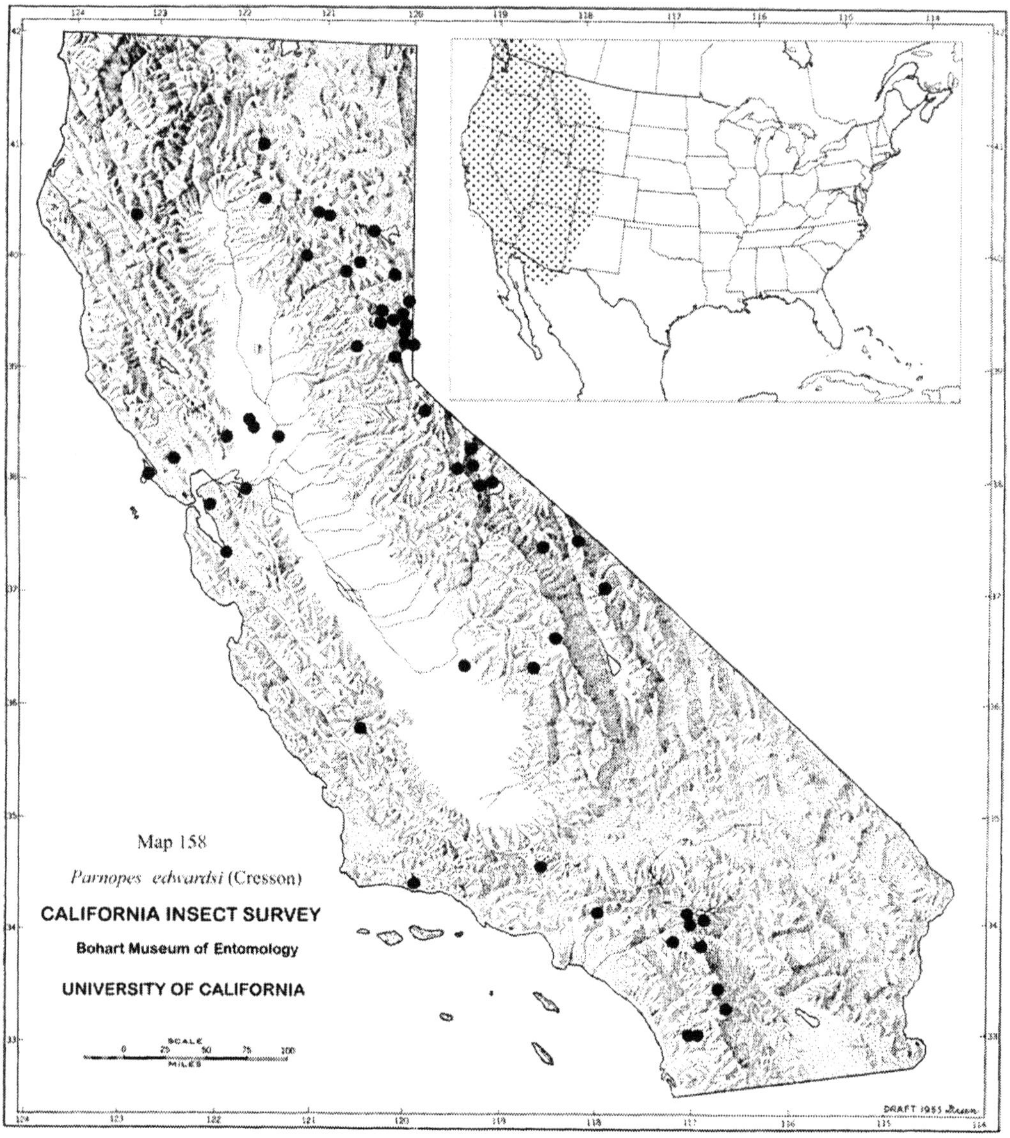

Map 158. California distribution of *Parnopes edwardsi*. Inset: overall distribution.

Parnopes fulvicornis Cameron

Figs. 7, 91; Map 159

Parnopes fulvicornis Cameron 1888:466. Holotype female; Mexico: Presidio (BMNH).

Parnopes westcottii Melander and Brues 1902:39. Lectotype male; USA: California (WSU). **New Designation**.

Parnopes diadema Viereck 1904:245. Holotype female; USA: Arizona, Florence (ANSP).

Parnopes taeniatus Viereck 1904:249. Holotype female; eastern Washington (ANSP).

Parnopes excurvatus Viereck 1904:250. Holotype male (not female); USA: Arizona, Florence (ANSP).

Geographic Range (Map 159). North America mostly west of the 100th meridian, also in Mexico, Guatemala and Canada (British Columbia and Ontario); 452 specimens were studied.

California Records. Collected June through August; **Calaveras Co**.: 2 mi. s Copperopolis; **Contra Costa Co**.: Antioch; **Imperial Co**.: San Felipe Creek; **Inyo Co**.: Eureka Valley Dunes; Keeler; **Riverside Co**.: Magnesia Canyon; **Yolo Co**.: Davis.

Hosts. This species reportedly parasitizes *Microbembex nigrifrons* Provancher (given as *aurata* J. Parker) and tentatively *M. monodonta* (Say) (Bohart and MacSwain 1940, Krombein 1958).

Discussion. *Parnopes fulvicornis* is most similar to the non-Californian species *excurvatus*. Both species can be extensively red-colored. A distinguishing feature of *fulvicornis* is the dense, silvery setae covering the face and almost completely obscuring the underlying punctation, bordered dorsally by the V-shaped brow. Specimens from desert regions tend to be strongly red-colored, particularly in females. Those from wetter areas are bluer or greener with less red coloration. Individual *fulvicornis* tend to be smaller than other *Parnopes* in California, ranging from 5-8 mm in length.

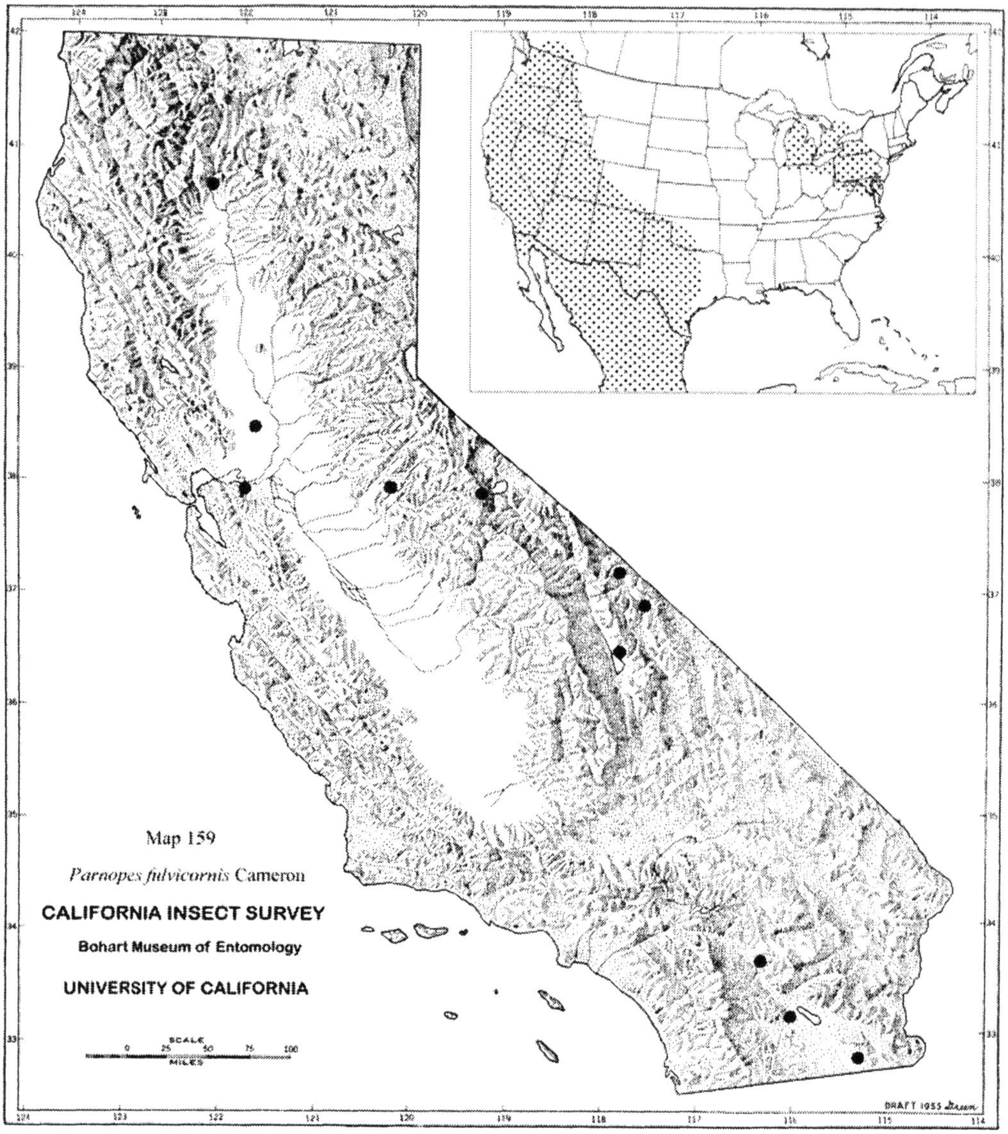

Map 159. California distribution of *Parnopes fulvicornis*. Inset: overall distribution.

LITERATURE CITED

Aaron, S. F. 1885. The North American Chrysididae. Transactions of the American Entomological Society 12:209-248.

Bischoff, H. 1910. Die Chrysididen des Königlichen Zoologischen Museums zu Berlin. Mitteilungen Museum für Naturkunde Berlin 4:427-493.

Bodenstein, W. G. 1939a. A new *Holopyga* from the western United States. Entomological News 50:19-21.

_____. 1951. Superfamily Chrysidoidea, pp. 718-726. *In*. Muesebeck, C. F. W., et al. Hymenoptera of America north of Mexico, synoptic catalogue. 1420 pp. U. S. Gov. Printing Office, Washington, D. C.

Bohart, G. E. and J. W. MacSwain. 1940. Notes on two chrysidids parasitic on western bembicid hosts. Pan-Pacific Entomologist 16:92-93.

Bohart, R. M. 1962. A review of the hexadentate species of *Chrysis* of America north of Mexico. Acta Hymenopterologica 1:361-375.

_____. New species of *Chrysis* in the *lauta, propria* and *venusta* groups from North America. Proceedings of the Biological Society of Washington 77:223-236.

_____. 1966a. New species of *Ceratochrysis* from North America. Journal of the Kansas Entomological Society 53:132-136.

_____. 1966b. New species of cuckoo wasps bred from twigs in western North America. Proceedings of the Biological Society of Washington 79:131-134.

_____. 1980. New genera and species of North American Chrysididae. Journal of the Kansas Entomological Society 53:132-136.

_____ and R. L. Brumley. 1967. Two new species of *Hedychridium* from California. Pan-Pacific Entomologist 43:232-235.

_____ and L. E. Campos. 1960. A review of the genus *Omalus* Panzer in North America. Annals of the Entomological Society of America 53:235-250.

_____ and L. S. Kimsey. 1978. A revision of the New World species of *Hedychridium*. Proceedings of the Biological Society of Washington 91:590-635.

_____ and L. S. Kimsey. 1982. A synopsis of the Chrysididae in America north of Mexico. Memoirs of the American Entomological Institute (33):1-266.

_____ and J. D. McLaughlin. 1979. Evidence indicating *Ammophila* as host of *Spintharosoma*. Pan-Pacific Entomologist 54:310

_____ and J. A. Powell. 1956. Observations on the nesting habits of *Eucerceris flavocincta* Cresson Pan-Pac. Entomologist 32:143-144.

Bridwell, J. C. 1919. Some notes on Hawaiian and other Bethylidae with descriptions of new species. Proceedings of the Hawaiian Entomological Society 4:35-38.

Brullé, A. 1846. *In* Lepeletier de Saint-Fargeau, Histoire Naturelle des Insectes. Hyménoptères. Vol. 4, 680 pp. Roret, Paris.

Buysson, R. du. 1891. Contribution aux Chrysidides du globe. Revue Entomologique (Caen) 10:29-47.

_____. 1898. Contribution aux Chrydidides du globe. Revue Entomologique (Caen) 17:125-147.

_____. 1901. Sur quelques Chrysidides du Musèe de Vienne. Annalen der Naturhistorisches Hofmuseum, Wien 16:97-104.

_____. 1904. Contribution aux Chrysidides du globe. Revue Entomologique (Caen) 23:253-275.

_____. 1908. Hymènoptéres nouveau. Revue Entomologique (Caen) 27:207-219.

Byers, G. W. 1978. Nests, prey, behavior and development of *Cerceris halone*. Journal of the Kansas Entomological Society 51:818-831.

Cameron, P. 1888. Family Chrysididae, pp. 1-147. *In*: Biologia Centrali-Americana, 1883-1900. Hymenoptera, vol. 1.

Carrillo, J. L. and L. E. Caltagirone. 1970. Observations on the biology of *Solierella peckhami*, *S. blaisdelli* and two species of Chrysididae. Annals of the Entomological Society of America 63:673-681.

Cockerell, T. D. A. 1894a. Description of a new *Chrysis*. Entomological News 5:125.

_____. 1894b. Descriptions of new Hymenoptera. Entomological News 5:328-329.

_____. 1897. Contributions from the New Mexico Biological Station, V. Some new Hymenoptera from the Mesilla Valley, New Mexico. Annals and Magazine of Natural History (6)19:394-403.

_____. 1903. *Euthrips* and *Hedychridium* in New Mexico. Canadian Entomologist 35:262.

Cooper, K. W. 1952. A remarkable new species and subgenus of chrysidid wasp from North America. Transactions of the American Entomological Society 78:137-148.

Cresson, E. T. 1865a. On the Hymenoptera of Cuba. Proceedings of the Entomological Society of Philadelphia 4:103-110.

_____. 1865b. Catalogue of Hymenoptera in the collection of the Entomological Society of Philadelphia. Proceedings of the Entomological Society of Philadelphia 4:242-313.

_____. 1879. Four new species of Hymenoptera. Monthly Proceedings. Transactions of the American Entomological Society 7:iv.

_____. 1916. The Cresson types of Hymenoptera. Memoirs of the American Entomological Society 1:1-141.

_____. 1928. The types of Hymenoptera in the Academy of Natural Sciences of Philadelphia other than those of Ezra T. Cresson. Memoirs of the American Entomological Society 5:1-90.

Dahlbom A. G. 1854. Hymenoptera Europea praecipue borealia etc. Vol. 2, xxiii+411 pp. Lundbergiana, Lund.

Dahlsten, D. L. 1961. Life history of a pine sawfly, *Neodiprion* sp., at Willits, California. Canadian Entomologist 93:182-195.

_____. 1967. Preliminary life tables for the pine sawflies in the *Neodiprion fulviceps* L. complex. Ecology 48:275-289.

Eickwort, G. C. 1973. Biology of the European mason bee, *Hoplitis anthocopoides* in New York State. Search Agriculture, Cornell Univerisity 3(2):1-31.

Evans, H. E. 1966. Nests and prey of two species of *Philanthus* in Jackson Hole, Wyoming. Great Basin Naturalist 26:35-40.

_____. 1970. Ecological-behavioral studies of the wasps of Jackson Hole, Wyoming. Bulletin of the Museum of Comparative Zoology, Harvard University 140:451-511.

_____. 1973. Further studies on the wasps of Jackson Hole, Wyoming. Great Basin Naturalist 33:147-155.

_____ and J. E. Gillaspy. 1964. Observations of the ethology of digger wasps of the genus *Steniolia*. American Midland Naturalist 72:257-280.

Fabricius, J. C. 1775. Systema entomologiae, etc. xxviii+832 pp. Kortii, Flensburgi et Lipsiae.

_____. 1787. Mantissa insectorum etc. Vol. 2, 382 pp., C. G. Proft, Hafniae. C. Richard, Brunsvigae.

_____. 1804. Systema Piezatorum, xiv + 439 pp. + 32 pp. Brunsvigae, C. Richard.

Ferguson, W. E. 1962. Biological characteristics of the mutillid subgenus *Photopsis* Blake and their systematic values. University of California Publications in Entomology 27:1-92.

Grandi, G. 1961. Studi di un entomologa sugli imenotteri superiori. Bolletino dell'Instituto di Entomologia dell'Università di Bologna 25:1-659.

Gribodo, G. 1874. Diagnosi di alcune specie nuove del genere *Chrysis*. Annali Museo Civico Storia Naturale, Genova 6:358-360.

_____. 1879. Note Imenotterologiche. Annali Museo Civico Storia Naturale, Genova 14:325-347.

Grigarick, A. A. and L. A. Stange. 1968. The pollen-collecting bees of the Anthidiini of California. Bulletin of the California Insect Survey 9:1-113.

Haldeman, S. S. 1844. Descriptions of insects presumed to be undescribed. Proceedings of the Academy of Natural Sciences, Philadelphia 2:53-55.

Halsted, J. A. 1978. On the rearing of *Microchridium minutum* and its probable host *Ammoplanellus* (*Ammoplanellus*) *umatilla*. Pan Pacific Entomologist 63:256-257.

Hicks, C. H. 1929. *Pseudomasaris edwardsii* Cresson, another provisioning wasp, with further notes on *P. vespoides* (Cresson). Canadian Entomologist 61:121-5.

_____. 1932. Notes on *Sphex aberti* (Hald.). Canadian Entomologist 64:145-151.

_____. 1934. Observations on a chrysidid parasite and its host. Entomological News 44:206-209.

Horning, D. S. Jr. 1971. Two new species of *Chrysura* from western North America. Pan-Pacific Entomologist 47:26-32.

Huber, J. T. and D. H. Pengelly. 1978 (1977). A revision of the genus *Elampus* Spinola (*Notozus* of authors). Proceedings of the Entomological Society of Ontario 108:75-137.

Hungerford, H. B. 1937. *Pseudomasaris occidentalis* (Cresson) in Kansas. Journal of the Kansas Entomological Society 10:133-4.

Kimsey, L. S. 1981. The Cleptinae of the Western Hemisphere. Proceedings of the Biological Society of Washington 94:801-818.

_____. 1982. A new North American chrysidid genus and redescription of the *Pseudolopyga* Krombein. Pan-Pacific Entomologist 57:351-358.

_____. 1987. New species of *Cleptes* from Asia and North America. Pan-Pacific Entomologist 63:56-59.

_____ and R. M. Bohart. 1991 (1990). The chrysidid wasps of the world. Oxford University Press, ix + 652.

Krombein, K. V. 1958a. Biology and taxonomy of the cuckoo-wasps of coastal North Carolina. Transactions of the American Entomological Society 84:141-168.

_____. 1958b. Additions during 1956 and 1957 to the wasp fauna of Lost River State Park, West Virginia with biological notes and description of new species. Proceedings of the Entomological society of Washington 60: 49-64.

_____. 1959. Biological notes on *Chrysis (Ceratochrysis) enhuyki* Cooper, and its host, *Leptochilus republicanus zendaloides* (Robertson). Entomological News 70:17-23.

_____. 1963. A new *Chrysura* from Plummers Island, Maryland. Entomological News 74:149-152.

_____. 1967. Trap-nesting wasps and bees. Smithsonian Press, Washington D.C., vi + 570 pp.

_____. 1979. Superfamily Bethyloidea, pp. 1203-1251. In Krombein et al. Catalogue of Hymenoptera in America north of Mexico. vol. 2, xvi + 1199-2209. U. S. Gov. Printing Office, Washington.

Kurczewski, D. W. 1967. *Hedychridium fletcheri*, a probable parasite of *Tachysphex similes*. Journal of the Kansas Entomological Society 40:278-284.

Kurczewski, F. E. and E. J. Kurczewski. 1979. An annotated list of cuckoo-wasps from Erie County Pennsylvania. Proceedings of the Entomological society of Washington 72:190-201.

Linnaeus, C. 1758. Systema naturae, etc. 10th ed. Tomus 1. 823 pp. Laurentii Salvii, Holmiae.

Linsenmaier, W. 1959. Revision der Familie Chrysididae. Mitteilungen der Schweizerischen Entomologischen Gesellschaft 32:1-232.

_____. 1984. Das subgenus *Trichrysis* Lichtenstein in Nord und Südamerika. Mitteilungen der Schweizerischen Entomologischen Gesellschaft 57:194-224.

_____. 1994. Grundriss der *Chrysis ignita* – gruppe von Nordamerika. Entomofauna, Zeitschrift für Entomologie 42:481-500.

Medler, J. T. 1964. A note on *Rynchium leucomelas* (Saussure) in trap-nests in Wisconsin. Entomological News 75:26-27.

Melander, A. L. and C. T. Brues. 1902. New species of *Gasteruption, Trigonalys, Parnopes* and *Psammophila*. Biologists Bulletin 3:33-42.

Mocsáry, A. 1887. Studia synonymica. Természetrajizi Füzetek 2:12-20.

_____. 1889. Monographia Chrysididarum orbis terrestris universi. Budapest: Academia Scientarum Hungarica, 643 pp.

_____. 1911. Species Chrysididarum novae. Annales Museum Nationale Hungarici 9:443-474.

_____. 1912. Species Chrysididarum novae. Annales Museum Nationale Hungarici 10:549-592.

_____. 1914. Chrysididae plerumque exoticae novae. Annales Museum Nationale Hungarici 12:1-72.

Moczar, L. 1996. Additions to American Cleptinae. Memoirs of the Entomological Society of Washington 17:153-160.

Moore, C. G. 1966. Taxonomy of the *coerulans* group of the genus *Chrysis* in North America. Annals of the Entomological Society of America 59:1125-1131.

_____ and F. D. Parker. 1962. A host of *Pyria inaequidens* (Dahlbom). Pan-Pacific Entomologist 38:14.

Norton, E. 1879. On the chrysides of North America. Transactions of the American Entomological Society 7:233-242.

Parker, F. D. 1962. A host of *Chrysis (Trichrysis) mucronata* Brull, and an additional host of *Chrysis (Chrysis) coerulans*. Pan-Pacific Entomologist 38:140.

_____. 1967. Notes on the nests of three species of *Pseudomasaris* Ashmead. Pan-Pacific Entomologist 43:213-16.

_____. 1968. Host-parasite associations in some twig-nesting Hymenoptera from western North America, II. Pan-Pacific Entomologist 44:1-6.

_____. 1970. New North American *Microdynerus* with notes on the nests of two species. Pan-Pacific Entomologist 46:241-53.

_____ and R. M. Bohart. 1966. Host-parasite association in some twig-nesting Hymenoptera from western North America. Pan-Pacific Entomologist 42:91-98.

Patton, W. H. 1879. Descriptions of several new Proctotrupidae and Chrysididae. Canadian Entomologist 11:64-68.

Powell, J. A. 1964. Biology and behavior of nearctic wasps of the genus *Xylocelia*, with special reference to *X. occidentalis* (Fox). Wasmann Journal of Biology 21:155-176.

_____. and C. Hogue. 1979. California insects. University of California Press, Berkeley, x + 388 pp.

Provancher, A. 1881. Fam. IX des Chrysides. Chrysididae. Faune Canadienne. Les Insectes Hymènoptéres 12:300-304.

Radoszkowski, O. 1881. Hymènoptéres d'Angola. Jornal de sciencias mathematicas, physicas, e naturaes da Academia Real das Sciencas de Lisboa 8:197-221.

Rohwer, S. A. 1909. Some new chrysidid wasps from western United States. Psyche (Cambridge) 16:87-92.

Rosenheim, J. A. 1987a. Host location and exploitation by the cleptoparasitic wasp *Argochrysis armilla*: the role of learning. Behavioral Ecology and Sociobiology 21:401-406.

_____. 1987b. Nesting behavior and bionomics of a solitary ground-nesting wasp, *Ammophila dysmica*: influence of a parasite pressure. Annals of the Entomological Society of America 80:739-749.

_____. 1988. Parasite presence acts as a proximate cue in the nest-site selection process of the solitary digger wasp, *Ammophila dysmica*. Journal of Insect Behavior. 1:333-342.

_____. 1989. Behaviorally mediated spatial and temporal refuges from a cleptoparasite, *Argochrysis armilla*, attacking a ground-nesting wasp, *Ammophila dysmica*. Behavioral Ecology and Sociobiology 25:335-348.

_____. 1993. Single-sex broods and the evolution of nonsiblicidal parasitoid wasps. The American Naturalist 141:90-104.

_____ and J. K. Grace. 1987. Biology of a wood-nesting wasp, *Mimumesa mixta* (W. Fox) and its parasite *Elampus viridicyaneus* Norton. Proceedings of the Entomological Society of Washington 89:351-355.

_____, T. Meade, I. G. Powch and S. E. Schoenig. 1989b. Aggregation by foraging insect parasitoids in response to local variations in host density: determining the dimensions of a host patch. Journal of Animal Ecology 58:101-117.

Say, T. 1824. Natural History. In W. H. Keating, Narrative of expedition to the source of St. Peter's River, etc., vol. 2, appendix part 1, pp. 253-459. H. C. Carey and I. Lea, Philadelphia.

_____. 1828. Descriptions of new species of Hymenoptera of the United States. *In:* Contributions of the Maclurian Lyceum to the Arts and Sciences 1:66-83.

_____. 1836 (1835). Descriptions of new North American Hymenoptera and observations on some already described. Boston Journal of Natural History 1:210-305.

Smith, D. R. 1962. Parasites reared from a species of *Neodiprion* found on Douglas-fir in Idaho. Pan-Pacific Entomologist 38:189.

Stage, G. I. 1960. First North American record of the adventive wasp *Chrysis fuscipennis* Brullé. Pan-Pacific Entomologist 36:191-195.

Taylor, L. H. 1924. Chrysididae from lower California. Proceedings of the California Academy of Sciences (4) 13:325-332.

Telford, A. D. 1964. The nearctic *Parnopes* with an analysis of the male genitalia in the genus. University of California Publications in Entomology 36:1-42.

Thorpe, R. W. 1968. Ecology of a *Proteriades* and its *Chrysura* parasite, with larval descriptions. Journal of the Kansas Entomological Society 41:324-331.

Tsuneki, K. 1970. Bemerkungen und Beschreibungen über de Japanischen Heteronychinen. Life Study (Fukui) 14:27-34.

Viereck, H. L. 1904. The North American cuckoo wasps of the genus *Parnopes*. Transactions of the American Entomological Society 30:245-250.

_____. 1906. Notes and descriptions of Hymenoptera from the western United States. Transactions of the American Entomological Society 32:173-247.

Walker, F. 1866. New species in: Lord, J. K., The naturalist in Vancouver Island and British Columbia. Vol. 2, 375 pp. London.

Plates

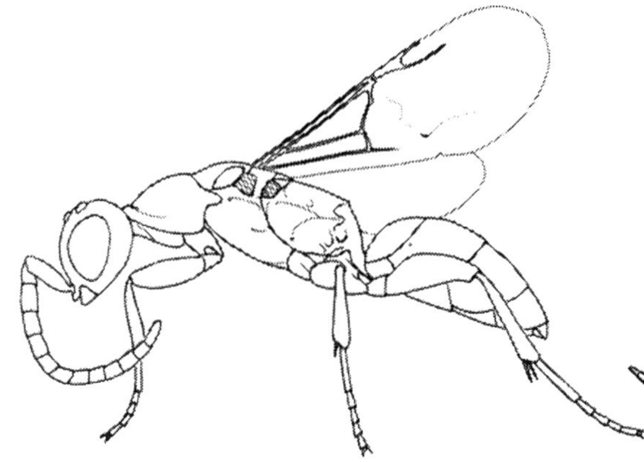

2. Cleptes speciosus ♂

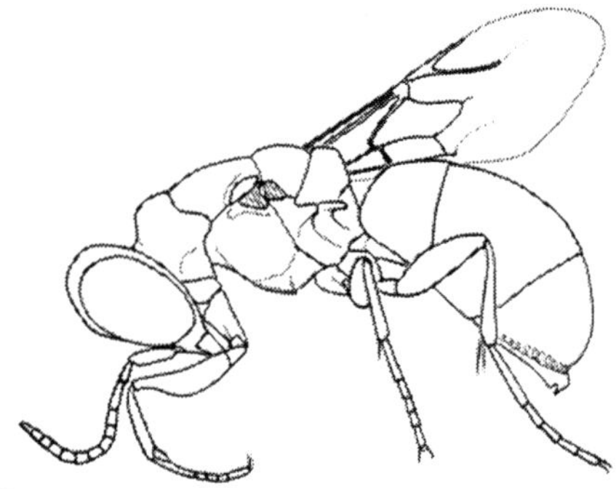

5. Argochrysis mesillae

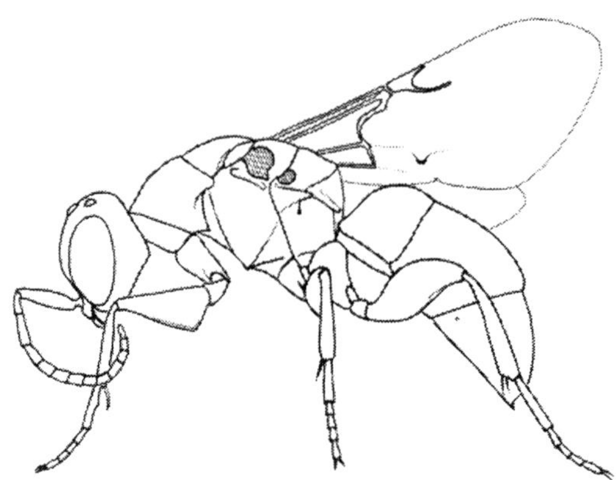

3. Holopyga ventralis

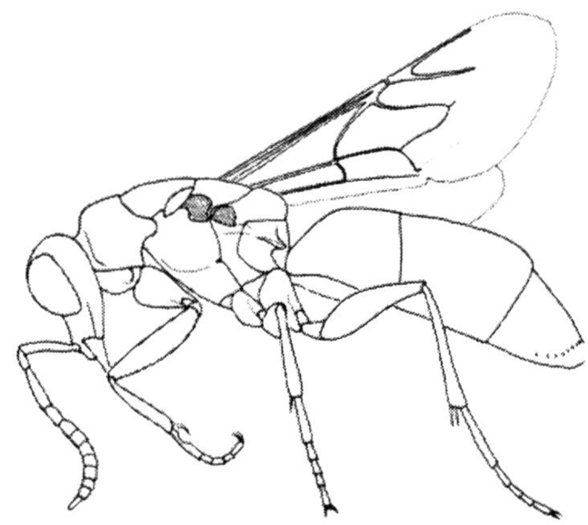

6. Chrysura tota ♂

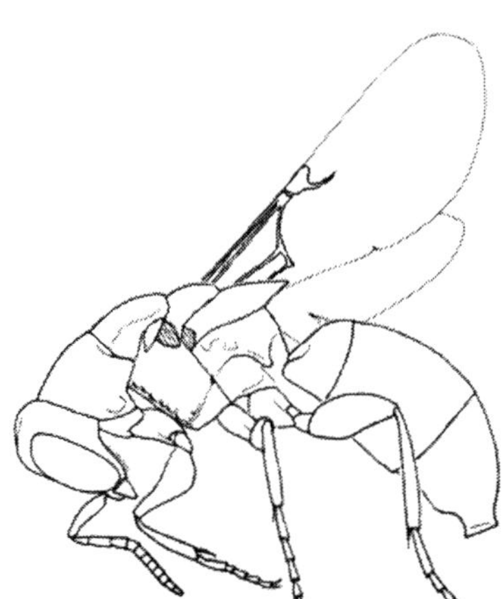

4. Elampus hyalinus ♀

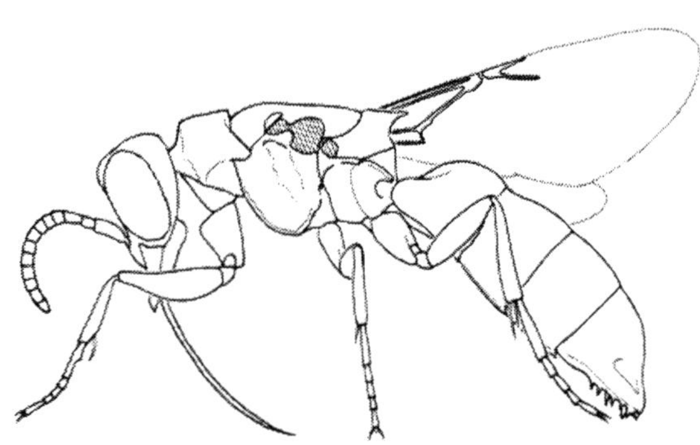

7. Parnopes fulvicornis ♂

Plate 1. Figs. 2-7. Lateral view of body.

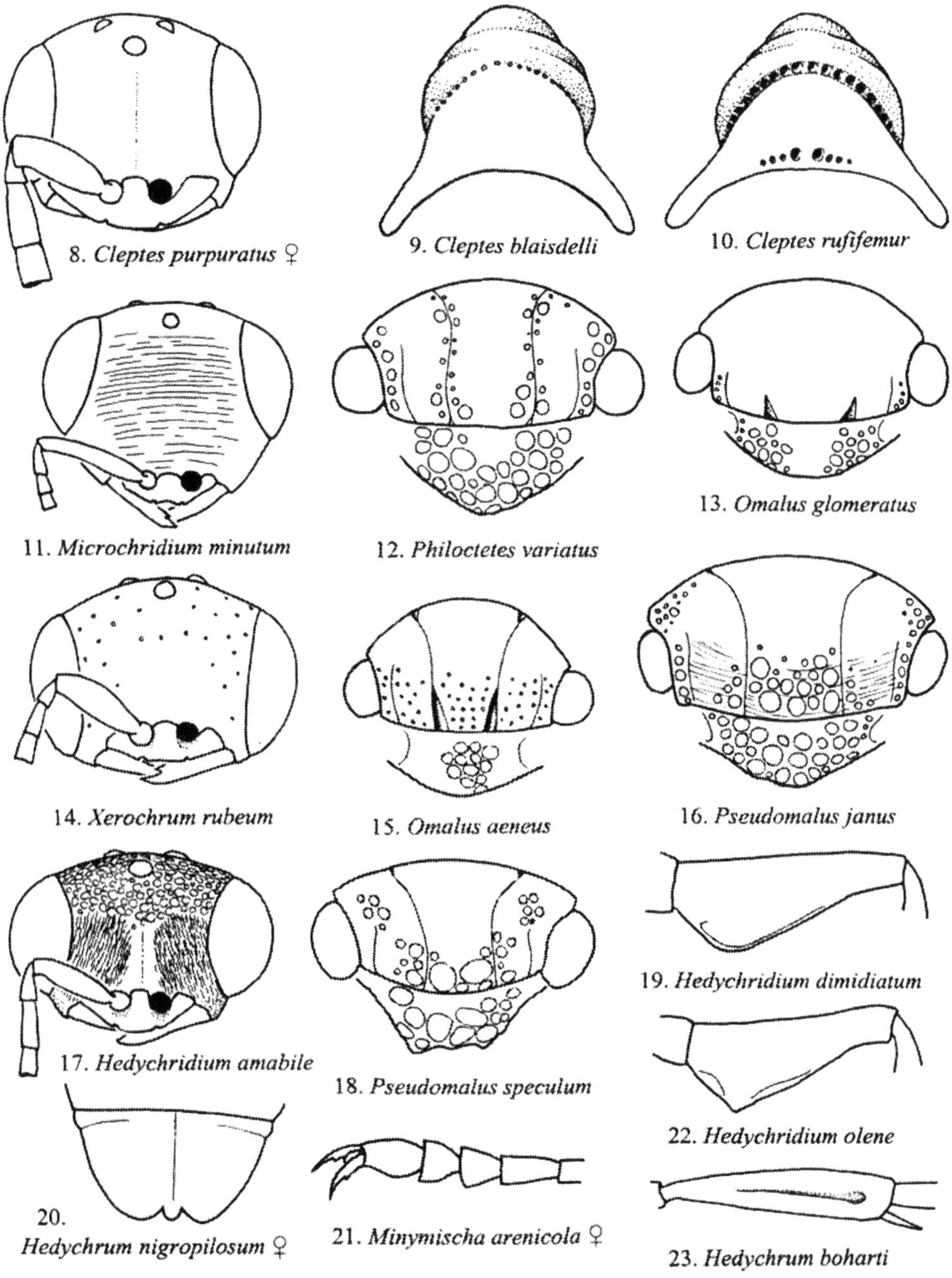

Plate 2. Figs. 8, 11, 14, 17. Front view of head, with right antennae removed. Figs. 9, 10. Dorsal view of pronotum. Figs. 12-16, 18. Figs. 19, 22. Lateral view of forefemur. Fig. 20. Ventral view of metasomal sternum III. Fig. 21. Lateral view of foretarsomere. Fig. 23. Inner surface of hindtibia.

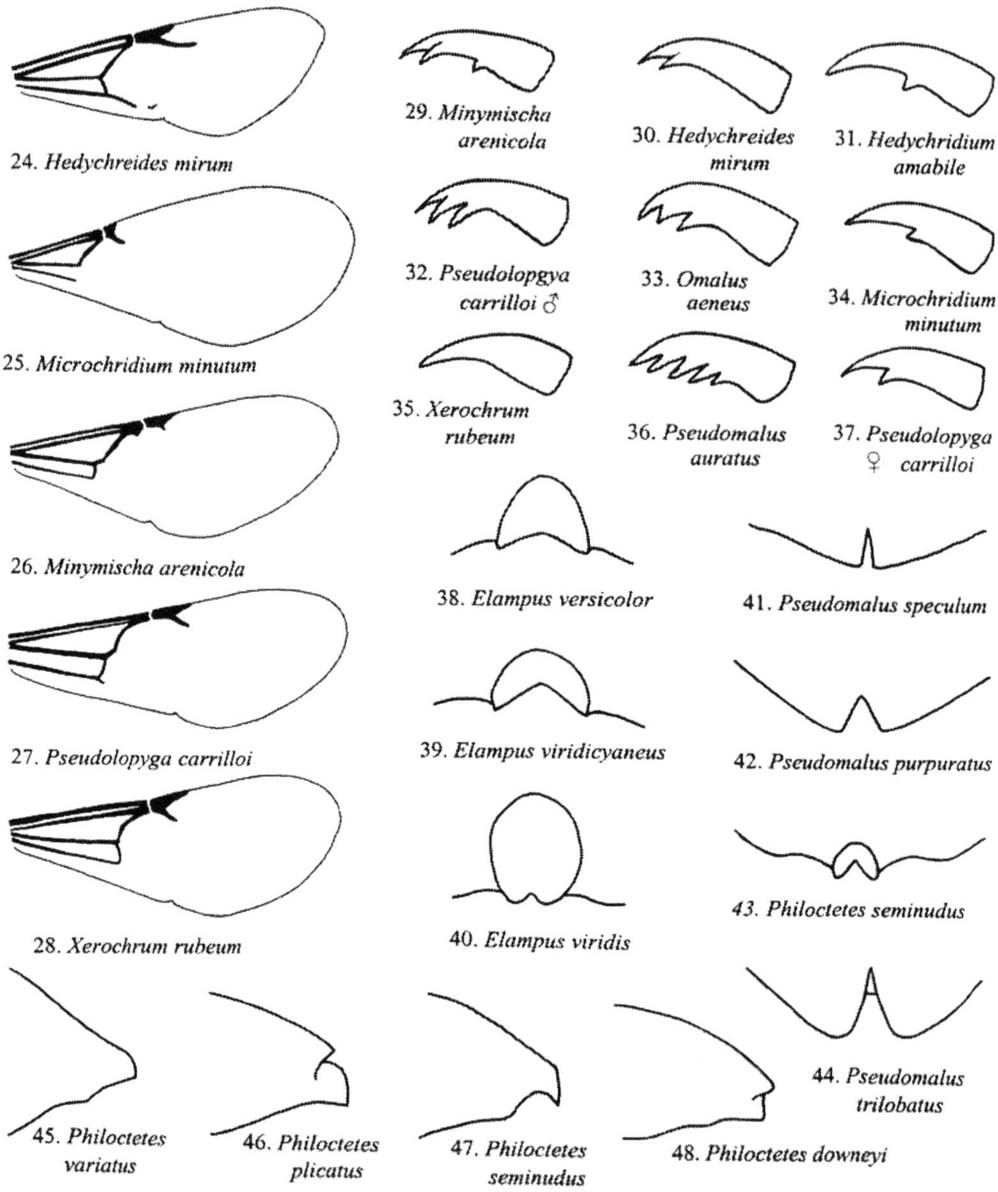

Plate 3. Figs. 24-28. Forewing. Figs. 29-37. Hindtarsal claw. Fig. 38-44. Posterior view of apical metasomal tergum. Figs. 45-48. Lateral view of apical metasomal tergum.

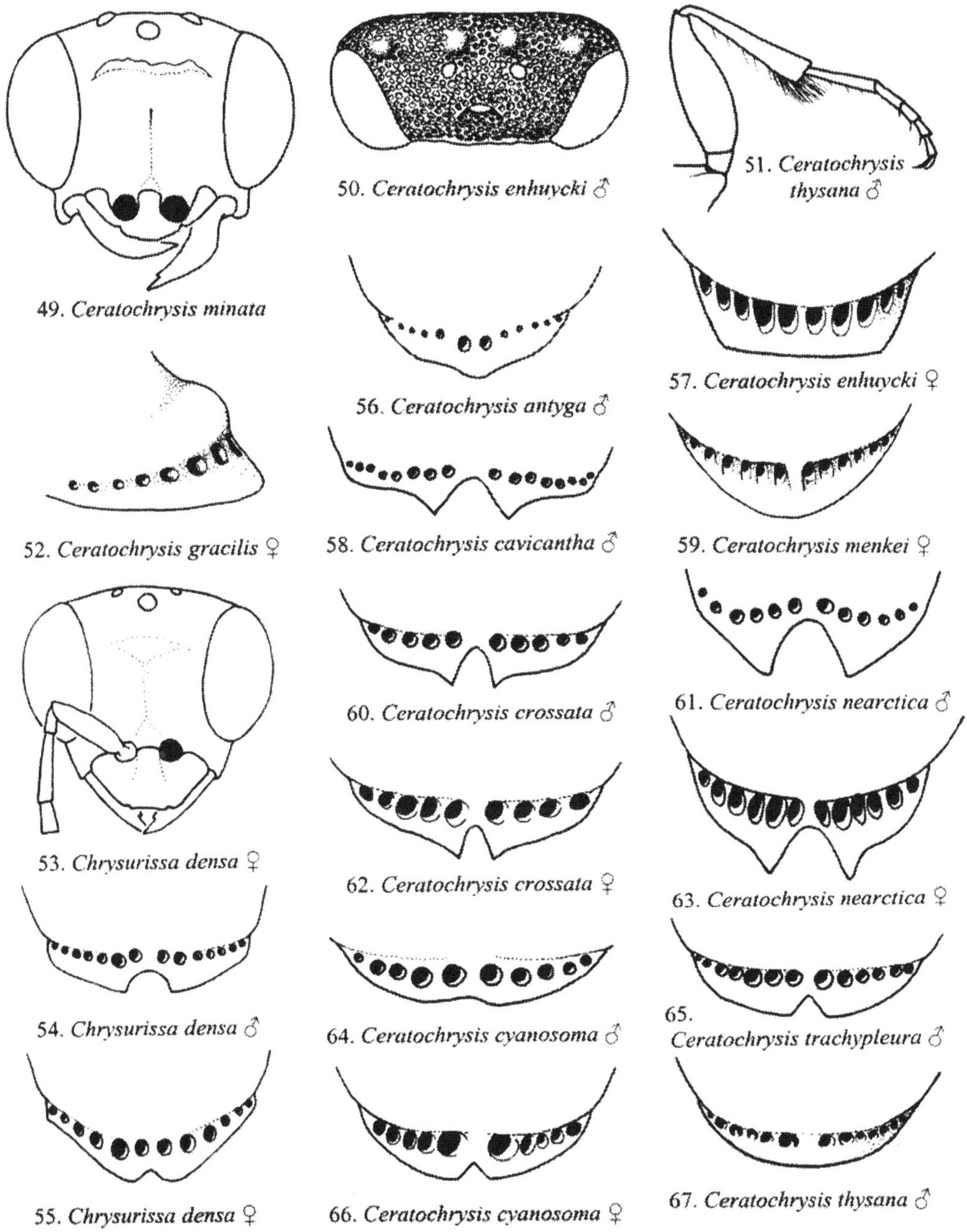

Plate 4. Figs. 49, 53. Front view of head, with both antennae removed. Fig. 50. Dorsal view of vertex. Fig. 51. Lateral view of midleg. Fig. 52. Lateral view of apical metasomal tergum. Figs. 54-67. Posterior view of apical metasomal tergum.

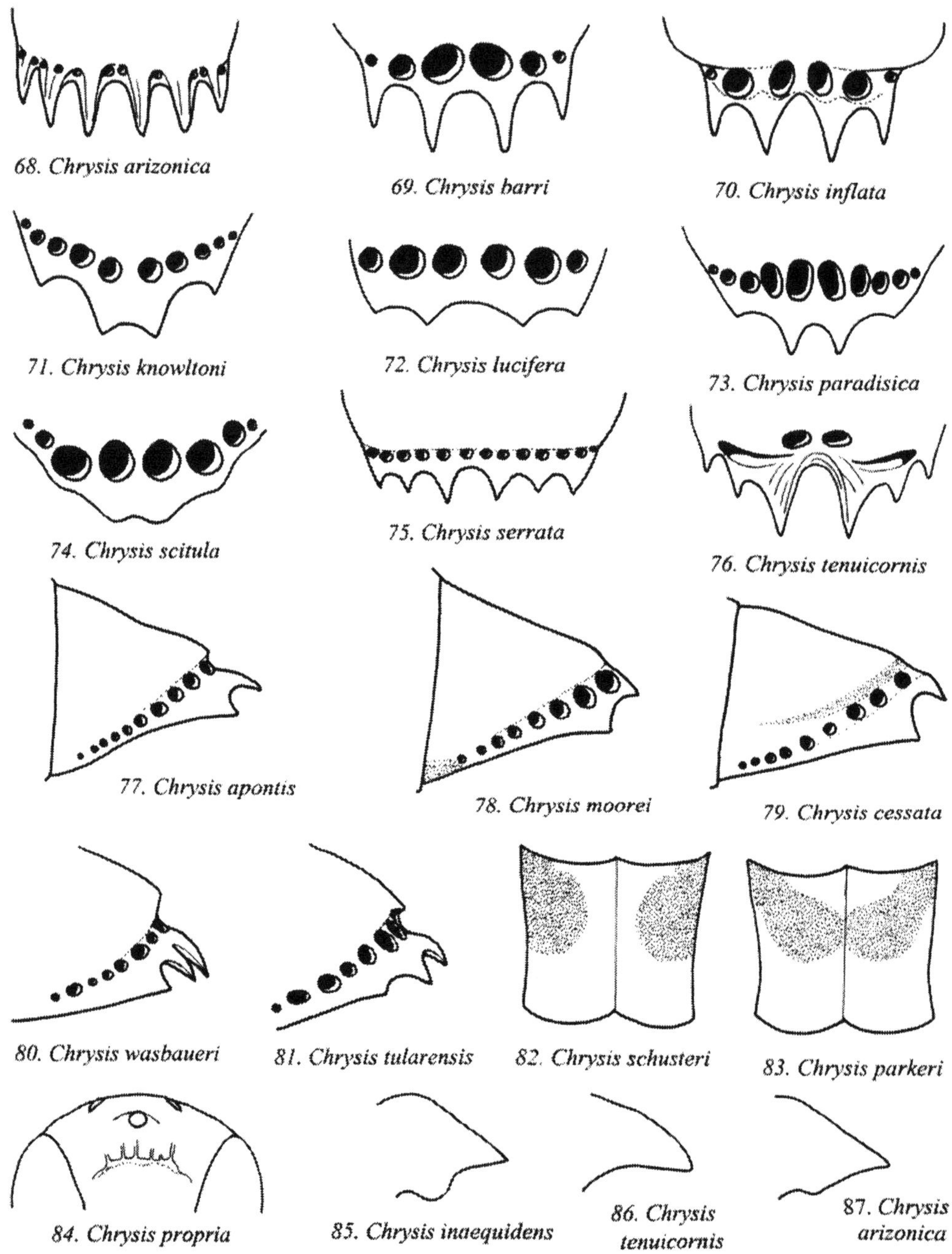

Plate 5. Figs. 68-76. Posterior view of apical metasomal tergum. Figs. 77-81. Lateral view of apical metasomal tergum. Figs. 82, 83. Ventral view of metasomal sternum II. Fig. 84. Front view of vertex. Figs. 85-87. Lateral view of propodeal projection.

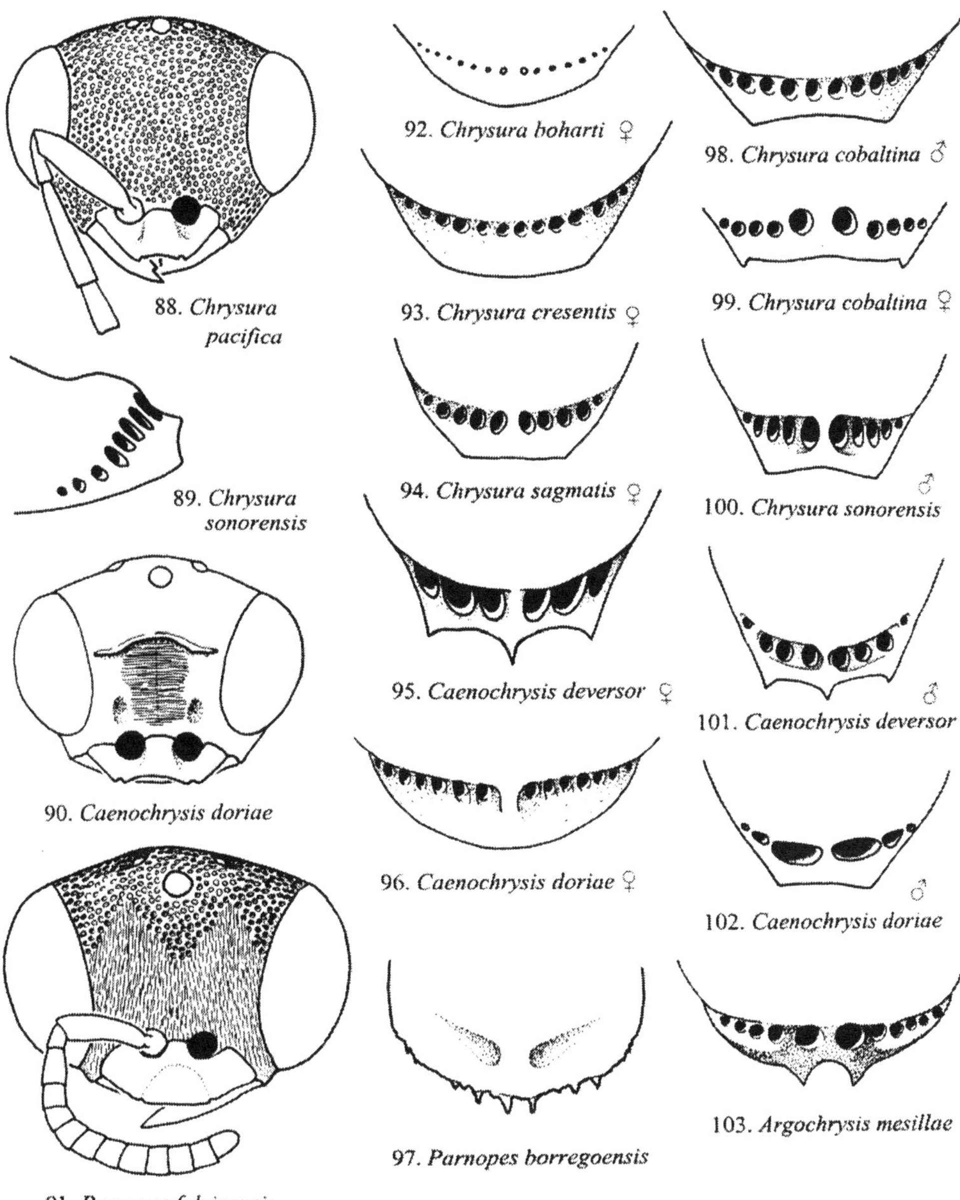

Plate 6. Figs. 88, 90, 91. Front view of head, with one or both antennae removed. Fig. 89. Lateral view of apical metasomal tergum. Figs. 92-103. Posterior view of apical metasomal tergum.

www.ingramcontent.com/pod-product-compliance
Ingram Content Group UK Ltd.
Pitfield, Milton Keynes, MK11 3LW, UK
UKHW050450190926
475586UK00010B/102